AF815801

THE ECONOMICS OF SUSTAINABLE DEVELOPMENT

For other titles published in this series, go to
www.springer.com/series/6360

NATURAL RESOURCE MANAGEMENT AND POLICY

EDITORIAL STATEMENT

There is a growing awareness to the role that natural resources such as water, land, forests and environmental amenities play in our lives. There are many competing uses for natural resources, and society is challenged to manage them for improving social well being. Furthermore, there may be dire consequences to natural resources mismanagement. Renewable resources such as water, land and the environment are linked, and decisions made with regard to one may affect the others. Policy and management of natural resources now require interdisciplinary approach including natural and social sciences to correctly address our society preferences.

This series provides a collection of works containing most recent findings on economics, management and policy of renewable biological resources such as water, land, crop protection, sustainable agriculture, technology, and environmental health. It incorporates modem thinking and techniques of economics and management, Books in this series will incorporate knowledge and models of natural phenomena with economics and managerial decision frameworks to assess alternative options for managing natural resources and environment.

The Series Editors

THE ECONOMICS OF SUSTAINABLE DEVELOPMENT: THE CASE OF INDIA

SURENDER KUMAR
TERI University, New Delhi, India

SHUNSUKE MANAGI
Yokohama National University, Japan

Springer

Surender Kumar
TERI University
Lodhi Road
New Delhi-110003
IHC Complex
India
skumar@teri.res.in

Shunsuke Managi
Yokohama National University
Fac. Business Administration
79-4 Tokiwadai
Yokohama
Hodogaya-ku
240-8501 Japan
managi.s@gmail.com

ISBN 978-0-387-98175-8 e-ISBN 978-0-387-98176-5
DOI: 10.1007/978-0-387-98176-5
Springer Dordrecht Heidelberg London New York

Library of Congress Control Number: 2009926030

Printed on acid-free paper

Springer is part of Springer Science+Business Media (www.springer.com)

Foreword

This book highlights methodological approaches for the economics of sustainable development and brings together recent empirical work done in India, especially by Dr. Surender Kumar and Dr. Shunsuke Managi. Various chapters in this book use Indian data to show the very wide applicability of methodologies in the theory of production for dealing with many empirical issues of environmentally sustainable development in a developing country. I congratulate the authors for the time and effort devoted to compiling this very useful reference on the subject and the publishers for publishing this volume.

The methodologies of cost functions, distance functions, and production functions have been used in many recent studies and in the studies reported in this book for environmental valuation. Environmental valuation is required for designing policy instruments like pollution taxes for sustainable development and for measuring green GDP. The UN methodology of integrated environmental and economic accounting provides ways of measuring the cost of maintaining environmental resources at sustainable levels or the maintenance cost for estimating green GDP. Some of the chapters in this book show that the methodology of distance functions could be used for estimating the cost of environmentally sustainable development.

It has long been recognized that services offered by environmental resources as productive inputs are as important as conventional inputs of labor, capital, and land in various developmental activities. However, only recently have some studies, including ones reported in this book, attempted to estimate total factor productivity in industry and agriculture, taking into account the contribution of environmental resources as productive inputs. Some recent studies, including one reported in this book, use directional distance function for measuring the technical efficiency and environmental efficiency of an industry to show, respectively, the extent a firm could simultaneously increase output and reduce environmental degradation or pollution given the resource constraints.

This book provides an interesting analysis of testing the much studied environmental Kuznet curve hypothesis by comparing estimated environmental productivity for different states in India with their per capita incomes. The methodology described for estimating environmental productivity provides good insights for pursuing it further. In addition, the methodologies and empirical analysis given in chapters on environmentally sensitive productive growth and the macroeconomic effects of oil price shocks raise awareness of the need for more studies along these lines.

In all, I found this book to be very rich in applied methodologies and empirical analysis for studying the very important problem of sustainable development in a developing country context. It would be a good reference book for students and teachers of environmental studies and sustainable development in universities and research institutions in both developed and developing countries.

Professor M N Murty
Professor Rtd. and Senior Consultant
Institute of Economic Growth
Delhi, India

Acknowledgments

In writing this book, we have benefited from the wisdom and generosity of many, but we owe by far the greatest debt to Professor M.N. Murty, who has been a constant source of inspiration. Dr. R.K. Pachauri, Chancellor TERI University and Chair IPCC, Dr. Leena Srivastava, Executive Director TERI, Mr. Rajiv Seth, Registrar TERI University, and Dr. M. Govinda Rao, Director of the National Institute of Public Finance and Policy provided enormous inspiration, encouragement and full cooperation throughout writing of this book. We are very grateful to them.

We thank M.N. Murty, D.N. Rao, and Pradyot Ranjan Jena, our coauthors, who all not only provided insightful comments but invested themselves in our scholarly development. It has been our pleasure to work with them. Chapters 6, 8, 9, and 11 are a result of joint work with them. Appreciation is also due to participants in various conferences, workshops, and seminars for their comments and suggestions.

We express thanks to Madhu Khanna, B.N. Goldar, Shreekant Gupta, and Gopinath Pradhan for having discussion at the various stages of writing.

Chapters 8 and 9 formed part of the work of the research project on "Accounting and Valuation of Industrial Pollution: Environmentally Corrected GDP for India," supported by the Ministry of Environment and Forests (MoEF), Government of India under the World Bank aided Environmental Management Capacity Building Technical Assistance Project in India. Grant-in-Aid for Scientific Research from the Japanese Ministry of Education, Culture, Sports, Science, and Technology (MEXT), and financial support from the Japan Society for Promotion of Sciences (JSPS). We express our sincere thanks to all these agencies. The results and conclusions of this research do not necessarily represent the views of the funding agencies.

We are especially grateful to series editors Professors D. Zilberman, Alberto Garrido, and Renan Goetz, and we owe a great debt to Jon Gurstelle, our editor at Springer.

A large part of the credit for making this book happen goes to Mr. Pawan Kumar. Surender could work for this book because he did everything else. Gratitude also goes to our wives, Rekha, and Keiko, and to our parents for their encouragement and support. Finally, we should mention our kids, Shashwat, and Yui, with whom we love to play, but with whom we have not been able to play lately. With the book finished, we can now catch up with them.

This monograph contains some revised versions of previously published papers. Permission to reproduce the materials published in the following papers was granted by Emerald, Kluwer (Springer), The International Research Center for Energy and Economic Development (ICEED), Cambridge University Press, Elsevier, and IWA Publishing.

Kumar, S. (2006). A decomposition of total factor productivity growth: a regional analysis of Indian industrial manufacturing growth. International Journal of Productivity and Performance Management 55(3/4): 311–331.

Kumar, S. (2006). Analyzing industrial water demand in India: an input distance function approach. Water Policy 8(1): 15–29.

Kumar, S. (2006). Environmentally sensitive productivity growth: a global analysis using Malmquist-Luenberger index. Ecological Economics 56(2) 280–293.

Kumar, S., Rao D.N. (2001). Measuring benefits of pollution abatement using health production function: a case study of Panipat thermal power station. Environmental and Resource Economics 20: 91–102.

Kumar, S., Rao D.N. (2003). Environmental regulation and production efficiency: a case study of thermal power sector in India. Journal of Energy and Development 29(1): 81–94.

Managi, S., Jena P.R. (2008). Environmental productivity and Kuznets curve in India. Ecological Economics 65: 432–440.

Murty, M.N., Kumar S. (2002). Measuring cost of sustainable industrial development in India: a distance function approach. Environment and Development Economics 7: 467–486.

Murty, M.N., Kumar S. (2003). Win-win opportunities and environmental regulation: testing the porter hypothesis for Indian manufacturing industries. Journal of Environmental Management 67(2): 139–144.

Contents

Chapter 1
Introduction

1.1 The Emerging Giant

Economic growth in India will reshape the world. In 1991, India began a series of economic reforms in response to a severe balance of payments crisis. Many of the reforms led to a substantial liberalization of the corporate sector, directly or indirectly, by easing restrictions on firms' activities and enhancing overall competition. Now, India's real gross domestic product (GDP) has increased by an annual average of nearly 9% in the past 5 years.

Initially, the premodern economy is characterized by activities in agriculture- and resource-based activities, low levels of income, and little industrialization. These are fairly low-tech operations, thriving largely on low labor costs. The economic transition to a modern economy is the shift in early industrialization from agriculture to factories. In addition, there is a shift to the large-scale industrial corporation with professional management. Then Indian firms start up a large organization for outsourcing business processes to serve companies around the world. Over time, as have its peers, India has moved into higher-value businesses. As a result, firms in India, as a new champion with China, are becoming increasingly innovative in their business models and in their products.

Goldman Sachs invented the BRICs acronym in 2001, which refers to the fast-growing developing economies of Brazil, Russia, India, and China. The countries are developing so rapidly that Goldman Sachs has argued the four nations as a group will overtake the G7 in 2032. In the revised 2007 figures of the Second Follow-Up Report by Goldman Sachs, based on increased and sustaining growth and more inflows into foreign direct investment, Goldman Sachs predicts that "from 2007 to 2020, India's GDP per capita in US$ terms will quadruple," and that the Indian economy will surpass the Japanese and German economies and be almost the same as the United States (in US$). India's GDP will be the world's third largest, following China and the United States in 2050 (see Table 1.1). India will also overtake China to become the most populous country in the early 2030s.

S. Kumar and S. Managi, *The Economics of Sustainable Development*,
Natural Resource Management and Policy 32,
DOI: 10.1007/978-0-387-98176-5_1,

Table 1.1 BRICs in 2050: gross domestic product

Gross domestic product (2007)			Gross domestic product (2050)[a]		
Rank	Country	GDP (millions of USD)	Rank	Country	GDP (millions of USD)
1	United States	13,843,825	1	China	70,710,000
2	Japan	4,383,762	2	United States	38,514,000
3	Germany	3,322,147	3	**India**	37,668,000
4	China	3,250,827	4	Brazil	11,366,000
5	United Kingdom	2,772,570	5	Mexico	9,340,000
6	France	2,560,255	6	Russia	8,580,000
7	Italy	2,104,666	7	Indonesia	7,010,000
8	Spain	1,438,959	8	Japan	6,677,000
9	Canada	1,432,140	9	United Kingdom	5,133,000
10	Brazil	1,313,590	10	Germany	5,024,000
11	Russia	1,289,582	11	Nigeria	4,640,000
12	**India**	1,098,945	12	France	4,592,000
13	South Korea	957,053	13	South Korea	4,083,000
14	Australia	908,826	14	Turkey	3,943,000
15	Mexico	893,365	15	Vietnam	3,607,000

[a]*Source*: Goldman Sachs (2007)

1.2 The Problems

Many problems still remain to be solved. About one-third of the total population of the country survives on less than US$1 per day. Subsequently, environmental problems are threatening India's sustainable future. Rapid economic growth tends to be detrimental to the environment due to the greater use of natural resources and the higher level of emission of pollutants. Hence, the issue arises of a potential conflict between economic policies and environmental quality.

Policy makers in India are facing tradeoffs between economic growth and environmental protection (see Khanna and Zilberman, 1999, 2001, for recent analyses). Clearly, although local pollution matters to citizens in India, India nevertheless plays a key role in climate change because of the future potential of emissions as a byproduct of economic activity. Additionally, India is considered as environmentally important in other ways. For example, India is recognized as 1 of the 17 "megadiversity regions" of the world and accounts for 67% of the world biodiversity. India has elaborate statutes, regulations, institutional frameworks, and policies on almost every conceivable topic – from hazardous waste to public liability to forests and wildlife. However, monitoring and enforcement capabilities are weak. The complexity and magnitude of environmental problems are increasing at a very high pace. In the control of local pollution, the government relies on traditional command and control policy to reduce the pollution instead of market mechanisms such as tax and emission trading. These potential benefits need to be discussed in future policy implementation in India.

Though India's GDP will be the world's third largest, GDP capita is now ranked outside the world's top 15 countries (see Table 1.2). This indicates potential for economic growth and for potential byproducts from environmental emissions.

Table 1.2 BRICs in 2050: gross domestic product per capita

Gross domestic product per capita (2007)			Gross domestic product per capita (2050)[a]		
Rank	Country	GDP per capita (in USD)	Rank	Country	GDP per capita (in USD)
1	United States	45,790	1	United States	91,683
2	United Kingdom	44,693	2	South Korea	90,294
3	France	41,523	3	United Kingdom	80,234
4	Canada	40,222	4	Russia	78,576
5	Germany	40,079	5	Canada	76,002
6	Italy	35,494	6	France	75,253
7	Japan	34,254	7	Germany	68,253
8	South Korea	19,983	8	Japan	66,846
9	Russia	9,115	9	Mexico	63,149
10	Turkey	8,893	10	Italy	58,545
11	Mexico	8,486	11	Brazil	49,759
12	Brazil	6,859	12	China	49,650
13	Iran	3,815	13	Turkey	45,595
14	China	2,485	14	Vietnam	33,472
15	Indonesia	1,918	15	Iran	32,676

[a]*Source*: Goldman Sachs (2007)

For example, India's per capita energy consumption is very low in comparison to its counterpart country China. In India the per capita energy consumption was only 439 kg of oil equivalent (Kgoe) in 2003 as compared to 1,090 Kgoe in China, 7,835 Kgoe in the United States and the world average of 1,688 Kgoe. Therefore, it is essential to understand the process for reducing emissions while keeping efficiency and activity in a market economy (i.e., the importance of understanding economic growth and efficiency in managing environment).

Economic growth has long been a central issue in modern economics. Relatively little attention, however, was given to the relationship between economic growth and the environment until recent years. Modern growth theories suggest that in a world of finite resources – either manmade or natural – environmental sustainability is potentially not compatible with continuous positive economic growth. Failure to achieve environmental sustainability even becomes an obstacle in achieving long-term economic growth. Given the tradeoffs between environment and development, the issue is not to achieve the maximum economic growth or total maintenance of environment, but is one of arriving at the optimality both in economic progress and environmental protection; the concept of sustainable development may be the guiding force.

The neoclassical growth model assigns little significance to natural resources. In the aggregate production function specification, output (e.g., GDP) is considered as a function of capital and labor, constrained by the prevailing level of technology. The model shows that the rate of economic growth is controlled by the rate of capital accumulation. The phenomenon may continue in the medium term (50–100 years), but the long-term growth is limited by the growth rate of labor force and diminishing marginal returns to capital in the absence of technological progress (Auty, 2007). The recent literature shows that the endowment of two additional forms of capital, natural capital (Sachs and Warner, 1995) and social capital (Acemoglu et al., 2002), play a significant role in a country's economic performance.

Economists are interested in the application of technology in the economy for analyzing the cause of long-run economic growth over time. Endogenous growth theory has been used to analyze economic growth and the environment. There is a set of necessary conditions under which it is optimal to sustain both economic growth and environmental conservation (e.g., Aghion and Howitt, 1998). An industrial sector that does not cause environmental degradation needs to be an engine of economic growth. Optimal pollution regulations become stricter along a sustainable growth path. This increases industry regulations and environmental expenditures and lowers its net marginal productivity of capital. Until this marginal productivity declines to the level of the discount rate of a representative household, economic growth continues. However, if the economy is supported by a clean industry that is a growth engine, then the results are different. The regulated industry can keep productivity higher than the discount rate by increasing input from the clean industry. Therefore, the economy sustains growth on an optimal path accompanied by environmental conservation.

If human capital and knowledge accumulate, long-term economic growth can be sustainable. This is because human capital and knowledge generally do not damage the environment. However, empirical analyses analyzing technological improvements for environmental protections are missing in the literature. Therefore, this book analyzes the performance of the industry in India by considering not only the importance of economic activity, but also the importance of environmental pollution abatements.

1.3 What We Do

This book describes the current status and future prospects for India. We discuss macroeconomic developments, regional disparity and poverty situation, the trend in natural resource depletion and environmental degradation, the trajectory of economic development, and conventional wealth. Then we provide a history of environmental regulations in India and discuss the state of the environment in India and possible reasons for noncompliance of environmental standards in the country. These are discussed in Chapters 2 and 3.

Provision of environmental services involves spatial externalities. The costs of provision are borne at the level of provision, but the benefits are realized on a larger scale. Mismatch between the decision-making responsibilities and costs and benefits have been considered a cause for the underprovision of services. Environmental policy debate predominately focuses on negative externalities and favors pollution taxes, fees, charges, among others, sink for pollutants and ignores positive externalities offered by the natural resources. These externalities can be internalized by compensating the providers of the services. Fiscal transfers are an innovative way of compensating the local and state public actors, i.e., decentralized jurisdictions in federal systems for the environmental services they provide beyond their own boundaries. In Chapter 4 we discuss the possibility of fiscal transfers.

This book studies many different aspects of industries in India from supply and demand sides. Efficiency and productivity analyses especially are provided in detail. Before development to liberalization of the economy started in 1991, India had been one of the most overregulated and closed economies in the world. Up to this point the Central Government's control over industrial development was maintained through public ownership and a license-permit-quota system. Planned industrialization took place in a highly protected environment, which was maintained by high-tariff, nontariff barriers and controls on foreign investment, together forming a set of policy tools that impeded rather than facilitated the growth process of the economy. The new industrial policy introduced in 1991 is considered a watershed event for the Indian economy that shattered this old order. Trade liberalization and deregulation became the central elements. Here it should be noted that the pickup in India's industrial growth precedes the 1991 liberalization by a full decade. In Chapter 5, we examine market productivity and test whether the post-reform period shows any improvement in productivity and efficiency in comparison to the pre-reform period.

The Indian economy today is more prone to industrial pollution and is making compliance decisions to meet environmental standards that involve investments of billions of rupees, i.e., environmental regulations impose significant costs upon industry. These costs are fairly high and therefore require economic justification. The justification can be given by estimating the benefits associated with these costs. We estimate the economic value that people in an urban area in India place on improving air quality in Chapters 6 and 7.

The development of the power sector in India has proceeded so far with little attention paid to its environmental implications. Such a course of development, however, seems difficult to continue in the face of growing degradation of environmental quality and the increasing public awareness of environmental problems in the country. The share of the thermal-power sector is about two-thirds of India's total electricity production. In the thermal-power sector, coal contributes the largest share of fuel consumption. We use the output distance function in Chapter 8 to examine the impact of environmental regulation and pollution abatement on the production efficiency of the Indian thermal-power sector. Also, the factors responsible for inefficiency are estimated by enabling explicitly the impact of environmental regulations and pollution abatement on production efficiency.

Sustainable industrial development requires the preservation of the environment. Industries create a demand not only for waste receptive services from the environmental media – air, forests, land, and water – but also for some material inputs supplied by environmental resources. The demand for environmental services from various economic activities can exceed the natural sustainable levels of supply at a given time; if measures are not taken to reduce this excess demand to zero, there can be degradation of environmental resources. The cost of reducing the demand for environmental services to the natural sustainable level of supply is regarded as the cost of sustainable use of environmental resources and, in the case of industrial demand for environmental services, it is the cost of sustainable industrial development. In Chapter 9, we measure the cost of sustainable industrial development.

Environmental regulation makes firms internalize the costs of environmental externality that they generate. This may result in firms complying with the regulation becoming less competitive in the market than the noncomplying firms. This conventional view about the effects of regulation on the competitiveness of firms has recently been subjected to scrutiny, especially in the context of empirically testing the so-called Porter hypothesis. The objective of Chapter 10 is to study the effect of environmental regulation relating to water pollution by the manufacturing industry on the productive efficiency of firms. The panel data of 92 water-polluting firms are used to test the Porter hypothesis.

Use of water may be broadly classified into three consumption categories: agricultural, industrial, and domestic. While there is substantial literature dealing with the agricultural and domestic uses of water, relatively few have systematically analyzed industrial water use, especially in the context of developing countries. This may partly be due to the lack of reliable information on water consumption at the firm level. There is no consensus on the range of industrial water demand price elasticity and the sensitivity of water demand to other factors such as other input prices and output levels. The question of assessing the economic value (shadow price) of water still remains open. Chapter 11 contributes to the literature on industrial water use by estimating the industrial water demand for a panel of Indian manufacturing firms.

After reaching a 25-year low in February 1999, oil prices have been rising sharply. Given the macroeconomic developments that followed the oil shocks of the 1970s, the substantial rise in oil prices since 1999 generated concerns about the prospects for growth and inflation and integrally related questions about the appropriate way for monetary and energy policies to respond. Much of the empirical literature is concerned with the developed countries, particularly the United States and Western Europe. Chapter 12 is intended to analyze the oil price – macroeconomy relationship by means of applying the vector autoregressive (VAR) approach for Indian economy.

It has been increasingly recognized that technological progress can play a key role in maintaining a high standard of living in the face of these increasingly stringent environmental regulations. However, the extent of the contribution of technological progress depends on how well environmental policies are designed and implemented. Successful environmental polices can contribute to technological innovation and diffusion, while poor policy designs can inhibit innovation. Chapter 13 measures technological/productivity change for environmental (nonmarket) outputs of data of sulfur dioxide (SO_2), nitrogen dioxide (NO_2), and suspended particular matter (SPM) in India using state-level industry data. Then the changes in environmental productivity in different states are linked with their respective per capita income to find an Environmental Kuznets Curve (EKC) type relationship. It is also interesting to see how the performances of environmental managements are changed in India. The comparison to other countries is estimated and discussed in Chapter 14.

Finally, technological progress plays a crucial ameliorating role in reducing energy consumption for combating climate change. Energy economists often cite market-based instruments such as energy taxes for encouraging energy-saving technological progress. Energy policy interventions may change the constraints and incentives that affect technological change. In the earlier literature on energy and environmental policy models, technological change is incorporated as an exogenous variable, i.e., technological developments are autonomous and do not depend upon on policy or economic variables and there is very little empirical evidence on induced technological developments. However, recently some attempts are made to model policy-induced technological changes in the climate–economy models. Chapter 15 extends the literature on induced technological progress by measuring both innovations and diffusion.

Part I
Macroeconomic Development and the Environment

Chapter 2
Economic Development and Environment

2.1 Introduction

The Indian economy is characterized by extraordinary contrasts. On the one hand, it is the fourth largest economy in the world in terms of purchasing power parity. It has been growing at an average rate of more than 8% per annum since 2003–2004, and the annual per capita income is increasing at the rate of about 7%. On the other hand, about one-third of the total population of the country survives on less than US$1 per day. Both sides of the picture are leading to degradation and depletion of the environment and natural resources. Similarly, the country has elaborate statutes, regulations, institutional frameworks, and policies for environmental conservation and preservation. The complexity and magnitude of environmental problems are increasing at a very high pace due to weak monitoring and enforcement and lack of capabilities (Gupta, 2001). These contrasts are posing a question about the sustainability of the present growth trajectory from both economic and environmental points of view.

This chapter intends to provide a critical account of India's development history from independence and, more particularly, after the early 1990s. In 1991, India altered its development strategy from an inward-oriented development path to a path that integrates the economy with the global economy. Note that, though the development strategy has been changed drastically, the regulatory and institutional framework for environmental protection and conservation has not been updated according to the changed scenarios. As a prelude to examining alternatives for environmental regulations, this chapter analyzes recent economic and environmental trends.

The chapter is organized as follows. Section 2.2 shows a picture of the country. Section 2.3 provides an account of the macroeconomic developments in the country since independence. Regional disparity and the poverty situation in the country are discussed in Section 2.4. Section 2.5 depicts the trend in natural resource depletion and environmental degradation that is occurring in the country. Section 2.6 questions the present trajectory of economic development followed in India and

S. Kumar and S. Managi, *The Economics of Sustainable Development*,
Natural Resource Management and Policy 32,
DOI: 10.1007/978-0-387-98176-5_2, © Springer Science+Business Media, LLC 2009

provides estimates of growth rate of per capita genuine and conventional wealth. The last section (Section 2.7) offers some concluding remarks.

2.2 Country Profile

2.2.1 Geographical Profile

Indian civilization is one of the oldest in the World, spanning more than 4,000 years and projecting a unique assimilation of various cultures and heritage. It is a land of spiritual integrity and philosophy. Unity in diversity is its magnificent facet, which had been fused by the feeling of national fervor. Religious tolerance and cultural amalgamation shape its unique national character.

2.2.2 Physiographic Conditions

India, with an area of only 2.4% of the world's total land area, supports around 16.7% of the world's human population and around 18% of the world's livestock population (ESPASSA, 2008). It is the second most populous country in terms of humans and first in terms of cattle population. Its total geographic area is about 329 million hectares. It is situated to the north of the equator. It lies between 8°04′ and 37°06′ N latitude and 68°07′ and 97°25′ E longitude. The Indian Ocean in the south, the Arabian Sea in the west, the Bay of Bengal in the east, and the Himalayas in the north bound it. India's total land area is about 3.3 million km^2. The coastline, encompassing the mainland, Lakshadweep Islands, and the Andaman and Nicobar Islands, is 7,516.6 km. It occupies a major portion of the South Asian subcontinent.

The country's mainland is broadly classified into four regions: the Northern Mountains, which include the great Himalayas; the vast Indo-Gangetic plains; the Southern (Deccan) Peninsula bounded by the Western and Eastern Ghats; and the coastal plains and islands. About 69% of its total geographic area is dry-land (arid, semiarid, and dry subhumid), and the country is divided into 10 biogeographic zones.

India is primarily a tropical country but due to great altitudinal variations almost all climatic conditions from hot deserts to cold deserts exist. There are four seasons: (1) winter (December–February), (2) summer (March–June), (3) southwest monsoon (June–September), and (4) postmonsoon (October–November). The southwest monsoon is the principal rainy season for almost the entire country and contributes almost 80% of the precipitation. The distribution of the southwest monsoon ranges from over 2,500 mm in the western coast and extreme northeastern sector to within 25–50 mm in the extreme tips of the peninsular region. Most of central India receives rainfall of over 1,000 mm, and in the northern plains the rainfall varies between 500 and 750 mm. The mean annual temperature varies from 10 to 28°C. The mean summer and winter temperatures show significant variation in the northern sectors (<10°C); the southern sectors however, show <5°C variation in mean summer and mean winter temperature.

2.2.3 Sociocultural Conditions

The total population of India is 1,027,015,247 persons, comprised of 531,277,078 males and 495,738,169 females as per the census of March 2001. The population has grown at the rate of 1.93% per annum during 1991–2001. The crude birth and death rates according to the 2001 census are 24.8 and 8.9, per 1000 respectively. The sex ratio (i.e., number of females per thousand males) of population was 933, rising from 927 in the 1991 census. The total literacy rate was 65.38%.

The country includes various ethnic groups, religions, and languages. All the five major races – Australoid, Mongoloid, Europoid, Caucasian, and Negroid – find representation among the people of India. As per the 2001 census regarding religion, about 81% of population of the country is Hindu followed by 13.4% Muslims. The followers of other religions such as Christians, Sikhs, Buddhists, Jains, and others also live in the country with dignity. There are 22 national languages that are constitutionally recognized, and Hindi is the official federal language of the country. Besides these languages, about 844 different dialects are practiced in different areas of the country.

2.2.4 Indian Polity and Governance

India obtained freedom on 15 August 1947. India is the largest democracy in the world. It has a multiparty political system. It is governed by a written constitution. Its constitution came into force on 26 January 1950 and since then democracy has been flourishing. The fact is that India has repeatedly been able to mount general elections since it gained freedom from British rule in 1947, on a scale never before been witnessed in history. It has a strong and independent judiciary and press. The Supreme Court of India is the apex body of the Indian legal system, followed by other high courts and subordinate courts. India retained civil society and state institutions that have provided stability.

India is a federal country. It has 28 states and 7 union territories. It has a three-tier federal system of governance, and the responsibilities of governance are shared between the union government, the state governments, and the local governments (rural and urban local bodies). The Indian Constitution allocates the division of responsibilities on all matters between the different tiers of government.

2.3 Macroeconomic Growth

On the eve of independence, 14 August 1947, Pt. Jawaharlal Nehru, the First Prime Minister of India, reminded the nation that the tasks ahead included "the ending of poverty and ignorance and disease and inequality of opportunity." These were the basic foundations that laid down India's development strategy, which has been articulated through the five-year and annual plans put together by the Planning Commission. India's development strategy, prior to 1991, can be summarized in the following points (Srinivasan, 2006a):

1. The commanding heights of the economy were entrusted to the public sector, although a large share of GDP and even larger share of employment were generated by the private sector. Import substitution across the board and industrialization were identified as core strategies to achieve the given objectives of economic growth and poverty alleviation.
2. Given the scarcity of resources, it was thought that state could better utilize its resources. The government appropriated a large share of the savings of the economy for its own use, largely for public investment before 1980s and for public consumption thereafter.
3. India followed the strategy of steering the private sector to conform the priorities and targets of 5-year plans through direct controls. The instruments of controls were quantitative rather than of taxes and subsidies. Moreover, these controls were exercised on a discretionary, case-by-case basis, rather than through a set of rules. Producers were insulated from domestic and international competition.

During the first three decades of planning during 1950–1980 the GDP increased at the rate of 3.75% per year. In that period per capita GDP increased slightly more than 1.5% per year from a situation of literary stagnancy during the colonial period. A closer look at the performance of the economy reveals continuing growth in terms of GDP during the first three five-year plans. The growth was interspersed during the period of 1965–1980 in general and during the decade of the 1970s in particular. GDP grew 4.1% between 1950–1951 and 1964–1965, 3.1% between 1964–1965 and 1980–1981. Table 2.1 explains the macroeconomic growth story of India in terms of selected indicators since 1950–1951.

A massive balance of payment crisis emerged in 1966, headed by a drought in 1965 and resulting need for heavy imports of food. India approached the IMF and World Bank for financial assistance, and as usual their condition was liberalization of the economy. But due to internal political compulsion, the controls were intensified rather than liberalized. Government intervention in agriculture to support the green revolution through various subsidies and price controls increased and was perpetuated by vested interests.

The 1980s witnessed relatively higher growth rates of GDP and per capita GDP.

During the period of 1980–1981 to 1990–1991, the GDP grew at the rate of 5.6% per year. The population growth rate declined from about 2.2% per year from 1950–1980 to 2% per year during the 1980s. Per capita GDP growth rate during the

Table 2.1 Selected macroeconomic indicators for India (%)

	1950s	1960s	1970s	1980s	1990–1991	1991/1992 to 1997/1998	1996–1997 to 2003/2004	2002/2003 to 2006/2007	2007–2008 (AE)
Real GDP growth	3.6	4.0	2.9	5.6	5.3	5.7	5.2	8.7	8.7
Agriculture and allied	2.7	2.5	1.3	4.4	4.0	3.7	0.9	4.9	2.6
Industry	5.8	6.2	4.4	6.4	5.7	7.0	4.1	8.3	8.6
Manufacturing	5.8	5.9	4.3	5.8	4.8	7.5	3.9	9.1	9.4
Services	4.2	5.2	4.0	6.3	5.9	6.4	7.8	10.2	10.6
Real GDCF/GDP	12.5	16.9	19.4	20.2	24.4	22.5	24.1	31.4	NA
ICOR	3.5	4.3	6.6	3.6	4.6	4.0	4.6	3.6	NA
Nominal GDCF/GDP	10.8	14.3	17.3	20.8	26.0	23.9	24.5	33.0	NA
GDS/GDP	9.6	12.3	17.2	19.0	22.8	22.7	24.1	32.7	NA
Saving-investment gap/GDP	−1.2	−2.0	−0.1	−1.8	−3.2	−1.2	−0.4	−0.3	NA
WPI inflation (average)	1.2	6.4	9.0	8.0	10.3	9.6	4.6	5.5	4.1

Source: Mohan (2008)

decade doubled relative to the last three decades. Moreover, in the 1980s, all major sectors saw higher growth rates. Industrial growth was 6.3% per year, manufacturing growth was 6.7% per year, agriculture grew by 3.2% per year, and the growth rate in services was 6.7% per year.

Note that during the 1980s, and more particularly in the second half of the decade, the process of relaxation of controls started, i.e., a process of relaxation of controls here and an increase in incentives of subsidies there was the strategy of the decade (Srinivasan, 2004). The government deserted the fiscal prudence of the previous three decades and, borrowing at home and abroad, financed the fiscal deficits. The gap in public sector savings and expenditure widened from about −3.7% of GDP in the first three decades to more than −8.2% of GDP in 1990–1991. This debt-led growth story ended in a macroeconomic and balance of payment crisis in 1991. At the time of crisis, the foreign exchange reserves were not enough to support imports for even two weeks, the external debt was several times that of the reserves, and the fiscal deficit of the central government and inflation were in double digits.

In 1991, a new economic policy was introduced with a view to bring about major changes in the economy's structure and policy regime. Ray (1997a) articulates the main objective of the new economic policy as (1) integration of India's local economy with the global economy through a predominant and dynamic private sector free from unnecessary state controls in most spheres of economic activities, (2) having a dynamic external sector where there is virtual free trade and free flow of investments to and from India, and (3) state intervention in the production of goods and services only where externalities make private sector operations very inefficient in addition to the normal range of public goods like defense, law, and order, among others. However, the state must play an active role in relieving extreme poverty and ensure access to basic needs like, health and education by the poor.

As a result of the new economic policy in 1991, the growth impulses appeared to have gathered momentum. The rate of growth in GDP rebounded from 1.5% in 1991–1992 to about 7.8% in 1996–1997. Subsequently the growth rate fluctuated until 2002–2003. Since 2003–2004, there has been a distinct strengthening of the growth momentum, and in the last two years it has averaged to about 9.5% per year (Mohan, 2008).

To understand the macroeconomic growth story it is interesting to analyze saving and investment trends. From the data one finds that gross domestic savings (GDS) have increased continuously from an average of 9.6% of GDP during the 1950s to about 35% of GDP in 2006–2007. Similarly, over the same period, gross domestic capital formation (GDCF, investment) has increased from 10.8% of GDP during the 1950s to about 36% of GDP in 2006–2007. The noticeable feature of these trends is that Indian economic growth is financed predominately by domestic savings (Mohan, 2008). The share of foreign savings in the financing of India's growth, which also can be termed as current account deficit, is very modest, and when it tried to reach about 2% of GDP, the economy faced a severe balance of payment crisis, as can be observed from the experience of 1960s and 1980s.

However, the long-term upward trends in savings and investment have been scattered with phases of stagnation. In India's growth history, it is also worth noticing that capital resources have been employed productively. Except for the decade of 1970s, the incremental capital output ratio (ICOR) has stayed close to about four (Mohan, 2008).

Post-reform macroeconomic performance can be attributed to again adopting fiscal prudence. During that period, public investments were used to try to keep the fiscal deficit under control. The data on public and private investments reveal that public investment crowds private investment in India, thereby raising questions about the long-run sustainability of the growth process. Note that public investment has started to increase since 2003–2004, reversing the declining trend that started in mid-1980s.

So far we have discussed the aggregate story. It would be interesting to see the sectoral composition of growth dynamics in this context. In India, the agriculture sector employs about 56% of the total labor force. The share of the sector in terms of employment was about 70% at the time of independence. However, the share of the sector in terms of GDP has been continuously declining from more than 50% in 1950–1951 to less than 20% in 2006–2007. Table 2.1 also provides the annual growth rates of different sectors of the economy since 1950–1951.

In the preceding discussion, we observe that both before reforms and after reforms the major policy initiatives were limited to the industrial sector. The industry-first approach was taken throughout the period of India's development. The desired objective of higher and sustained GDP growth rates can be achieved by reversing the balance between public and private ownership, and by opening up industry to foreign investment, international trade, and competition so that industry can perform its expected role as the leading sector. It is also thought that higher growth in the industrial sector can help in transferring labor from agriculture to industry. In this scenario, the notion that the agricultural or rural sector can play the leading role is not entertained (Kalirajan and Sankar, 2001).

Note that the government recognized the need for agricultural reform and its importance from the very beginning of reform. The Ministry of Finance's Discussion Paper on economic reforms (1993) proclaims that: "No strategy of economic reform and regeneration in India can succeed without sustained and broad-based agricultural development." It sets out the critical areas for reform, which include reduction of input subsidies, restructuring of public investment on agriculture, upgrading of quality of research and extension services, resurrection of private investment in the sector, strengthening of the institutional credit system, and land reform in several states. However, no major policy reform initiatives were taken for the sector. For example, subsidized supply of chemical fertilizers encouraged farmers to substitute chemical fertilizers for organic manures. As a result, the share of organic manure in the value of intermediate inputs, at 1980–1981 prices, fell from 8.7% in 1960–1961 to 3.1% in 1995–1996. This substitution, along with overuse of groundwater for irrigation (because of extremely low prices for electricity with zero marginal prices for kilowatt hour of energy in many states), heightened the environmental problems (Kalirajan and Sankar, 2001).

The historical review of India's growth story also reveals that periods of slow overall growth have invariably been characterized by slow agricultural growth even in the years when the weight of agriculture in GDP has decreased remarkably. Despite the recent improvements in the agriculture sector, given the dependence of Indian agriculture on monsoons, the immediate need is to make improvement in irrigation facilities through public investment coupled with institutional support (Mohan, 2006).

Moreover, the structural composition of the economy shows that the share of the industrial sector in GDP was about 24% and the share of the service sector was about 56% in recent years. This skewed structural growth should not be considered an indicator of the maturity of the economy, but mainly a lack of industrialization and infrastructural development (Sengupta, 2006).

India's per capita energy consumption is very low in comparison to its counterpart country China. In India the per capita energy consumption was only 439 kg of oil equivalent (Kgoe) in 2003 as compared to 1,090 Kgoe in China, 7,835 Kgoe in the United States and the world average of 1,688 Kgoe. The *National Action Plan on Climate Change* (Government of India, 2008) shows that over the period of time energy intensity measured in terms of energy requirement to produce one unit of GDP was continuously declining, and in 2005 it was 0.16 – much lower than the developed and counterpart developing countries (see Fig. 1.3.2 in the Action Plan document). However, the figure of energy intensity should not be taken at face value. The figures are an indicator of inadequate access to commercial energy by a large section of the population and inadequate development of infrastructure and industrial sectors (Sengupta, 2006). Table 2.2 presents the data on selected energy indicators for 2003.

The other area that requires immediate attention in India is the infrastructure sector. According to the Planning Commission, to achieve 9% per year growth during the 11th Five Year Plan the infrastructure investment ought to grow at the rate of 8% per year from the prevailing level of about 4.6% per year. This implies that the government should take utter care so that investment in infrastructure from both public and private sectors comes forward in the desired direction and magnitude.

2.4 Poverty and Regional Disparities

Notwithstanding its recent macroeconomic performance, India is still among the poorest countries in the world. A quick look at the figures for population below the poverty line using the head count ratio, a most important development indicator, shows that in India about 300 million people survive on less than US$1 per day in 2000. About 25% of the national population earns less then US$0.40 per day. Per the report of the National Commission of Enterprises in the Unorganized Sector (NCEUS), 77% of Indians live on less than Rupees 20 per day. Moreover, India has very high rate of malnutrition among children under the age of 3.

Table 2.2 Selected energy indicators for 2003

Countries	GDP per capita (PPP $2,000)	Poverty ratio (national poverty line)[a]	TPCES intensity of GDP (Kgoe/PPP 2,000$)	Electricity intensity of GDP (Kwh/PPP 2,000$)	Share of industry in GDP (%)	TPCES per capita (Kgoe)	Electricity consumption per capita (Kwh)
China	4,838	4.6	0.23	0.29	53	1,090	1,379
India	2,732	28.6	0.16	0.20	26	439	553
USA	35,487	–	0.22	0.37	23	7,835	13,066
World	8,180	–	0.21	0.31	28	1,688	2,429

TPCES total primary commercial energy supply
Source: Sengupta (2006)
[a]The figure for India is from National Sample Survey 1999–2000 and for China the figure for 1998

Poverty can be defined as a state in which individuals or groups of people are unable to satisfy the basic needs of life. Since poverty is a very contentious issue, different countries have varied definitions and approaches for measuring it. As per the Planning Commission in India, the poverty line in rural areas is drawn with an intake of 2,400 calories in rural areas and 2,100 calories in urban areas. Moreover, there are issues relating to price indices for updating the poverty lines. Note that in the 1990s the designs of household expenditure surveys changed, and as a result of that a problem of potential noncomparability over the surveys has emerged. Given all these issues, the researchers have had to make several strong assumptions in the measurement of poverty; they obviously differed in their methodologies and reached varying conclusions about the estimates of the population below the poverty line in the country.

The Planning Commission has estimated that 27.5% of the population was living below the poverty line in 2004–2005, compared to a poverty rate of 51.3% in 1977–1978 and 36% in 1993–1994. As noted earlier, researchers have provided differing estimates of the population below the poverty line; however, it is fair to say that all of them agree that the poverty ratio did not increase in the 1990s and differ only on the rate of decline and whether the rate of decrease was higher in 1980s or 1990s. For example, two recent papers by Dev and Ravi (2007) and Himanshu (2007) have analyzed recent trends in poverty and inequality and have come to broadly similar conclusions. That is, the pace of poverty reduction accelerated (sharply according to Himanshu) between 2000 and 2005 relative to the reduction between 1994 and 2000, but Sundram (2007) found that in terms of persons, with the Planning Commission poverty lines, in rural India, the head count ratio (HCR) declined by 4.8% points or 0.8 points per year or at 2.7% per annum between 1994 and 2000 and by 0.9 points per year or at 3.4% per annum between 2000 and 2005, indicating a small increase in the pace of poverty decline in the first 5 years of the 21st century. In urban India, however, in terms of HCR for persons, Sundram finds a clear slowdown – from 0.78 points per year between 1994 and 2004 to just 0.3 points per year between 2000 and 2005. But using an alternative poverty line, Sundram (2007) finds that a slightly faster pace of poverty reduction between 2000 and 2005 is reversed, with a small reduction in the pace of poverty reduction from 2.8% per annum to 2.5% per annum. His estimates with alternative poverty lines also reinforce the result of a slower reduction in urban poverty between 2000 and 2005 relative to that between 1994 and 2000. Table 2.3 presents the poverty estimates provided by Sundram. His estimates are based on the household expenditure survey data related to 55th and 61st round of National Sample Survey Organization (NSSO).

Poverty in terms of income or consumption does not express the true picture of destitution. In addition, poverty can be looked at as having different dimensions, viz., UNDP's Human Development Index (including health, access to nutrition and water, life expectancy, and education, among other factors), social exclusion, marginalization, etc. These all in one way or other are linked with the environment and natural resources. Unaccounted for benefits, which singly or in combination ecosystems provide, are the means for obtaining adequate nourishment, avoiding

Table 2.3 Estimate of head ratios of households and persons with planning commission and alternative poverty lines: all-India: 1993–1994 to 2004–2005 [head count ratio (%)]

	Households			Persons		
	1993–1994	1999–2000	2004–2005	1993–1994	1999–2000	2004–2005
Planning commission's definition of poverty lines						
Rural	28.0	23.3	18.8	31.8	27.0	22.7
Urban	22.7	18.1	16.6	28.1	23.4	21.9
Alternative definition of poverty lines						
Rural	30.3	25.1	21.7	34.2	28.9	25.5
Urban	21.3	17.8	17.4	26.4	23.1	22.8

Planning commission poverty lines: 1993–1994: rural, 205.84; urban, 281.33, 1999–2000: rural, 327.56; urban, 454.11; and 2004–2005: rural, 356.30; urban, 538.60

Alternative poverty lines: 1993–1994: rural, 211.30; urban, 274.88; 1999–2000: rural, 335.46; urban, 451.19; and 2004–2005: rural, 371.29; urban, 546.20

Note 1. Alternative poverty lines have been updated by reference to CPI for agricultural laborers for rural India and CPI for industrial workers for urban India

Note 2. All estimates for 1993–1994 are on mixed reference period and estimated from Unit Record Data

Source: Sundram (2007)

Table 2.4 High poverty states of India

State	% of India's poor in 1999–2000	% of population in 2001
Uttar Pradesh (including Uttaranchal)	20.4	17
Bihar (including Jharkhand)	16.4	10.7
Madhya Pradesh (including Chhattisgarh)	11.5	7.9
Maharashtra	8.8	9.4
West Bengal	8.2	7.8
Orissa	6.5	3.6
Total	71.8	56.4

Source: ESPASSA (2008)

diseases, acquiring clean and safe air, water, and shelter and many other sociocultural activities.

One striking fact in the post-reform growth of India is that there was significant widening of regional disparities in growth and poverty reduction across states. For example, in the early 1960s the per capita Gross State Domestic Product (GSDP) of richer states such as Punjab, Maharashtra, and Gujarat was, on average, about 80% higher than the average per capita GSDP of the bottom four states, viz., Bihar, Uttar Pradesh, Orissa, and Madhya Pradesh. This disparity increased to 125% by the early 1970s. During the 1980s, the disparity between the states marginally declined to 100%; however, it jumped to 200% toward the end of the 1990s (Rao, 2008).

The majority of India's poor continue to be located in rural areas despite a declining trend in official income-based poverty estimates. State-wise, nearly 72% of India's poor and half of her population are located in the following six states: Uttar Pradesh (including Uttaranchal), Bihar (including Jharkhand), Madhya Pradesh (including Chhattisgarh), Maharashtra, West Bengal, and Orissa (see Table 2.4).

In the official data on the below-poverty-line population for the year 1993–1994, seven states – Bihar, Orissa, Uttar Pradesh, Madhya Pradesh, Maharashtra, Assam, and West Bengal – had a poverty ratio (% of population in poverty) in excess of the all-India average for rural areas (37.2%). Not only is the distribution of poverty spatially uneven in India, but the gap in terms of poverty incidence between the poor and the affluent states in the country is growing over time (ESPASSA, 2008).

Table 2.5 provides data on the incidence of rural and urban poverty at three points in time for the five highest and lowest poverty states. This table reveals that there is considerable stability over the three points of time in which states happened to have a high or low poverty ratio (Srinivasan, 2004).

Sengupta (2006) describes regional disparities in terms of development indicators, including energy consumption. In his paper, annexure tables 2.2–2.6 show the variation in the pattern of energy consumption in terms of the share of household with electrical connections, the share of fuel in average monthly per capita expenditure, the share spent on efficient commercial energy for lighting and cooking across the states, and how the variation is linked to the poverty incidence in a state.

Table 2.5 Regional disparity in poverty

High poverty states				Low poverty states			
Rural		Urban		Rural		Urban	
	1987–1988				1987–1988		
Orissa	58.7	Bihar	51.9	Punjab	12.8	Himachal Pradesh	7.2
Bihar	53.9	Karnataka	49.2	Haryana	15.3	Assam	11.3
West Bengal	48.8	Madhya Pradesh	47.3	Himachal Pradesh	16.7	Punjab	13.7
Tamil Nadu	46.3	Uttar Pradesh	44.9	Andhra Pradesh	21.0	Delhi	15.1
Madhya Pradesh	42.0	Orissa	42.6	Gujarat	28.6	Haryana	18.4
	1993–1994				1993–1994		
Bihar	58.0	Madhya Pradesh	48.1	Punjab	11.7	Assam	7.9
Orissa	49.8	Orissa	40.6	Andhra Pradesh	15.9	Himachal Pradesh	9.3
Assam	45.2	Tamil Nadu	39.9	Gujarat	22.2	Punjab	10.9
Uttar Pradesh	42.3	Karnataka	39.9	Kerala	25.4	Delhi	16.1
West Bengal	41.2	Andhra Pradesh	38.8	Haryana	28.3	Haryana	16.5
	1999–2000				1999–2000		
Orissa	47.8	Orissa	43.5	Punjab	6.0	Himachal Pradesh	4.6
Bihar	44.0	Madhya Pradesh	38.5	Haryana	7.4	Punjab	5.5
Assam	40.3	Bihar	33.5	Himachal Pradesh	7.5	Assam	7.5
Madhya Pradesh	37.3	Uttar Pradesh	30.8	Kerala	9.4	Delhi	9.2
West Bengal	31.7	Andhra Pradesh	27.2	Andhra Pradesh	10.5	Haryana	10.0

Source: Srinivasan (2004)

These tables not only explain the disparities across states but also indicate the challenge the country will face in the future if the transition from traditional to commercial and cleaner fuels is to be completed in the next decade or so. The *National Action Plan on Climate Change* shows that there is a positive and significant relationship between per capita commercial and clean energy fuels and the human development index using the cross-country figures for the year of 2004 (Fig. 1.2.1 in the Action Plan, Government of India, 2008).

This spatial variation in incidence and depth of income poverty is the outcome of a highly uneven performance by the states of India in reducing poverty over time. The factors identified as having contributed to poverty reduction include favorable initial conditions of human and physical resource development as well as equitable access to physical and human infrastructure (Datt and Ravallion, 1998).

2.5 Depletion of Natural Resources and Environmental Degradation

Inclusion of the environment in conventional two-sector macroeconomic models comprising production and consumption activities reflects interaction between economic activities and the environment. The environment contributes to economic activities in three ways: by providing direct consumption goods such as some food items for consumption and raw material for various production activities; by providing sink facilities to absorb the waste generated during the various economic activities; and by providing various amenity values that add to human welfare or utility. Note that these functions performed by the environmental sector are interlinked. For example, deforestation for the purpose of consumption and production activities reduces the availability of the resource for other purposes. The sink facilities provided by the forests get reduced and amenity values are also decline. That is, like manmade resources the environment is also a scarce resource.

In Section 3, it is observed that since the inception of the development planning process, India has been able to maintain some continuing positive growth rate in aggregate as well per capita income and has been progressing at an impressive pace since 1991, when it changed its development strategy. In a world of finite resources, the present pattern of growth is consistent only with the abundant availability of natural resources, but as growth puts more strain on resources, environmental scarcity increases and raises questions about the sustainability of the growth trajectory.

Recognizing that environment is an important factor input both in consumption and production activities, it is useful to provide an account of the resource damage that is occurring in the country over the period of time in terms of monetary values. The presentation in monetary values helps in formulating rational economic policy.

The *World Development Indicators* (WDI) provides estimates of the damage caused since 1970 by the depletion and degradation of various exhaustible and renewable resources in terms of monetary values and as a proportion of national income. The estimates are available for the extraction of various minerals and metals, energy resources such as oil and gas, coal and lignite, deforestation, and for atmospheric degradation. In the last category the estimates are available for the damages caused by CO_2 emissions and by particulate matter.

Figure 2.1 summarizes estimates of the damages since 1970 as a proportion of gross national income from the depletion of exhaustible resources such as mineral, oil and gas, and coal resources, depletion of forest resources and damages due to the proxy of global pollutant CO_2 emissions produced in the country and a local pollutant particulate matter. The figure shows that the rate of depletion of mineral resources is increasing in the twenty-first century. Similarly, the rate of depletion of energy resources is also increasing. With respect to both of the exhaustible resources, damages as a proportion of gross national income (GNI) are increasing, suggesting that the intensity of resource use of the economy is increasing.

The WDI provides data on the damages due to particulate matter (PM) since the early 1990s. In Figure 2.1, it may be observed that damages as a proportion of GNI are diminishing over the period of time. This may be because in India there is a comprehensive program of environmental regulation which focuses generally on atmospheric pollution related to local pollutants; in addition, environmental awareness in the public in general and the urban public in particular is increasing. A similar trend may be observed with respect to CO_2 emissions, and this may be due to declining energy intensity in the economy (Table 2.2).

In India, the land under forest cover is about 18% of total land area; actual forest cover with crown density is only about 11%. Over a period of time, there has been considerable depletion of forest cover in the country. Between 1995 and 1997, more than 17,000 km^2 of forestland was lost. Concerned about these losses, the Supreme Court in 1996 directed that all ongoing activities in any forest area in any state should be stopped forthwith (Mandal and Rao, 2005). It is interesting to note that the depletion in forest resources measured as a proportion of GDP is declining over

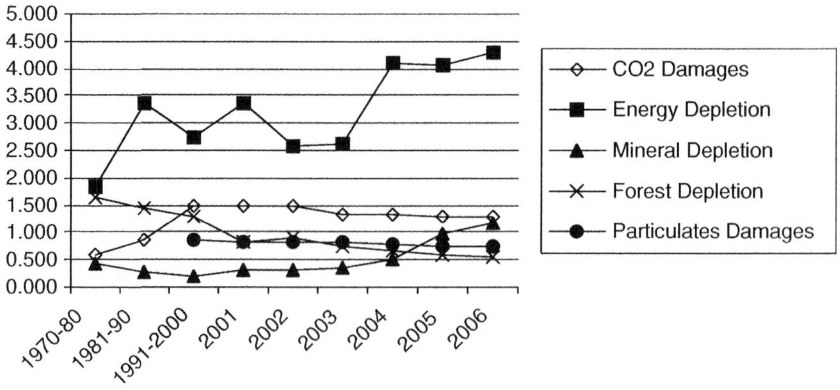

Fig. 2.1 Natural resource damages (% of GNI)
Source: authors' calculations based on WDI data

the period of time and increases momentum in the 1990s. This may be attributed to judicial intervention and subsequent state alertness with respect to the national wealth.

Figures 2.2–2.6 present resource damage as a proportion of GNI with respect to per capita income. It is thought that the fundamental reason behind the depletion and degradation of natural resources is economic activity; thus per capita income may be considered as the best proxy for production and consumption activities. As the economy is growing, the relative damage with respect to particulates and deforestation is declining (Figs. 2.5 and 2.6).

Figure 2.2 shows that there is an inverted U-shaped relationship between damages as a proportion of GNI and per capita income in India, an environmental Kuznets curve (EKC) type of relationship. The EKCs explain that in the beginning of economic development, environmental damage increases, but as the economy

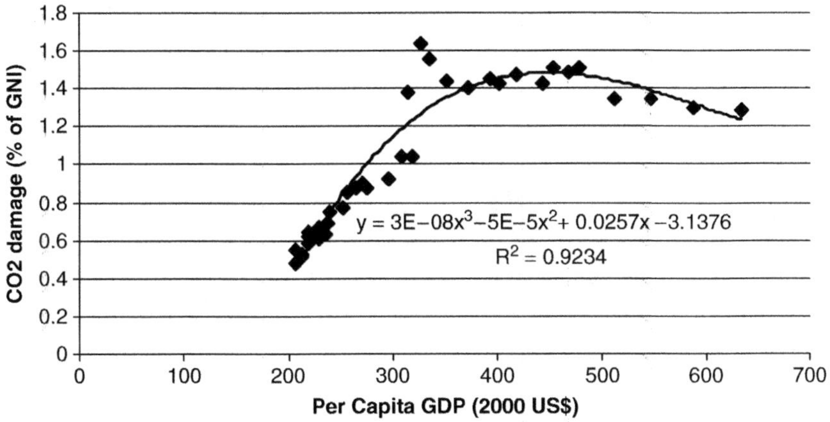

Fig. 2.2 CO_2 emissions damage versus per capita GDP
Source: authors' calculations based on WDI data

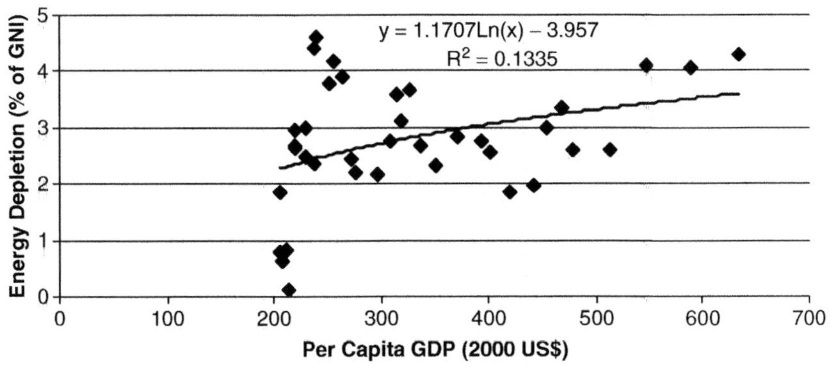

Fig. 2.3 Energy depletion versus per capita GDP
Source: authors' calculations based on WDI data

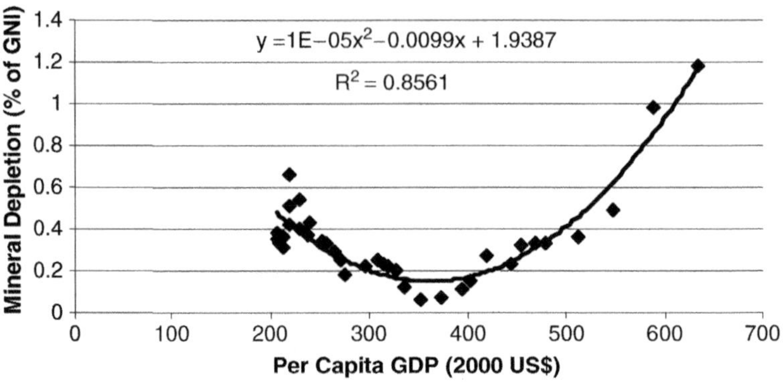

Fig. 2.4 Mineral depletion versus per capita GDP
Source: authors' calculations based on WDI data

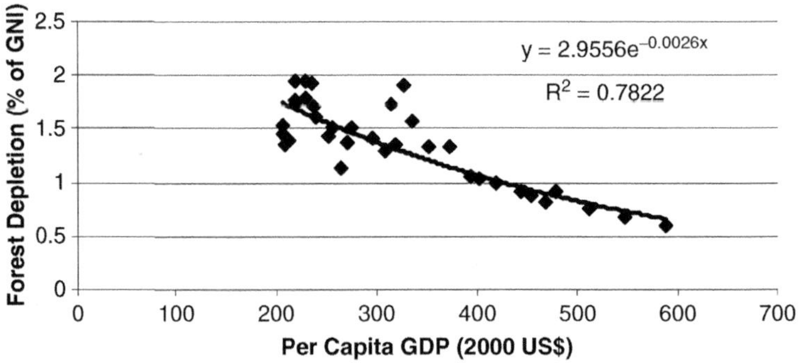

Fig. 2.5 Forest depletion versus per capita GDP
Source: authors' calculations based on WDI data

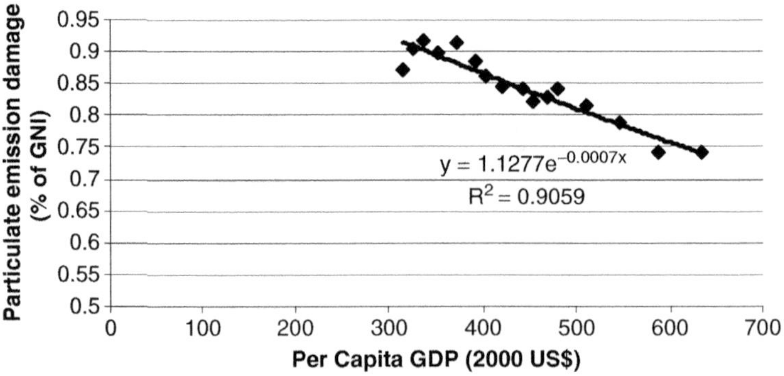

Fig. 2.6 Particulate emissions damage versus per capita GDP
Source: authors' calculations based on WDI data

matures the damage starts to decline. Note that the EKC are tried to observe between absolute damages or pollution/emission levels and per capita income. On the basis of the figure it may be argued that in India the carbon intensity of the economy increased at the initial levels of development; that is, the carbon intensity of the economy increases until the per capita gross income reaches about US$450 (at year 2000 prices) and after that it starts to decline. India attains the turning point at a much lesser level of per capita income, not only in comparison to her counterpart developing countries but even to most of the developed countries. Note that it is possible with the increasing CO_2 emission in absolute terms, however, it reflects a responsible behavior of the Indian economy toward the global problem of climate change given her development priorities.

As discussed earlier, the resource damage intensity in the economy, measured as a ratio of depletion of mineral and energy resources to GNI, is increasing not only over the period of time but with respect to per capita GDP also, as can be observed from Figures 2.3 and 2.4. This implies that as the per capita income is increasing, the depletion of exhaustible resources is increasing at a faster rate than the economy.

Regarding WDI data related to resource damages, two points are worth mentioning. First, these estimates do not include water resources and their degradation, forests as agents of carbon sequestration, fisheries, land degradation, and biodiversity loss. Moreover, the data include only two air pollutants and ignore all others. In developing countries like India, indoor air pollution is much more damaging. Second, estimates of damages were measured at the market price of the resource and, as is well established, the market prices of the resources do not reflect their true shadow prices. As a result the damage estimates understate the true damages due to resource depletion and degradation.

Land degradation is one of the major environmental problems in India. It occurs through the natural and manmade processes of wind erosion, water erosion, and water-logging. The result of such degradation is the loss of invaluable nutrients and lower food production. Poor land use practices and management are prime factors in rapid land degradation. It is estimated that about 57% of the total land is experiencing some form of degradation. The business-as-usual scenario estimates that India would lose about 40 million tonnes of major soil nutrients annually (Pachauri, 2004).

India is recognized as one of the 17 "megadiversity regions" of the world and accounts for 67% of the world biodiversity. Loss of biodiversity is a significant issue to India, since many plant and animal species are severely threatened due to destruction of their habitats and an overexploitation of resources. A large number of species are either endangered or on the verge of extinction. According to SACEP, India has 47,000 species of flowering and nonflowering plants representing about 12% of the recorded world's flora. Of these, 5,150 are endemic, 2,532 species are found in the Himalayas and adjoining regions, and 1,782 are found in peninsular India.[1]

[1] http://www.sacep.org/html/mem_india.htm, as accessed on 14 July 2008.

The availability of fresh water is going to be the most pressing problem in India over the coming decades. Urban growth, increased industrial activities, intensive farming, and the overuse of fertilizers and other chemicals in agricultural production have put more stress on water resources. Untreated water from urban settlements and industrial activities and runoff from agricultural land carrying chemicals are primarily responsible for the deterioration of water quality and the contamination of lakes, rivers, and groundwater aquifers. India receives an average annual rainfall equivalent of about 4,000 billion cubic meters (BCM). This sole source of water is unevenly distributed both spatially and temporally. The rivers of India face serious pollution problems. The quality of surface and groundwater has deteriorated significantly over the last two decades. The water quality of most of the rivers in India is not even fit for bathing, recreation, or the other social uses that have endured for thousands of years. High arsenic concentrations have been recorded from a large number of rural wells in West Bengal, India.

Increasing amounts of untreated hazardous waste are becoming a serious environmental issue in India. The waste is generated by various industrial processes, mining extraction, tailings from pesticide-based agricultural practices, and urban households. The largest quantities of hazardous waste are generated by the following industries: petrochemicals, pharmaceuticals, pesticides, paints and dyes, petroleum, fertilizers, asbestos, caustic soda, inorganic chemicals, and general engineering. The rate of generation of solid waste in urban centers has outpaced population growth in recent years with the wastes normally disposed of in low-lying areas of the city's outskirts. Daily waste generation in India varies between 0.45 and 0.89 kg/capita. According to SACEP, at present, around 7.2 million tonnes of hazardous waste is generated in the country, of which 1.4 million tonnes are recyclable, 0.1 million tonnes are incineratable, and 5.2 million tonnes are destined for disposal on land.

The increasing resource intensity of economy activities associated with environmental degradation is creating doubts about the sustenance of present growth trajectory of the economy.

2.6 Sustainability of Growth

The distortions created through the degradation and damage of our natural resource wealth perhaps impose a higher burden than any other form of distortion and present a serious challenge in achieving healthy and sustainable progress (Pachauri, 2004). Apprehensions about this trend have been further fueled by concerns related to the adverse impacts of climate change.

Modern growth theories suggest that in a world of finite resources – either manmade or natural – environmental sustainability is potentially not compatible with continuous positive economic growth. Failure to achieve environmental sustainability even becomes an obstacle in achieving long-term economic growth. Given the tradeoffs between environment and development, the issue is not to achieve maximum economic growth or total maintenance of environment, but is to achieve optimality both in economic progress and environmental protection. The concept of sustainable development may be the guiding force.

The neoclassical growth model, which has dominated mainstream economic growth theory since the second half of the last century, ignores the role of natural resources. In the aggregate production function specification, output (e.g., GDP) is considered as a function of capital and labor, constrained by the prevailing level of technology. The model shows that the rate of economic growth is controlled by the rate of capital accumulation. The phenomenon may continue in the medium term (50–100 years), but long-term growth is limited by the growth rate of the labor force and diminishing marginal returns to capital in the absence of technological progress (Auty, 2007). The recent literature shows that the endowment of two additional forms of capital, natural capital (Sachs and Warner, 1995) and social capital (Acemoglu et al., 2002), play a significant role in a country's economic performance.

Though a complete operationalization of sustainability or achieving optimality is not possible, adopting wealth, which comprises all forms of capital – physical, social, and natural – as indicators of economic well-being for an economy, implies that sustainable development requires the creation of wealth, or at a minimum, requires that the economy's wealth does not decline over the period of time (Dasgupta, 2001). Wealth or capital is an accounting value of a country's assets. Change in capital stock is known as investment and genuine investment, I_t is defined as follows. Generally investment is equal to saving:

$$I_t = \sum (P_{it} \cdot \mathrm{d}M_{it}/\mathrm{d}t) + \sum (h_{jt} \cdot \mathrm{d}H_{jt}/\mathrm{d}t) + \sum (n_{kt} \cdot \mathrm{d}N_{kt}/\mathrm{d}t), \qquad (2.1)$$

where M_{it} is the quantity of ith manufactured asset; H_{jt} is the jth form of human capital; N_{kt} is the kth form of natural capital, and P_{it}, h_{jt}, and n_{kt} are, respectively, the accounting prices of manufactured, human, and natural capital.

Since 1999, the World Bank has been publishing estimates of genuine savings. Genuine savings are the adjusted estimates of saving adjusted not only for depreciation of manufactured capital, but also for the depletion of exhaustible resources such as minerals and hydrocarbons, renewables such as forests coupled with atmospheric resources such as emissions of carbon and particulate matter. Adjustments are also made for education expenditures. In making these estimates some crude assumptions are made with respect to prices of natural capital, estimation of natural resource rents, etc.[2] Genuine saving is calculated as gross saving plus education expenditure minus the value of depletion of natural capital and damage due to atmospheric pollution.

Figure 2.7 shows the various estimates of savings in India since 1970. The gross saving in 1970 was about 15% of GNI, and the estimates of genuine saving were about 8%. The estimates of gross saving were increasing during the 1970s and 1980s. At the beginning of the 1990s these estimates were around 22% of GNI, but the estimates of genuine savings remained almost unchanged, hovering around 8%

[2]For details on these assumptions see Hamilton and Hassan (2003).

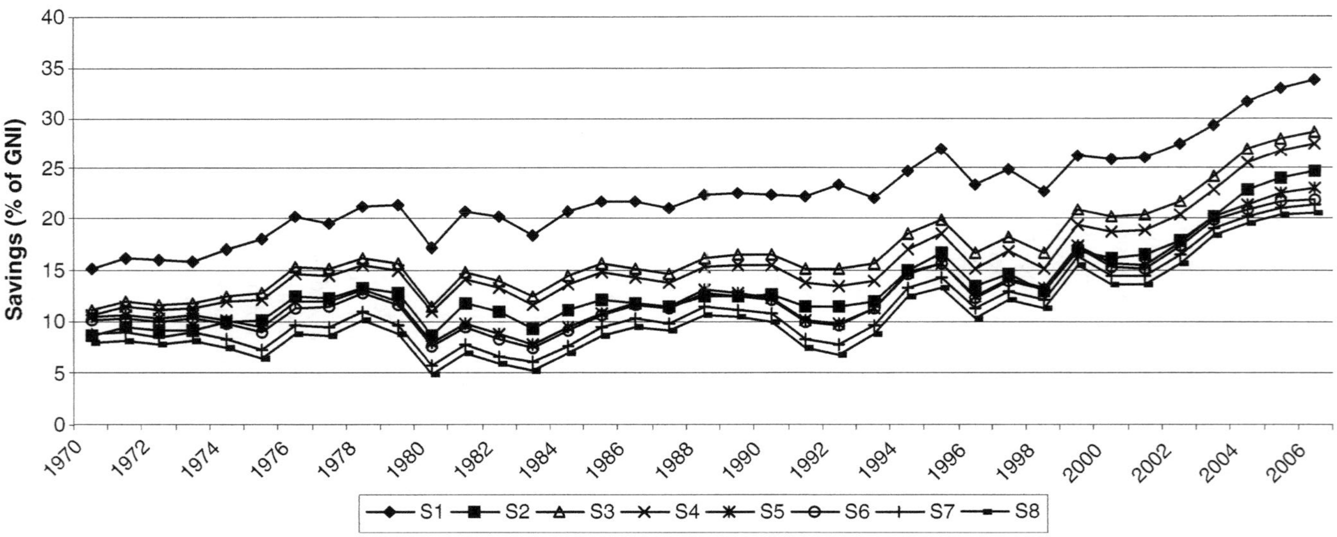

Fig. 2.7 Measures of genuine savings or investment as percentage of GDP
Note: S1: gross saving; S2: S1 – depreciation of manufactured capital; S3: S2 + education expenditure; S4: S3 – CO_2 emissions damages; S5: S4 – energy depletion; S6: S5 – mineral depletion; S7: S6 – forest depletion; and S8: S7 – particulate emission damages
Source: authors' calculations based on WDI data

of GNI. As explained earlier, we find that in 2006 the estimates of gross saving reached at the level of 33% of GNI and genuine saving estimates were about 21%. The growth in genuine savings may be caused by various factors, such as change in the development strategy in 1991 that caused an increase in gross savings and an increase in education expenditures, a complete ban on green felling in 1996 by the Supreme Court of India, the declining carbon intensity of the economy and reduction in particulates due to the introduction of CNG in public transport in Delhi, improvement in environmental regulatory performance, and increasing environmental awareness. But the issue of concern is the increasing resource and energy use intensity of the economy. Ayres (2008) calls for a radical change in the development trajectory. He says that nations should change their development path from one which favors increasing energy and resource use to increase productivity of manufactured capital and labor to one that concentrates on increasing resource productivity.

Figure 2.8 scatters genuine saving as a percent of national income against income measured as GDP per capita. The first point to note is that India never observed a negative genuine savings rate; however, the savings rate experienced a downturn during some years when compared with the adjunct previous year. Second, there is a clear upward trend in the scatter; as the economy's health improves, genuine saving increases. This result is very striking, as Hamilton and Hassan (2003) find that many countries under US$1,000 per capita income have negative genuine saving rates.

According to Hartwick's rule, known as "invest resource rents," a nation should invest all rent earned from exhaustible resources currently extracted in productive assets in order to have a sustainable consumption path. Figure 2.9 explores the question of whether India is consuming or investing natural resource rents by scattering genuine saving rates against the share of exhaustible resources, viz., mineral and hydrocarbons in GNI. If the country is investing all rents earned from

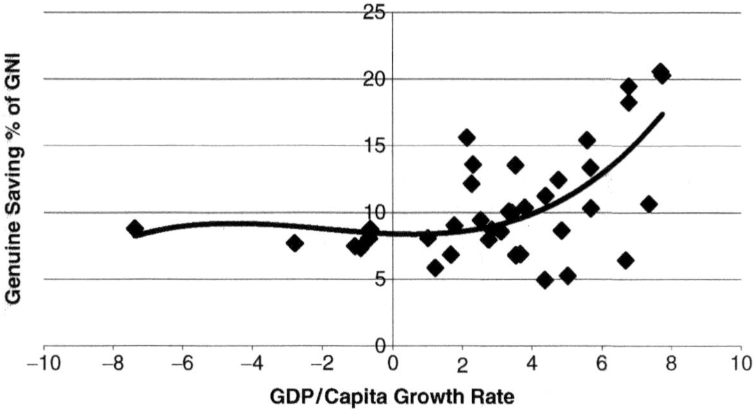

Fig. 2.8 Genuine saving versus GDP per capita growth rate
Source: authors' calculations based on WDI data

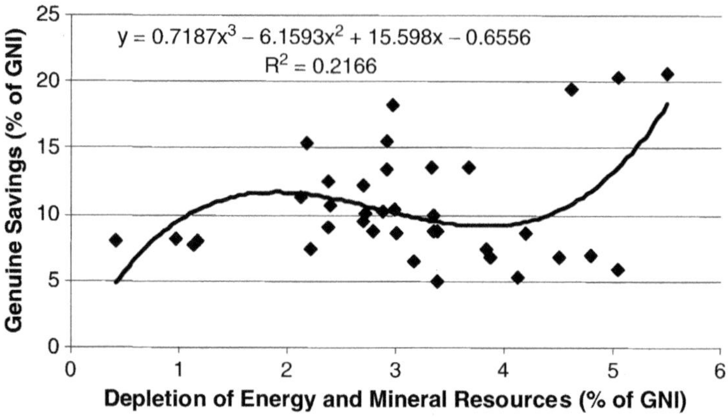

Fig. 2.9 Genuine saving versus depletion of energy and mineral resources
Source: authors' calculations based on WDI data

the extraction of these resources, then scatter in the figure should exhibit no trend and the Indian economy, according to Hartwick's rule, is on the sustainable path.

Arrow et al. (2004) are of the view that a society can be on the sustainable consumption path if it is able to maintain or increase its productive base. They define productive base as the stock of all society's capital assets at time *t*, inclusive of manufactured capital assets, human capital, and natural capital. It also depends on the level of technological progress. As defined earlier, they consider genuine investment as change in productive base. Note that maintaining a productive base does not imply maintaining any particular set of resources at any given time since there is substitutability between different kinds of assets. The growth rate of genuine wealth can be computed by dividing the figures of genuine investment or savings by the ICOR in an economy. To compute the figures at a per capita level – that is, to make adjustments for population growth – the population growth rate is subtracted from the figures for genuine wealth. The figures of per capita growth rate in genuine wealth are adjusted for the growth rate in technology and/or institutions measured as the growth rate of total factor productivity (TFP).[3]

Note that in India capital accumulation is to a large extent financed by domestic savings; therefore, there is no major difference between figures for genuine saving and genuine investment. In the computation of genuine wealth figures, unlike Arrow et al. (2004), we use the actual figures of ICOR rather than presumed figures. In India, except for the decade of 1970s, the ICOR has hovered around 4, and this conventional measure of capital intensity includes only manufactured capital. To account for human and natural capital, therefore, we increase the observed ICOR by 1. Moreover, for making adjustments for TFP growth rate, we use the estimates

[3]For the methodology on estimation of genuine per capital wealth growth rate and its adjustment factors see Arrow et al. (2004).

provided by Kumar and Managi (2008). In their study, the estimates are produced over the period of 1963–2000 for a large number of countries. For estimating TFP, unlike other previous studies, their study considers three inputs – labor, capital, and energy to produce GDP and the emissions of carbon and sulfur.

Figure 2.10 shows the trend in the growth rate of per capita genuine wealth and conventional wealth (manufactured assets). This figure provides some important insights into the question of sustainability of the Indian growth trajectory. First, both per capita conventional and genuine wealth have continuously increased since 1970. Second, during the study period of 37 years, conventional wealth increased at the rate of mere 1.06% per year and the growth rate of per capita genuine wealth was virtually near zero – only 0.07% per year. For the period 1970–2000, the growth rates of per capita conventional and genuine wealth were 0.55% and −0.34% per year, respectively; however, Arrow et al. (2004) observed that the growth rate in per capita genuine wealth was 0.54% per year. The difference in these two estimates may be attributed to the use of different parameters for manufactured capital intensity and TFP growth rate.

Third, the growth rate of per capita conventional wealth was negative until 1977 and then became positive; however, it was negative in 1982, 1987, 1991, and 1992. The growth rate of genuine wealth was negative till 1992 and then it became positive. However, it was positive in 1988 and 1989. These preliminary estimates reveal that the development trajectory followed by the country before 1991 was not sustainable. During the 1980s, although the country observed a positive growth rate in manufactured assets, the growth rate of decline in human and natural assets was more than enough to offset the positive growth rate of manufactured capital.

Fourth, the discussion in Section 3 reveals that although India took four decades to come out of a situation that the economist Raj Krishna called Hindu rate of growth, the growth rate of GDP achieved in the 1990s was also not sustainable since

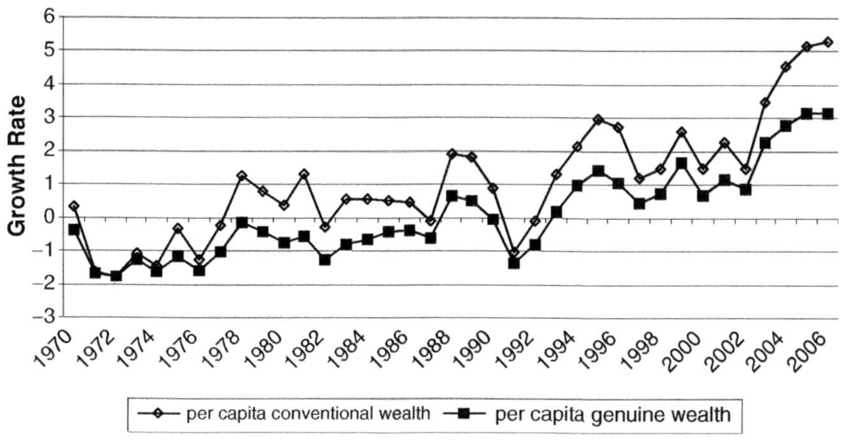

Fig. 2.10 Growth rate of per capita conventional and genuine wealth
Source: authors' calculations based on WDI data

the per capita genuine wealth was declining. It is only from the first quinquennium of the 21st century that the growth rate in income may be considered as sustainable. During 2001–2006, the growth rates of per capital conventional and genuine wealth were 3.7% and 2.23% per year.

Lastly, the point that deserves special attention is that the trend in the difference in the growth rates of these two measures of per capita wealth – conventional and genuine – is increasing over time. Until the mid-1970s, the difference in the growth rates was negligible, but it began to increase and reached about 2 by 1981. During the decades of the 1980s and 1990s it varied between 0.33 and 1.7. Since 2001, we observe that the difference in these two growth rates is continuously increasing and in 2006 it was 2.12. The increasing difference in the growth rates of the different measures of wealth implies that if both the growth rates are positive and there is substitutability between the two, the economy may continue to grow. If there is limited substitutability between natural capital and manufactured capital, it is doubtful that the present growth trajectory will be sustained.

Increasing populations are considered a major reason for the destruction and depletion of natural resources, and this consideration is relevant to the Malthusian approach to environmental accounting. It is true that since populations are increasing and aggregate genuine wealth is not increasing, wealth will be shared among more people. Hamilton (2002) examines the effect of population growth on genuine saving estimates. Figure 2.11 scatters the growth rate of per capita wealth against the population growth rate and shows that there is an inverse relationship between the two factors: as the population growth rate exceeds the rate of 2% per year, the growth rate of per capita genuine wealth becomes negative and the growth path becomes unsustainable. In India, the population growth rate is declining; at present it is about 1.37% per annum. The dependency ratio is also declining.

The preceding discussion on the sustainability of income or welfare should be read with caution. As indicated earlier, in the estimates of genuine saving or investment, all natural resources are not accounted for. For example, these estimates

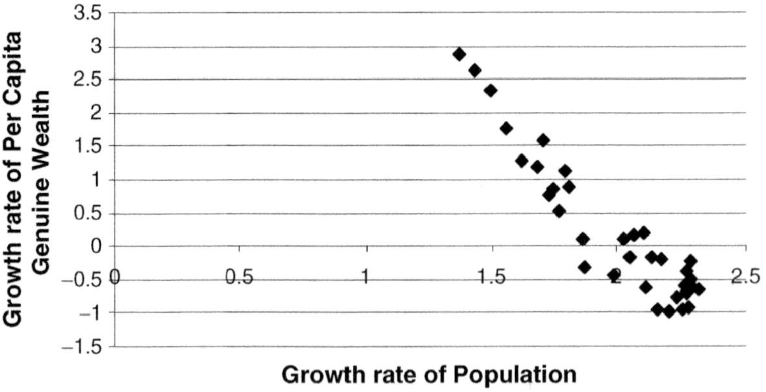

Fig. 2.11 Growth rate of per capita genuine wealth versus growth rate of population
Source: authors' calculations based on WDI data

do not account for the degradation of land resources–except for forests that are being logged. In India, agricultural land is subject to utter degradation. About 57% of the total land area is under some form of degradation, and this sector employs about 60% of the total labor force of the country. Similarly, the depletion and degradation of water resources are not accounted for in these estimates of genuine wealth. In the last two decades, the water quality of almost all of the rivers, large lakes, ponds, and streams throughout the country has been degraded to the extent that today that these water sources are not even of bathing quality. There are many other forms of natural resource degradation that must be accounted for in estimates of genuine wealth as they are lowering the amenities available to humans.

In the estimation of the growth rate of per capita genuine wealth, it is assumed that there is substitutability between different forms of assets. These estimates miss critical bottlenecks that limit the substitution possibilities. For example, in rural India it is often not possible for people to find an appropriate substitute if their water holes vanish and the local woodlands recede (Dasgupta, 1993).

Of course, due to economic activities, environmental degradation is taking place throughout the country. However, note that the regional distribution of natural resources and the level of economic development are not similar across states. The poverty distribution in India coincidently is linked with the distribution of ecosystems and their health in the country (ESPASSA, 2008). As noted earlier, economic regional disparity in the postliberalization regime is increasing. Most of the manufactured capital formation is taking place in those states which are economically better off than poor states and that house most of the natural resources.

2.7 Conclusions

India is the largest democracy in the world and is the fourth largest economy in the world in terms of purchasing power parity. However, about one-third of the total population of the country survives at less than US$1 per day. These two facts lead to degradation and depletion of the environment and natural resources. Similarly, the country has elaborate statutes, regulations, institutional frameworks, and policies on almost every conceivable topic from hazardous waste to public liability to forests and wildlife. However, monitoring and enforcement capabilities are weak. This chapter overviews the complexity and magnitude of environmental problems in addition to general economic performance. These contrasts raise questions about the sustainability of the present growth trajectory from both economic and environmental points of view.

Chapter 3
Environmental Regulations and Compliance in India

3.1 Introduction

In the last two and a half decades, India witnessed a rise in the scale of economic activities. The growth rate of per capital income was 1.7% per annum during the period of 1951–1980 and increased to about 7% in 2006–2007. The incidence of poverty (population below the poverty line) has declined from about 51% in the 1970s to about 27% in 2004–2005. India has also succeeded in reducing infant mortality and in increasing school enrollments. However, challenges remain in areas such as child malnutrition, primary and secondary education completion rates, maternal mortality, and gender balance in education and health. The resurgence of tuberculosis and the threat of HIV/AIDS are also a cause for concern. Degradation of the environment is a significant barrier to the achievement of the Millennium Development Goals (MDGs) related to reduction of poverty, hunger, and disease. Therefore, the problem is more acute in low-income countries like India that are struggling for "development" and focusing on increasing levels of economic activity on one side while on the other side facing the negative impacts of degraded environmental quality.[1]

The environment provides private as well as public goods type of services. It provides private goods like food, fresh water, wood, fiber, and fuel, which people can buy from the market. In addition to supporting all life and regulating natural systems, the environment supplies public goods type of services like fresh air, biodiversity, nutrient cycling, soil formation, control of diseases and floods, avoiding climatic change problems, and aesthetic, spiritual, and recreational benefits. Markets for public good services are absent; everybody receives the benefits of conservation of the environment. Equally, everybody receives the damages from

[1] Dasgupta (2007) warns that if nothing substantial is done to prevent the degradation of ecosystems; the average per capita consumption level at the world level may decline. He finds that economic development during 1970–2000 in the Indian subcontinent was either unsustainable or barely sustainable when the productive base of the countries is taken into account.

S. Kumar and S. Managi, *The Economics of Sustainable Development*, Natural Resource Management and Policy 32,
DOI: 10.1007/978-0-387-98176-5_3,

the degraded environment. While both rich and poor gain from the conservation of the environment, the poor are relatively more affected by the degradation.

A recent WHO report, based on analysis of data available from national health authorities and review of the scientific literature and expert surveys, shows that people in developing countries lose 20 times more healthy years than people in developed countries from environment-related health factors. Lack of water, sanitation, and hygiene results in the loss of 0.4 million lives, while air pollution contributes to the death of 0.52 million people annually in India. Environmental factors contribute to 60 years of ill health per 1,000 population in India compared to 54 in Russia, 37 in Brazil and 34 in China.

The first real impetus for developing a framework for environmental protection in India came after the UN Conference on the Human Environment in 1972. Environmental policy in the 1970s and 1980s recognized the need for an institutional identity to environmental policy making, resulting in the setting up of the Ministry of Environment and Forests (MoEF) as a full-fledged ministry in 1985 and a spurt in environmental legislation leading to an extensive framework of environmental laws in the country. In addition, informal regulations in the form of an active judiciary and citizen participation have been playing a considerable role in environmental protection.

The remainder of this chapter is organized as follows: Section 3.2 describes environmental regulations in India; Section 3.3 presents in brief the state of the environment in India; Section 3.4 lists possible reasons for noncompliance of environmental standards in the country. The last section offers some concluding remarks.

3.2 Environmental Regulations in India

Market failures provide a justification for government intervention in markets. Markets do not exist for most of environmental services, such as fresh air and water in rivers, because these services possess the characteristics of public goods. Hence, environmental regulations are inevitable to achieve economic efficiency in production and consumption. Governments can play a significant role in environmental protection by assigning property rights to resources and in undertaking measures to reduce the transaction costs to facilitate bargaining between private parties (Coase, 1960). The Coasean approach also presupposes a court system to deal with cases when bargaining between two or more parties does not result in a mutually beneficial solution (Sankar, 1998). The court system is also required to enforce the reached agreement, so that it cannot be breached so easily. Pigou foresees state intervention in the form of a tax on the polluting units. The Pigouvian prescription has now taken the form of the polluter pays principle.

Pollution is an environmental externality created by the production and consumption of goods and services in the economy. The externality of pollution could be of different types: local, national, and global; public or private. Different approaches are required to deal with these different types of externalities. For example, groundwater pollution by an industrial estate is a local externality, which could be tackled by an institutional arrangement involving industry and local community. River pollution could be an externality requiring the involvement of federal and provincial governments for control, as is the case for the project Ganga Action Plan in India. Pollution of an international river or coastal waters could result in transnational pollution requiring international cooperation, as happened with an agreement among riparian countries (France, Germany, and Netherlands) for cleaning the river Rhine in Europe. Also, it is important to consider different measures of pollution: pollution at source and ambient pollution. Further, a distinction has to be made between point and nonpoint sources of pollution. Measurement of the pollution in these different dimensions is important from the standpoint of designing environmental policy for controlling and preventing it (Murty, 2008).

Public policy/response is an intended action of the public/state/civil society to alter individual behavior in a manner that minimizes the difference between social net benefits and private net benefits. Environmental policies are designed to alter the behavior of economic agents, either individuals or groups of individuals, in a manner that internalizes environmental externalities generated during the course of individual actions. As shown in Figure 3.1, policy responses can be classified into two categories: formal and informal. Legislative responses require policy responses mandated by the state. These policy responses may originate from the government to achieve the objective of maximizing social welfare or from society itself. As a society feels the heat of externalities, it exerts pressure on governments, and the government brings legislation to control externalities. Actions by the state to control externalities without public pressure can be put into the category of formal regulations. On the other hand, civil society pressure to control individual behavior in the social interest is known as informal regulation. Environmental regulations do not remain confined within the purview of governments in modern times because now firms are not individually governed units but must depend on markets to get funds and to sell their products. Markets also help alter individual behavior in a socially desirable manner. In India, we find both formal and informal kinds of regulations in the area of environmental externalities.

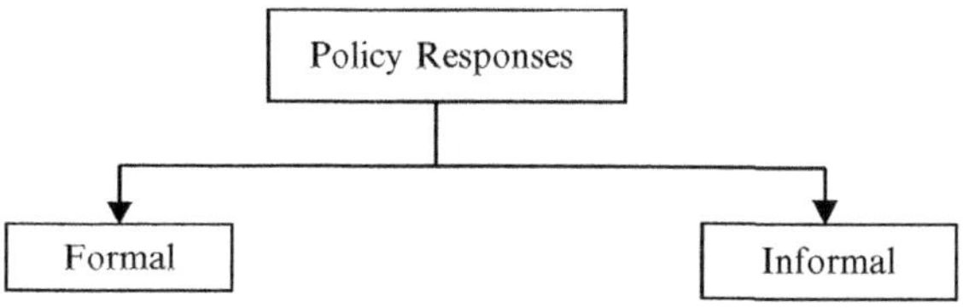

Fig. 3.1 Environmental regulations in India

3.2.1 Formal Regulations

Two international conferences on environment and development – one at Stockholm in 1972 and another at Rio de Janerio in 1992 – have influenced environmental policies in most countries, including India. Many countries and international agencies have accepted the polluter pays principle, the precautionary principle, and the concept of intergenerational equity as guidelines for designing environmental policies (Sankar, 1998).

Historically, policy responses for preventing and controlling environmental degradation in the country started slowly during the 1970s and gradually picked up speed in later years. Many of the legislative responses are more than 30 years old and form the foundations for current environmental policies. New environmental policies recognize the importance of the role of incentive-based policy instruments in controlling and preventing environmental pollution. That is, formal regulations may be classified into two categories (Fig. 3.2): The state intervenes in the form of legislation and policies and through public investments for environmental cleaning activities, such as the Ganga Action Plan (GAP) and the Yamuna Action Plan.

The overall framework of environmental legislation in India is set by the National Conservation Strategy and Policy Statement on Environment and Development issued by the MoEF in June 1992 (Datt et al., 2004). The Indian constitution enjoins the "States to take measures to protect and improve the environment and to safeguard the forests and wildlife in the country" (Article 48A). It also makes it a "fundamental duty of every citizen to protect and improve the natural environment including forests, lakes and rivers and wildlife and to have ecological compassion for the living creatures" (Article 51A(g)).

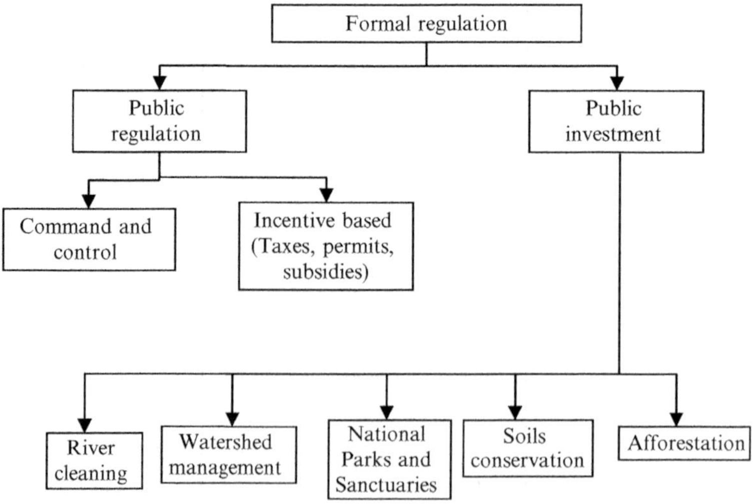

Fig. 3.2 Formal environmental regulations in India

Recognizing the severe problems related to the issue of pollution, both air and water, the Policy Statement for Abatement of Pollution, 1992, identifies the following steps in order to integrate environmental considerations into decision making at all levels:

- Prevent pollution at the source and encourage, develop, and apply the best available practicable technical solutions.
- Ensure that the polluter pays for the pollution and control arrangements.
- Focus on protection of heavily polluted areas and river stretches.
- Involve the public in decision making.

In order to ensure that the projects are adequately monitored the following requirements have been put in place:

- Investors are required to report every 6 months on the implementation of the environmental safeguards stipulated in the clearance by the MoEF.
- Field visits by MoEF and its regional offices to collect samples and data on the environmental performance of the cleared projects.
- In cases of inadequate compliance, the issue is taken up with concerned state governments and nodal ministries.

3.2.2 The Institutional Framework for Environmental Management: A Brief History

The UN Conference on the Human Environment in Stockholm in 1972 is a landmark in the evolution of environmental policy in India. A National Committee on Environmental Planning and Coordination (NCEPC) was formed in February 1972 in the Department of Science and Technology. This committee was the forerunner of the Department of Environment (DoE), which eventually became the present MoEF in 1985. The main responsibility of the NCEPC was to plan and coordinate, while the various ministries and agencies of the government were supposed to carry out the actual implementation. In January 1980, the central government set up a committee to recommend legislative measures and administrative machinery for environmental protection (known as the Tiwari Committee). The committee made extensive recommendations, including, inter alia, the establishment of a Department of Environment as part of the central government. This department came into existence on November 1, 1980. It was envisaged both as a coordinating and administrative agency. Its mandate was to coordinate national policies for environmental protection and resource management, as well as to have administrative responsibility for pollution monitoring and regulations. In 1985 DoE was converted into the MoEF. At present this ministry is the nodal agency in the administrative structure of the central government "for the planning, promotion and coordination of environmental and forestry programmes" (Government of India, 1995).

In tandem with these developments at the center, and at the urging of NCEPC, almost all states and union territories established environmental boards with terms similar to those of the national committee. Most of these have since been converted into environment departments (Gupta, 2001).

There is another important set of environmental institutions in India (particularly with regard to pollution control) that were established even before DoE. These are the central and state pollution control boards (CPCB and SPCBs). These boards were created under the Water (Prevention and Control of Pollution) Act to implement the provisions of the Act and were initially known as central/state water pollution control boards (PCBs). After the passage of the Air (Prevention and Control of Pollution) Act in 1981, these boards started addressing air pollution issues also and were given their current name.

These pollution control boards are statutory bodies. Their mandate is to implement and enforce the major pollution control laws (Jasanoff, 1986). State pollution control boards have been constituted in all states. The central board coordinates the activities of the state boards and serves as the state board for the federally administered union territories. It is supposed to compile and publish data on air and water pollution, and more importantly, to lay down ambient standards for air and water, as well as emission standards for these media.

The division of powers between the central and the state governments with respect to environmental legislation is not entirely clear. In general, it appears that while the central government is the legislating authority, the state governments are the implementing agencies. Specific differences are however discernible with respect to the different Pollution Control Acts as outlined below. In addition, governments may, according to their political mandates, provide more or less power to the state governments.[2]

3.2.3 Environmental Laws

There is a spate of environmental legislation in India starting from early 1970s.[3] Chronologically, these legislative acts are the Wildlife Protection Act of 1972, the Water (Prevention and Control of Pollution) Act of 1974, the Water Cess Act of

[2]For environmental federalism in India, see Gupta (2001) and Mandal and Rao (2005).

[3]Environmental laws in India for pollution control date back to the mid-nineteenth century. The Shore Nuisance Act, 1853, the Indian Penal Act, 1860, the Indian Easement Act, 1882, the Bengal Smoke Nuisance Act, 1905, the Bombay Smoke Nuisance Act, 1912, and the Motor Vehicle Act, 1839 were some of the pioneering legislations enacted before the independence. In the post-independence period the spate of legislations such as, the Factories Act, 1948, the Industries (Development and Regulation) Act, 1951, the River Board Act, 1956, the Atomic Energy Act, 1962, the Insecticide Act, 1968, the Merchant Shipping (Amendment) Act, 1970, and the Radiation Protection Act, 1971 also dealt with, to some extent, the problems of air and water pollution in India.

1977, the Forest Conservation Act in 1980, the Air (Prevention and Control of Pollution) Act in 1981, the Environment (Protection) Act of 1986, and the Public Liability Insurance Act of 1991. In 1988, the Water Cess Act of 1977 was amended as the Water (Prevention and Control of Pollution) Cess Act, along with the Air Act of 1981.[4] Of these, the acts that directly concern industrial production in India are the Water Act (1974), the Water Cess Act (1977 and 1988), the Air Act (1981 and 1988), and the Environment (Protection) Act or EPA (1986). While the first two are foundational legislation in the context of air and water pollution in the country, the EPA is designed to fill the gaps still remaining in the legal framework for the control of industrial pollution. The third is more revenue-generating legislation than a measure to restrict the consumption of water by industrial units. A list of environmental legislations in India is given in Appendix 1.

These laws constitute the foundations of domestic environmental regulation. As mentioned above, they provided for the setting up of pollution control boards at the central and the state levels, which were empowered to prevent, control, and abate air and water pollution and to advise governments on matters pertaining to such pollution. The CPCB is to coordinate the activities of the State Boards. The CPCB has also prepared a list of polluting industries in India (Appendix 2). The Acts also specify that industrial units have to provide on demand all information regarding their effluent and treatment methods.

3.2.4 Fiscal Instruments for Pollution Control in India

The government's approach to preventing pollution has been mostly in the nature of legislation-based command and control measures, while natural resource management has been largely carried out through programs supported by allocations from the central (e.g., programs of MoEF, Ministry of Nonconventional Energy Sources, Ministry of Agriculture, etc.) and state budgets. The use of fiscal instruments (other than expenditure policy) in environmental policy has been rather limited, even though the need to employ economic and fiscal policy instruments for the control of pollution and management of natural resources has gained steady recognition during the 1990s (Datt et al., 2004).

In the Policy Statement for the Abatement of Pollution, released in 1992, the MoEF noted the need for a mix of policy instruments in the form of regulations, legislation, agreements, financial incentives, etc., to address environmental concerns. In the Ninth Five-Year Plan (1997–2002), an important element of the environmental strategy was "integrating environment with decision making through valuation of environmental impacts; evolving market-based instruments as an alternative to the command and control form of environmental regulation;

[4]The Water (Prevention and Control of Pollution) Act and the Air (Prevention and Control of Pollution) Act are hereafter referred to as the Water Act and the Air Act, respectively.

appropriate pricing of natural resources based on their long-term marginal cost of supply; appropriate fiscal reform and natural resource accounting" (as cited in Datt et al., 2004).

A Task Force was constituted by the MoEF in 1995 to evaluate the scope for market-based instruments (MBIs) for industrial pollution abatement (Government of India, 1997). The Task Force recommended explicit incorporation of MBIs in pollution control laws, greater reliance on economic penalties in the short and medium term, and completely replacing criminal penalties by MBIs in the long run. It also recommended modification of the existing water cess to make it a genuine effluent-based tax based on the pollution load rather than the amount of water consumed, and also abolishing tax concessions on installation of pollution control equipment. It recognized the need for systematic data collection to estimate marginal abatement costs and the regulatory burden and called for the introduction of additional MBIs:

- Use of pollution taxes in accordance with the polluter pays principle, for small, dispersed sources of emissions/effluents
- Use of tradable permits for large firms provided there was an adequate number of firms in the market
- Levy of user fees differentiated according to the treatment cost imposed by each unit, to cover costs of common effluent treatment plants (CETPs) where individual treatment of waste discharge was not feasible because of the economies of scale

The MoEF again set up a Task Force in 2001 to expedite the implementation of pilot schemes on pollution charges in selected critically polluted areas (hotspots). The Terms of Reference drawn up for the pilot program noted that given the legal issues and impediments involved in implementing a typical tax-standard type of pollution charge scheme in India, the next best alternative needs to be adopted for providing a similar incentive to grossly polluting industries. The Terms of Reference suggested the following instruments as proxy pollution charges: (1) water cess charge, (2) legislation on water pollution and conservation, (3) bank guarantees based on marginal costs of abatement, and (4) pollution charges based on estimates of marginal costs of abatement. Project proposals have been called for from selected state pollution control boards (SPCBs) in this regard, and a detailed action plan for implementing the pilot program is slated to be drawn up (Datt et al., 2004).

The State of the Environment Report prepared for India in 2001 as part of a project supported by UNEP and the MoEF recommended that economic measures need to be put in place to encourage a shift from curative to preventive measures, internalization of the costs of environmental degradation, and conservation of resources. The revenue generated may be used for enforcement, collection, treatment facilities, and R&D. The Report also called for economic incentives for environmentally benign substitutes, technologies, and energy conservation. The need for evolving an appropriate tariff structure for water services to encourage wise usage and to generate funds for cash-strapped service providers was also recognized in the Report (Datt et al., 2004).

In order to encourage environmental conservation, donations given by the corporate sector for conservation of nature and natural resources are exempt from income tax. A depreciation allowance of 30% is also allowed on devices and systems installed in industrial units for minimizing pollution or for conservation of natural resources. In order to encourage the shift of polluting industries from congested urban areas, capital gains made in moving from urban to other areas are exempt from taxes if these are used for acquiring land and building production facilities in nonurban areas. Excise and custom duty exemptions or reductions are given for the use of environmentally friendly raw materials.

The actual use of fiscal incentives in the country has, however, been rather limited. These take the form of tax concessions for the adoption of pollution control equipment and a somewhat more structured policy for the promotion of renewable energy technologies. Tax incentives are usually specified for identified abatement technologies and activities, not providing dynamic incentives for technological innovation and diffusion. Also, since most of these are end-of-the-pipe treatment technologies, these incentives do not promote more efficient use of resources. There are some provisions for the use of levies, cess, fines, penalties, etc., for polluters, though their implementation and effectiveness could do with improvement.

The comprehensive environmental legislations and policy statements described above provide for the use of various environmental policy instruments. However, India has not used these legislation and policy statements so far for choosing a right mix of instruments for environmental regulation. It still uses command and control instruments. There are several empirical studies exploring the possibility of using economic instruments and the institutions that facilitate people's participation in the management of environmental resources (see Murty et al., 1999; World Bank, 1999; Murty and Kumar, 2004). These studies argue for the use of economic instruments for the control of pollution by the industries, especially by the big factories, and the use of institutions facilitating collective action to control industrial pollution by the small-scale industries in an industrial estate, and the management of forest resources.

3.2.5 Review of Some Recent Studies[5]

Predictions of environmental changes and the estimation of monetary values of them are needed for designing policy instruments for environmentally sustainable development (Murty, 2008). There are costs and benefits of environmental policy changes. For example, designing a water pollution tax requires estimates of marginal cost and marginal benefits of pollution reduction. The feasible policy changes in a country depend on the environmental laws in that country.

According to the taxes-standards method (Baumol and Oates, 1988), if taxes are designed and levied such that the tax on each pollutant is equal to the marginal cost

[5]This section, to a large extent, is based on Murty (2008).

of abatement corresponding to the standard, polluting firms will have incentives to comply with the standards. The designing of taxes for water pollution abatement requires information about the standards for different water pollution parameters and the estimates of water pollution abatement cost functions. Given this information, the pollution taxes required to make firms meet the standards can then be estimated.

Although it is widely known that command and control measures do not provide the necessary incentives to polluters for choosing least-cost methods of pollution control, the Government of India has so far resorted to such measures only for controlling industrial pollution in India. On the other hand, fiscal instruments such as pollution taxes or marketable pollution permits provide incentives for adopting least-cost pollution abatement measures. Ironically, there have been no serious attempts made in India so far for using such instruments for industrial pollution abatement. The currently levied tax on the consumption of water by industrial activities cannot be treated as a pollution tax, since its main objective is to raise revenue for the pollution control boards. As such, the tax collected is very nominal (Rs. 0.015–0.07 kL^{-1}), which normally does not have much effect on the industrial demand for water. Some of the recent research studies on water pollution abatement in India (Gupta et al., 1989; Mehta et al., 1995; Murty et al., 1999; Pandey, 1999; Misra, 1999; World Bank, 1999; Murty and Kumar, 2002, 2004) have found that the pollution tax on the industrial water use should be several times higher than the current rate of water cess to realize the prescribed water quality standards.

Some of the recent studies on industrial pollution abatement in India (see Gupta et al., 1989; Mehta et al., 1995; Murty et al., 1999; Pandey, 1999; Misra, 1999; World Bank, 1999; Murty and Kumar, 2004; Murty and Gulati, 2007) give some information about the rate of tax to be levied on industries for making them comply with the prescribed water and air quality standards. Mehta et al. (1995) considered an abatement cost function for effluent treatment plant (ETP) in paper and pulp units in India and concluded that marginal abatement costs of relatively high-cost producers should serve as the basis for setting charges/taxes so as to ensure that producers find it cheaper to abate than to pollute. They recommended four options for experimentation by policy makers: (a) abatement charges with the government undertaking cleaning up, (b) abatement charges with cleaning up contracted out based on competitive bidding, (c) a tax proportional to excess pollution on firms violating standards and subsidies for those going beyond the prescribed abatement standards, and (d) private permit trading system.

The water polluting firms in Indian industry are supposed to meet the standards set for the pollutants [35 mg/L for biological oxygen demand (BOD), 250 mg/L for chemical oxygen demand (COD), and 100 mg/L for suspended solid (SS)] by the CPCB. The air polluting firms are supposed to meet the stack emission standards of 115, 80, and 80 mg per Nm^3, respectively, for suspended particle matter (SPM), sulfur dioxide (SO_2), and nitrogen oxide (NO_x). A survey[6] of samples of water polluting

[6] A Survey of Water Polluting Industries in India, 1996 and A Survey of Water and Air polluting Industries in India, 2000, Institute of Economic Growth, Delhi.

industries in India shows that most of the firms have ETPs and that in addition some firms are using process changes in production and input choices to achieve the effluent standards. However, there is a large variation in the degree of compliance among the firms measured in terms of ratio of standard to effluent quality. The laxity of formal environmental regulation by the government and the use of command and control instruments could be regarded as factors responsible for large variations in firms' compliance to pollution standards. Using this data, Murty and Kumar (2004) provide estimates of taxes on 1 tonne of BOD, COD, and SS as Rs. 20,157, Rs. 48,826, and Rs. 21,444, respectively. Similarly, a recent study (Murty and Gulati, 2007) provides estimates of taxes on emissions of SPM, SO_2, and NO_x from thermal power generation in India as Rs. 2,099, 20,519 and 5,554 per tonne, respectively.

The MoEF has also commissioned several case studies to examine issues relating to economic instruments for pollution abatement. These studies estimated abatement costs of pollutants and recorded wide variations across different industries. The studies pointed out the inefficiency of the current legislation, which required all polluters to meet the same discharge standards, and called for the introduction of economic instruments for cost-effective pollution control. They emphasized the need for regulators to allocate their monitoring resources more efficiently by targeting industries characterized by relatively high discharges and low costs of pollution abatement. These studies also observed that taxes and incentives based on efficiency instruments more effectively align pollution control agencies with the polluters than does the command and control regime. Such instruments also facilitate the triple bottom line of economic efficiency, environment responsibility, and social relevance, entitling the industrial units to clean development mechanism (CDM) and other cleaner production benefits (Datt et al., 2004).

3.2.6 *Informal Regulation and People's Participation*[7]

Economic instruments and command and controls are the instruments of formal regulation. The designing and implementation of these instruments involve a top down, or centralized, approach. The success of these instruments in controlling pollution depends upon the quality of governance and its ability to incur high transaction costs. The governments of many developing countries cannot meet these requirements, which results in the failure of regulation. A bottom up, or decentralized, regulation involving civic society and local communities and a very limited role for government could save transaction costs and get rid of political and bureaucratic corruption.[8] Even without formal regulation by the government,

[7]This section, to a large extent, is based on Murty (2008).

[8]Coase theorem says that given the initial property rights to any resource either to the polluter or to the affected party, and if the cost of bargaining is zero, the bargaining between the two parties results in the optimal control of pollution. Even with the positive transaction costs, the bargaining could result in the reduction of externality though not to the optimum level (Coase, 1960).

informal regulations increase the expected penalties on the firms for noncompliance with pollution standards. The firms react by reducing pollution in the presence of informal regulations similar to formal regulations (Murty, 2008).

The management of environmental resources can no longer be taken as the responsibility of a single institution like market or government (Murty, 2008). Limitations of formal regulations have paved the way for having a mix of institutions and instruments. Various stakeholders, such as consumers, investors and producers, have incentives for pollution control. Consumers regulate the market for pollution-intensive commodities by expressing a preference for green products or commodities produced using cleaner technologies. Investors also have incentives to invest in industries using cleaner technologies. A higher level of observed pollution in a firm is an indication to investors that the firm uses inefficient technology, resulting in the loss of profits, and there may be a downward valuation of the firm's stocks in the capital market. On the other hand, a good environmental performance by the firm may result in upward valuation of its stocks (Murty, 2008).

In developed countries, informal regulations are an important component of enforcement of environmental laws. The U. S. Environmental Protection Agency (USEPA) encourages private litigants to bring suits against polluters. The European Commission in its 1993 Green Paper has clearly signaled its intent to beep up the rights of individuals to pursue polluters for compensation for environmental damages (as cited in Murty, 2008). Similarly, the Canadian government is involving private agents in every aspect, ranging from drafting to compliance of environmental pollution.

Some recent studies show that stock markets react to environmental performance of firms in both developed and developing countries (World Bank, 1999). The recent World Bank sponsored studies report that stock markets in developing countries like Argentina, Chile, Mexico, and Philippines are even more volatile in response to news about the environmental performance of firms. The average of gains in stock prices due to good news about environmental performance is found to be 20% in these countries.

There are cases of firms complying with environmental standards in developing countries even in the absence of formal regulation by the government. PT Indah Kiat Pulp and Paper (IKPP) in Indonesia is an interesting example (World Bank, 1999). IKPP is the largest and the cleanest paper-producing firm in Indonesia. Clean-up began in some of its mills in the 1990s with pressures from local communities. Furthermore, the need for going to the Western bond market for financing its expansion to meet growing export demand has made IKPP seek cleaner technologies. The good performance of the company in pollution management has resulted in the increase of its stock value in comparison to Jakarta's composite stock index.

Figure 3.3 describes the structure of informal environmental regulations in India.

Small-scale enterprises play a pivotal role in the industrial development of India. Nonavailability of economically viable technological options for complying with environmental standards under the command and control mechanism has been causing considerable hardship to these enterprises. The presence of scale

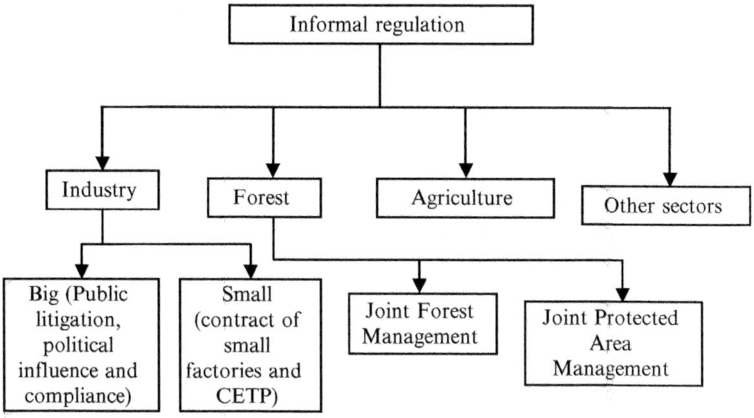

Fig. 3.3 Informal environmental regulation in India

economies in pollution abatement, especially in water pollution abatement, has compounded the problem. It is not economical for small-scale enterprises to have their own individual ETPs. Collective action involving all relevant parties for water pollution abatement (factories, affected parties, and government) is now seen as an institutional alternative to deal with the problem of water pollution abatement in industrial estates, especially in India (Murty et al., 1999). Collective action in industrial water pollution abatement is meant to bring about the necessary institutional changes that are compatible with the choice of cost-saving technologies. For example, a CETP can be adopted if necessary legislation is in place to define the property rights of the factories and the affected parties.

According to Murty (2008), three processes are involved in collective action for control of water pollution in an industrial estate. These are (a) collective action of affected parties, (b) collective action of factories, and (c) bargaining between a coalition of affected people and a coalition of factories. The collective action of affected people is possible if the damages from pollution are substantive enough to justify the transaction costs of coalition and bargaining. The factories in an industrial estate must have recourse to pollution abatement methods, taking into account possible collective action by the affected people. The available pollution abatement technologies may provide small factories with a broad spectrum of technological choices, out of which the CETP may be the least-cost technology. Therefore, collective action by the factories can be technology driven. Finally, bargaining between a coalition of affected people and a coalition of factories produces the end result of collective action that is the realization of prescribed environmental standards.

In India, there are also several examples of public litigation cases against factories for claiming damages from pollution by local people and resulting in big factories complying with the standards. The Pattancheru industrial estate in the Andhra Pradesh state of India is an example (Murty, 2008). Local opposition to pollution started in 1986 when about 3,000 villagers marched to the Chief Minister's office after suffering large-scale crop losses and health damages due to contamination of groundwater and the pollution of nearby river. In 1989, about 5,000 people held a demonstration before the state assembly, demanding an end to industrial pollution. In the same year farmers blocked the highway running through Patancheru for 2 days. The villagers had also filed court cases. These collective efforts of people forced the factories in the industrial estate to have a CETP for complying with the water pollution standards. Similar experiences are reported from many other industrial estates in the region.

Collective actions of local communities depend upon, among other factors, their affluence, the degree of political organization, education, and environmental awareness. Pargal and Wheeler (1996) find a negative relationship between BOD load in a factory effluent and per capita income and educational levels of local communities in a sample of 243 factories in Indonesia. Similarly, Murty and Prasad (1999; see Murty et al., 1999) observe a negative relationship between BOD effluent-influent ratio and a relative index of development of local community and the political activity of local community measured in terms of percentage of votes polled in the recent election to the Indian Parliament.

There are several empirical studies exploring the possibility of using of economic instruments and the institutions facilitating people's participation in the management of environmental resources (Chopra et al., 1990; Mehta et al., 1995; Murty et al., 1999; Pandey, 1999; World Bank, 1999; Murty and Kumar, 2002, 2004, 2006). These studies argue for the use of economic instruments for the control of pollution by the industries, especially by big factories, and the use of institutions facilitating collective action to control industrial pollution by small-scale industries in an industrial estate, and the management of forest resources (Murty, 2008). In the models of informal regulations, the governments play a catalytic role by providing information about environmental programs and available cleaner technologies and provide some financial incentives to local communities.

3.3 Current State of India's Environment

Despite the exhaustive formal and informal environmental regulations in India made for pollution control, the level of compliance is quite poor. According to the CPCB, as of June 2006, 73% of the 2,672 units under 17 categories of highly polluting industries were in compliance, which is a decrease from 2004, when the rate was 84% (OECD, 2006). Appendix 3 provides a summary of compliance status by industrial sector.

The quality of natural resources like water and air continues to deteriorate. According to one study, India loses about 6% of GDP due to pollution (Jha, 1999). Similarly, another study estimates that urban air pollution costs India US$ 1.3 billion a year and that water degradation leads to health costs amounting to US$ 5.7 million every year (Parikh, 2004).

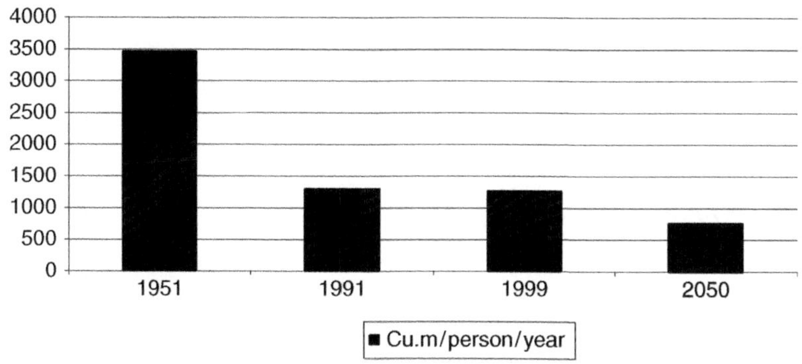

Fig. 3.4 Availability of per capita utilizable water (*source*: Parikh, 2004)

3.3.1 Water Pollution

Water pollution is a major cause of concern in India. It not only causes ecosystem damage, it also adversely affects health and thereby impairs the economic productivity of people. About 90% of surface water is polluted to the extent that it is not fit for bathing. Also, about 200 million people do not have access to safe drinking water, and utilizable water per capita is decreasing (Fig. 3.4). This level of pollution is set to create conflict over water and scarcity even in regions with abundant water (TERI, 2001).

The three major contributors towards water pollution are the domestic sector, the industrial sector, and the agricultural sector. Eighty percent of the effluents by volume are from the domestic sector. This is because only 20% and 2% of wastewater in Class I and Class II cities, respectively, is treated. Meanwhile, only 3.15% of the rural population has access to sanitation services, and 115 million homes have no access to toilets of any type. In the industrial sector only 59% of large and medium industries had adequate effluent treatment in 1995. In the agricultural sector, fertilizer use increased from 7.7 MT in 1984 to 13.4 MT in 1996 and pesticide use increased from 24 MT in 1971 to 85 MT in 1995 (Bhalla et al., 1999).

In the Ganga subbasin,[9] the anthropogenic demand on river waters is huge. In India, for instance, around 43% of the country's population (2001 census) belongs to the subbasin's catchment area, with a density of over 1,000 people per km^2. The demand for irrigation waters in the region is huge and growing, owing to the predominantly agrarian nature of the subbasin's economy. The Ganga and its

[9]The river Ganga runs over 2,500 km through four countries – China (Tibet), Nepal, India and Bangladesh – and forms one of the most populous as well as poverty-stricken river basins of the world. Along its length, large tributaries enter into the Ganga from both north and south, significantly affecting its flow and course. The total basin area of the Ganga is about 1,093,400 km^2, of which 79% is spread over eleven Indian states, 13% falls in Nepal, and both Bangladesh and China have 4% each (Murty, 2008).

tributaries, particularly the Yamuna, carry huge pollution loads that come from domestic as well as industrial sources. Increasing urbanization has led to huge additional demands on the rivers to meet domestic as well as nondomestic water needs of a rapidly growing urban population.

3.3.2 Water Pollution from Households

Household-borne effluents contribute 80% of water pollution in India. Untreated effluents from households pollute surface and groundwater sources. Local governments (city corporations, municipalities, and panchayats) having the responsibility for water supply and sanitation are supposed to treat the effluents as per the national water pollution standards or MINAS standards. However, a major portion of effluents, more than 75%, goes untreated to the environmental media. There are many states in India where household-borne effluents are not treated at all. There is 100% effluent treatment capacity in Haryana state, while Delhi state has the capacity to treat more than half of its effluent. Mumbai and Chennai have the treatment capacity to treat more than 90% of their effluents, while there are many cities having the capacity to treat 80% of the effluents they generate. Most rural households in India have no sanitation, and the local level village governments known as panchayats have no resources to provide it.

3.3.3 Water Pollution Loads from Industries

Effluents originating from households, industry, and agriculture contribute to surface and groundwater pollution. The CPCB, Government of India, provides source-specific pollution standards for industries with respect to pollution concentration of major water pollutants: BOD, COD, SS, and pH. The CPCB launched a water pollution control program in 1992 for the industries. It has identified 1,551 large and medium industries and given a time schedule for compliance with the prescribed standards. It has been found that many of these industries have ETPs and even those having ETPs do not comply with the prescribed pollution standards. There are 0.32 million small-scale industrial units in India, and due to the presence of scale economies in water pollution reduction, it is uneconomical for these units to have ETPs of their own (Murty et al., 1999). These small-scale units contribute almost 40% of industrial water pollution in India. However, small-scale units located in many industrial estates in India have gone for CETPs.

3.3.4 Water Pollution from Agriculture

Pollution by agricultural runoffs affects groundwater and surface water sources. The agriculture runoff contains pesticide and fertilizer residues. As for the fertilizers, they have an indirect adverse impact on the water resources. Indeed, by increasing the nutritional content of the watercourses, fertilizers allow organisms to proliferate. These organisms may be disease vectors or algae. The proliferation of algae may slow the flow in the watercourses, thus increasing again the proliferation of organisms and sedimentation. The WHO has defined a permissible limit of concentration of nitrates as 45 mg/L, which is also accepted by the Indian Council of Medical Research (ICMR). A relationship between N-fertilizers in several states and the respective concentration of NO_3 was found in tube wells during a survey carried out in 1986. It can be observed that in states such as Haryana, the NO_3 concentration was already exceeding by far the permissible limits in 1986.

3.3.5 Effects of Water Pollution

The socioeconomic costs of water pollution are extremely high: 1.5 million children under 5 years age die each year due to water-related diseases, 200 million person days of work are lost each year, and the country loses about Rs. 360 billion each year due to water-related diseases.

One of the reasons why environmental standards are ignored is because they are seen to be expensive. However, when one calculates the cost of lowering pollution one must compare it to the health and economic benefits of abating pollution. For example, Rs. 460 billion is needed to construct toilets in 115 million homes; wastewater treatment in 3,696 cities/towns would cost Rs. 180–600 billion depending on technology. Pollution abatement in industries would cost Rs. 140 billion (about 1.2% of total annual turnover). However, the loss from human health damages due to sanitation and water pollution is 360 billion rupees per year (Parikh, 2004).

Measurement of the economic values of environmental resources requires not only use of economic theory and technique but also physical biological links between the economic good being valued and the environmental media. For example, estimates of the value of a salt marsh in sustaining a marine fishery must be based on knowledge of the biological and ecological links between the marsh and the exploited fish species. Estimates of the health benefits from water pollution control must be based on scientific knowledge of the relationship between pollutant concentrations in water and humans, and estimates of the recreational fishing benefits stemming from water pollution control require knowledge of the relationships among pollutant levels, biological productivity, and anglers' activities. Lack of knowledge of these relationships may, in some instances, be a major barrier to empirical measurement of values.

The benefits from the improved fresh water quality comprise both private good and public good type of services provided by water resources. The private goods services include drinking water, use in the industrial processes, irrigation, fisheries, and navigation. The public goods services consist of recreation, aesthetic enjoyment, waste disposal, and biodiversity or aquatic life. There are markets for private

Table 3.1 Access to safe drinking water in households in India (%)

	Rural	Urban	Total
1981	26.5	75.1	38.2
1991	55.5	81.4	62.3
2001	73.2	90.0	77.9

Source: Economic Survey (2006–2007)

goods services from water resources, and market prices can be used for their valuation if markets are perfect. However, the markets for drinking water, water for industrial uses, and irrigation are imperfect for various reasons, especially in developing countries like India. The prices in imperfect markets normally understate the true values of these services. The estimation of the true values of even some of the private goods services of water therefore require some specially designed methods of valuation. In the case of public goods services from water resources, the markets are absent and we have to use specially designed methods to value them.

Water pollution is one of the main causes of public health risks in India. Table 3.1 shows that only 80% of households had access to safe drinking water in the year 2001. Some studies (e.g., McKenzie and Ray, 2004) show that India loses 90 million days a year due to water-borne diseases with the production losses and treatment costs worth Rs. 6 billion. Poor water quality, sanitation, and hygiene result in the loss of 30.5 million disability-adjusted life years (DALY) in India. The groundwater resources in vast tracts of India are contaminated with fluoride and arsenic. Fluoride problems exist in 150 districts of 17 states in the country, with the states of Orissa and Rajasthan the most severely affected. The high concentration of fluoride in drinking water causes fluorosis, which results in weak bones, weak teeth, and anemia. The presence of arsenic, a poison and a carcinogen, in the groundwater of the Gangetic delta causes health risks to 35–70 million people in West Bengal, Bihar, and Bangladesh.

3.3.6 Air Pollution

Air pollution has become a major cause of concern in India because most of the Indian urban pollution is exposed to some of the highest pollutant levels in the world (Smith, 2000), and there is a positive association between air pollution and mortality and morbidity, as is found in many studies (e.g., Kumar and Rao, 2001). Moreover, it is supposed that urban air pollution is associated with contamination from automobile exhausts and industrial effluents and that air pollution is an urban problem. However, in a developing country like India the problem of indoor air pollution far outweighs the problem of ambient air pollution. While in cities suspended particulate matter (SPM), sulfur dioxide (SO_2), and nitrogen oxide (NO_x) levels are much higher than permissible limits, in rural areas indoor pollution kills half a million people prematurely every year (Parikh, 2004).

Smith (2000) reports that the annual ambient concentration of pollutants in most cities, where monitoring is done for particulate matter, ranges between 90 and 600 $\mu g/m^3$, with a population mean of about 200 $\mu g/m^3$. Table 3.2 shows air quality in seven major cities during 2002. It can be observed that in Indian cities the level of SO_2 emissions is almost within the critical limits, but suspended particulate matter and respirable suspended particulate matter (RSPM) are at very high levels in both residential and industrial areas.

Increasing urbanization and industrialization are considered to be the driving factors for the problem of air pollution in India. During 1980–2000, due to urban growth about 600,000 hectares (ha) of land was transformed into urban centers. The transport sector is considered to be the highest contributor to air pollution, followed by refuse burning and industrial activities. For example, in Delhi the transport sector contributes about 70% of NO_x pollutants. Moreover, in energy consumption, India heavily depends on coal; demand for coal is expected to increase by 6.5% per annum and coal consumption is a major cause of air pollution in India.

Given the relationship between air pollution and health and exposure of the Indian population to air pollutants, the cost of air pollution in India is estimated to be very high. For instance, Brandon and Hommann (1995) estimate that in 1995 there were 19.8 million hospital admissions and 1,200 million minor sicknesses per year. In Delhi alone there are human health damages worth Rs. 1,170 million per year.

Approximately half of the world's population and up to 90% of rural households in developing countries still rely on unprocessed biomass fuels such as wood, dung, and crop residues. In rural areas, air pollution is primarily produced indoors from the use of firewood and other unclean sources of cooking fuel. Since 72% of India lives in rural areas this form of air pollution has a significant impact on the population as a whole. Indoor pollution causes 0.41–0.57 million premature deaths per year, and for each death there are about 6 person years of illness (Parikh, 2004). Making cleaner technologies available and affordable can aid in lowering pollution and improving health.

3.3.7 *Land and Forests*

In India about 57% of the total land area is under some form of degradation, and a greater part of the land is severely affected by soil and water erosion problems. Productive lands are essential to meet India's need for food, fuel, and fodder. In addition, they help conserve biodiversity and water. According to the National Wasteland Development Board about 175 million hectares (53% of the country's total geographic area) is degraded. This compromises life support systems and the livelihoods of poor and tribal people (Parikh, 2004).

There are 16 major forest types comprising 221 minor types in India. Of these, the tropical moist deciduous forest forms the major percentage (37%) of forest

Table 3.2 Standard for air quality and actual pollution in seven major cities during 2002

Pollution level	Standards set by CPCB of annual mean concentration range (μg/m^3) in industrial (I) and residential (R) zones				
	Industrial zones (I)			Residential zones (R)	
	SO_2 and NO_2	RSPM	SPM	SO_2, NO_2, RSPM	SPM
Low (L)	0–40	0–60	0–180	0–30	0–70
Moderate (M)	40–80	60–120	180–360	30–60	70–140
High (H)	80–120	120–180	360–540	60–90	140–210
Critical (C)	>120	>180	>540	>90	>210

	SO_2		NO_x		RSPM		SPM	
State/city	Ind	Res	Ind	Res	Ind	Res	Ind	Res
Hyderabad	L	L	L	M	M	H	M	H
Delhi	L	L	L	M	C	C	H	C
Ahmadabad	L	L	M	M	C	C	M	C
Bangalore	L	L	L	M	M	H	L	H
Mumbai	L	L	L	L	M	H	M	C
Chennai	M	L	L	L	M	M	M	M
Kolkata	L	L	H	H	H	C	M	C

Source: CPCB

Table 3.3 State-wise area of the arid zone in India

State	Area under arid zone	Percent area
Rajasthan	1,96,150	62
Gujarat	62,180	19
Punjab	14,510	5
Haryana	12,840	4
Maharashtra	1,290	0.4
Karnataka	8,570	3
Andhra Pradesh	21,550	7
Jammu and Kashmir	70,300	–
Total area	3,17,090	–

Source: ESPASSA (2008)

cover in India (ESPASSA, 2008). Tropical dry deciduous forest forms 28.6%, and the remaining types are scattered in minor proportions.

The arid and semiarid zones in India are spread over eight states, but 90% of the hot desert is located in the northwestern part of the country (Table 3.3). Of this, 62% is located in the state of Rajasthan. The Great Indian Desert, or the Thar, is situated on the eastern-most fringe of the Saharan-Rajasthan plain. This desert is by the far the most populated one in the world, the human population being 75 km^{-2} as compared to 3–5 in other deserts. The Indian Thar desert extends about 2.34 million km^2 covering parts of Rajasthan, Gujarat, southwestern Punjab, Haryana, and Karnataka. Many people and livestock depend on this desert. The soil of the land is fertile – full of dormant seeds of various species – and with a little precipitation it blooms with a wide range of vegetation and attracts animals and birds.

3.3.8 Valuation of Environmental Degradation in India

The preceding sections reveal that India suffers from a large number of environmental problems. Some studies have tried to estimate the value of environmental degradation at national and sectoral levels.

Murty and Kumar (2004) estimated the cost of industrial water pollution abatement and found that these costs account for about 2.5% of industrial GDP in India. Forests are associated with ecosystem services such as soil protection, water augmentation (recharging groundwater), flood control/regulation, carbon sequestration, and nutrient cycling. Manoharan (2000) provides a review of a large number of valuation studies that throw considerable light on the magnitude of intangible benefits or ecosystem services accruing from India's forests. Parikh (2004) provides estimates of how much India lost from land and forest degradation in 1990. These losses are very significant given the level of GDP in India (Table 3.4). The entire estimates are based on various studies that use different methodologies and thereby are not comparable, yet the estimates are good enough to understand the implication of losses.

Table 3.4 Estimates of economic value of environmental degradation in India (in billion rupees)

Human health damages due to water pollution and poor sanitation	360
Loss of crop productivity due to soil degradation	89–232
Loss of wood due to forest degradation	57
Human health damages due to air pollution	885–4,250

Source: Parikh (2004)

Table 3.5 Alternative estimates of costs of water pollution (Rs. millions per year at 1995 prices)

1. Damage costs	
(a) Value of annual loss of 30.5 million DALYs @ average per capita GDP of Rs. 12,000	366,000 3.95% of GDP (1995–1996)
2. Avoidance costs	
(a) Pollution abatement in organized industry	10,120
(b) Pollution abatement in small-scale industry	45,980
(c) Wastewater treatment in 3,696 cities/towns	3,620–10,540
(d) Provision of toilets to 115 million households	35,300–56,630
(e) Provision of safe drinking water	39,300
Annualized cost (assuming operation and maintenance costs of installed facilities at 20% of capital costs)	134,320–162,550 26,860–32,510
Annual costs (capital + O&M)	161,180–195,060
Annual cost as percent of GDP (1995)	1.73–2.1%

Source: Parikh (2004)
(a)–(d) at 15% discount rate and 15 years life

The potent question in valuation is that of avoidance of pollution costs versus damage due to pollution. Parikh (2004), again using the estimates of various studies, shows that the cost of avoidance is much lower than damage costs (Table 3.5). In 1995 alone, India lost about Rs. 366 billion, which accounts for about 3.95% of GDP due to water pollution and poor sanitation facilities. On the other hand, avoidance costs in terms of infrastructure and abatement costs are required to reduce the level of water pollution range from 1.73% to 2.2% of GDP. Moreover, these damage costs do not fully reflect the loss in social welfare. These estimates suggest that abatement of pollution is socially desirable and economically justified. It would be prudent to invest in water pollution management.

Parikh and Parikh (2001) estimate the costs of degradation of the four most important natural resources of India, namely, air, forests, water, and cultivable soils for the mid-1990s. Their findings are summarized in Table 3.6. These estimates alone account for about 3.58%–4.99% of GDP.

3.4 Causes of Poor Environmental Compliance

The degree of compliance with environmental regulations in a state depends on the probability of detection of noncompliance and the severity of punishment if detected and convicted. Firms treat fines as a cost of doing business, and firms

Table 3.6 Annual cost of environmental degradation in India 1994–1997 (% of GDP)

Resource	Range
Air	0.4
Forests	1.1–1.6
Soil	0.30–0.80
Water	1.70–2.1
Total	3.5–4.9

Source: Parikh and Parikh (2001)
*Does not include damage due to indoor pollution

minimize the significance of expected compliance costs and expected penalties. Given that, the effectiveness of environmental regulations could be increased either by increasing the penalties or by increasing the probabilities of detection through more monitoring or both. But in developing countries like India, it is difficult to do either, given the resource constraints and the political economy of environmental pollution.

Environmental compliance in a country is the function of the cost of pollution abatement, the comprehensiveness of the environmental laws in relation to the level of development of the country, the capacity of the industry to bear the costs of abatement, the costs of noncompliance and the probability of detection of noncompliance (Priyadarshini and Gupta, 2003).

The regulatory regime for environmental protection in India, like so much else in the country, is a picture of sharp contrasts (Gupta, 2001). On one hand, the country has elaborate statutes and regulations on almost every conceivable topic from hazardous waste to public liability to forests and wildlife. On the other hand, monitoring and enforcement capabilities are weak, and many of these statutes remain on paper. Similarly, while an extensive institutional framework and set of policies for environmental protection have evolved over the years, the complexity and magnitude of environmental problems has increased manifold.

Environmental legislation and policies in India are procedural and have typically taken punitive measures, including the extreme punitive measure of closure, without any clear policy guidelines. The approach adopted by the regulatory bodies is basically Command and Control, where laws exhibit an "end-of-pipe treatment" of pollution rather than prevention of pollution, i.e., proactive role. The command is the laying down of standards and pollution limits, while the control is the power to withdraw the water or power supply of noncomplying firms, the imposition of penalties and fines, or even imprisonment. Thus the present framework to address the problem of pollution control and prevention does not take into consideration incentives to control pollution and proves to be ineffective, as mentioned earlier.

The standards set and implemented in a state depend on the governance quality of the state. If the state is better administered, then implementation of environmental norms and standards may be better, though this is not necessarily the case. Infrastructure facilities, such as laboratories for testing samples, a thorough understanding of environmental problems, and good monitoring and enforcement capabilities, determine the effectiveness of legislation (Priyadarshini and Gupta, 2003).

Curmally (2002) points out that the work of the PCBs in India falls much short of the fulfillment of the above criteria, and so the effectiveness of the CAC method in controlling pollution is minimal. These problems lead to weak enforcement and poor monitoring.

Some provisions of environmental laws have, however, either not been implemented or have been interpreted so liberally as to defeat the very purpose of the legislation. For example, while the statutes of the Water, Air, and Environment Pollution Act tackle quite broad-based environmental problems and suggest punitive actions for the offenders, they are implemented by the SPCBs, which in general have poor track records for implementation. One of the reasons attributed for this is that members of the State Control Board have sometimes been political appointees and may not have the relevant environmental expertise or resources (Jha, 1999).

Notwithstanding the constraints to the implementation of various acts, 415 projects were appraised for environmental clearance using the prescribed EIA methodology for the year 1996. Of these 415 projects, only 170 were able to obtain environmental clearance. Of the remainder, 18 industrial projects were exempted from environmental clearance and the rest of the projects were rejected (Jha, 2004).

There are several examples of the inefficient functioning of these Boards. One important example often quoted by the press is that the Madhya Pradesh (one large state in India) SPCB had given a pollution control clearance to Union Carbide's pollution control equipment just a few weeks before the Bhopal gas accident. Further SPCBs may be slow to respond to community and NGO initiatives. Section 15(d) of the Environment Protection Act allows for community action against industries responsible for polluting the environment. However 60 days notice is required to be given to the SPCB, presumably to enable it to initiate action on its own. In several instances these community initiatives have not been acted upon. Also according to the Act, to convict a polluting industry, air and water samples have to be collected by the SPCB, and the latter has been known to delay collection indefinitely (Jha, 2004).

According to the Factories Act, submitting a detailed disaster management Plan and environmental impact assessment to the factory inspectorate and the environment ministry is mandatory for hazardous units. The act also stipulates that these documents have to be produced on demand by any citizen of India. These provisions have often been flouted by industries, including public sector industries. The pollution control board's technical capacity to carry out EIAs is also limited, leading to further difficulties in the implementation of the provisions of the Factories Act.

3.5 Concluding Remarks

The first real impetus for developing a framework for environmental protection in India came after the UN Conference on the Human Environment in 1972. Environmental policy in the 1970s and 1980s recognized the need for an institutional identity for environmental policy making, resulting in the setting up of the government agency in India. This chapter provides an explanation of environmental regulations and the state of environment in India. The degree of compliance with environmental regulations in a state depends on the probability of detection of noncompliance and the severity of punishment if detected and convicted. The noncompliance of environmental standards is a problem and possible causes are discussed.

Appendix 1: Key Environmental Legislation in India: An Illustrative List

Policies
1992 Policy Statement on Abatement of Pollution
1992 National Conservation Strategy and Policy Statement on Environment and Development
1998 National Forest Policy
2002 Wildlife Conservation Strategy
2006 National Environment Policy
Environment Acts
1927 The Indian Forest Act
1972 The Indian Wildlife (Protection) Act (amended 1993)
1973 The Water (Prevention and Control of Pollution) Act (amended 1988)
1977 The Water (Prevention and Control of Pollution) Cess Act (amended 1992)
1980 The Forest (Conservation) Act (amended 1988)
1981 The Air (Prevention and Control of Pollution) Act (amended 1987)
1986 The Environment (Protection) Act (amended 1992)
1988 The Motor Vehicles Act
1991 The Public Liability Insurance Act (amended 1992)
1995 National Environment Tribunal Act
1996 National Environment Appellate Authority Act
2002 The Wildlife (Protection) Amendment Act
2002 The Biological Diversity Act
2003 The Water (Prevention and Control of Pollution) Cess (Amendment) Act
Environment Rules
1986 The Environment (Protection) Rules
1989 Hazardous Wastes (Management and Handling) Rules
1990 Forest (Conservation) Rules (amended 1992)
1991 Chemical Accidents (Emergency Planning, Preparedness and Response) Rules
1998 The Biomedical Waste (Management and Handling) Rules
1999 The Recycled Plastics Manufacture and Usage (Amendment) Rules
2000 The Municipal Solid Wastes (Management and Handling) Rules
2000 The Hazardous Wastes (Management and Handling) Amendment Rules
2000 The Ozone Depleting Substances (Regulation and Control) Rules
2001 The Batteries (Management and Handling) Rules
2002 The Noise Pollution (Regulation and Control) (Amendment) Rules
2003 The Recycled Plastics Manufacture and Usage (Amendment) Rules
2003 Biomedical Waste (Management and Handling) (Amendment) Rules
2003 Forest (Conservation) Rules

(*continued*)

(continued)

2003 Draft Biological Diversity Rules
Environment Notifications
1994 Environmental Impact Assessment Notification 1994 (amended 2002)
1998 Constituting the Taj Trapezium Zone Pollution (Prevention and Control) Authority
1999 Fly Ash Notification
1985 The Vienna Convention/Montreal Protocol on substances that deplete the ozone layer
1972 The Rio Declaration on Environment and Development and the Agenda 21
International Agreements to which India is a Signatory
1975 The Convention on International Trade in Endangered Species of Flora and Fauna (CITES)
1991 The Convention on Wetlands of International Importance (the Ram Sar Convention)
1992 The Framework Convention on Climate Change
1992 The Convention for Conservation of Biological Resources

Source: World Bank (2007)

Appendix 2: Major Polluting Industries

Industry	Key environmental aspects
Aluminum	Disposal of red mud, bauxite tailings and other hazardous waste, dust emissions and high-energy consumption
Caustic	Water pollution due to disposal of brine mud, mercury and chlorine; chlorine emissions
Cement	Fugitive dust emissions from material handling and air emissions from stack; energy consumption
Copper	Sulfur dioxide and dust emissions; water pollution from electrolytic bath and other processes; disposal of slag from smelter
Distillery	Water pollution due to highly organic effluent from spent wash; soil contamination
Dyes and dyes intermediates	Water pollution due to toxic azo-dyes, highly organic colored and phenolic substances
Fertilizer	Water pollution due to heavy metal, ammonia- and fluoride-bearing effluent, ammonia emissions, fluoride-bearing dust and hazardous material
Iron and steel	Water pollution from cyanide, fluoride- and heavy metal–bearing effluent, dust emission from sintering, pelletization, pig iron plants; slag and dust disposal
Leather	Water pollution, particularly from hexavalent chromium and salt in discharge
Pesticides	Air pollution due to particulate and volatile organic compounds; effluent containing pesticides residues
Petrochemicals	Water pollution due to phenol- and benzene-containing effluent; fugitive emissions of toxic and carcinogenic and volatile organic compounds (VOC); hazardous material disposal
Pharmaceuticals	Water pollution due to organic residue-bearing effluent; VOC and particulate emissions; hazardous waste containing process sludge and spent catalyst
Pulp and paper	Water pollution from high organic and inorganic substance and chlorinated compounds in black liquor; highly malodorous emissions of reduced sulfur compounds and VOC

(*continued*)

(continued)

Refinery	Water pollution from effluent containing organic and inorganic material, oil, and solvent; air emission of particulate matter, sulfur dioxide, "benzene, toluene, and xylene," VOC
Sugar	Water pollution due to high biological oxygen demand (BOD) and chemical oxygen demand (COD) effluent and spillage of molasses; air pollution due to combustions of bagasse, coal, etc.
Thermal power plants	Air emissions from combustion, coal handling, water pollution due to discharge of boiler blow down, overflow from ash pond; land contamination due fly ash disposal practices
Zinc	Air pollution due to fugitive zinc dust, water pollution containing residues, disposal of solid and hazardous waste

Source: CPCB – List of "Red Category" Polluting Industries

Appendix 3: Sector-Wise Compliance Status of 17 Categories of Highly Polluting Industries (June 2006)

No.	Industrial category	Complying	Defaulting	Closed[a]	Total
1	Aluminum	6	1	0	7
2	Cement	198	16	20	234
3	Chlor-Alkali	24	10	0	34
4	Copper	3	1	0	4
5	Distillery	191	35	36	262
6	Dyes and DI	87	9	25	121
7	Fertilizer	104	10	21	135
8	Iron and steel	28	9	1	38
9	Oil refineries	17	3	1	21
10	Pesticides	95	9	11	115
11	Petrochemicals	73	7	1	81
12	Pharmaceuticals	351	124	59	534
13	Pulp and paper	118	32	37	187
14	Sugar	438	49	91	578
15	Tannery	97	13	17	127
16	Thermal power	129	51	8	188
17	Zinc	4	1	1	6
18	Total	1,963	380	329	2,672

Source: CPCB

[a]Some of the industries may have been shut down temporarily, often until corrective actions have been agreed upon

Chapter 4
Intergovernmental Fiscal Transfers and the Environment

4.1 Introduction

Provision of environmental services involves spatial externalities. The costs of provision are borne at the level of provision, but the benefits are realized on a larger scale.[1] Mismatch between the decision-making responsibilities and costs and benefits has been considered a cause for the underprovision of the services. Perrings and Gadgil (2003) suggest a number of measures to patch up the local costs and global benefits of biodiversity conservation. Intergovernmental fiscal transfers are an important instrument for internalizing the spatial externalities (Breton, 1965; Olson, 1969).

Ecological functions[2] performed by a subnational government are the classic case of fiscal externalities, and intergovernmental fiscal transfers should consider these functions like social and economic functions. Neglecting environmental services in the fiscal transfers causes twofold effects: inadequate incentives/compensation to those conserving natural resources on the one hand, and lack of disincentives to those frittering away such precious resources on the other hand. The implementation of the concept of sustainable development requires the inclusion of environmental services in intergovernmental fiscal relations. It calls for the consideration and appropriate financing of these services at any governmental level (Ring, 2002).

Environmental policy debate predominately focuses on negative externalities and favors pollution taxes, fees, among other instruments and ignores positive externalities offered by the natural resources. Conservation activities such as

[1] It also can be the other way round. The benefits are local but the costs are borne at the regional or national level. In general there is a mismatch if there is no fiscal equivalence.

[2] Environmental services or "ecological public functions consist of the protection and sustainable use of natural resources, living organism, ecosystems and landscapes." These also include negative impacts of human activities on the environment in the form of environmental pollution such as emissions, waste and contaminated sites, impaired or destroyed landscapes, etc. (Ring, 2002).

S. Kumar and S. Managi, *The Economics of Sustainable Development*, Natural Resource Management and Policy 32,
DOI: 10.1007/978-0-387-98176-5_4,

afforestation and protection of watersheds offer positive externalities, and surprisingly much less has been written about internalizing these externalities (Ring, 2008a). These externalities can be internalized by compensating the providers of the services, and fiscal transfers are an innovative way of compensating local and state public actors, i.e., decentralized jurisdictions in federal systems, for the environmental services they provide beyond their own boundaries.

Deteriorating environmental quality has been threatening the sustainability of the growth trajectory,[3] and Indian states have been raising concerns related to divergence between costs borne and benefits enjoyed by them for the conservation of natural resources.[4] The *12th Finance Commission* recognized the mismatch and allocated Rs. 10 billion for preservation of forests (Government of India, 2004). The *13th Finance Commission* has been given the mandate to make recommendations that help in managing ecology, environment, and climate change consistent with sustainable development (Government of India, 2008). This chapter intends to critically analyze the contribution of existing intergovernmental fiscal transfers to environmental sustainability in India.

India has a three-tier federal system of governance, and the responsibilities of governance are shared between the union government,[5] the state governments, and the local governments (i.e., rural and urban local bodies). Assignment of responsibilities on all matters, including environment, between the different tiers of government is governed by the Indian Constitution.[6] The assignment of functions related to environmental activities is fairly clear in India and tries to minimize the transaction costs by giving much scope to the decentralized systems of environmental governance. The central government is responsible for determining the overall policy frame, and state and local governments are involved in implementation. Environmental degradation cannot be avoided simply by assigning the task at appropriate level; genesis of degradation can be found in the incentives structure of governance (Mandal and Rao, 2005).[7]

Integration of environmental aspects in the fiscal transfers also helps the states in conserving natural resources and addressing poverty. Regional distribution of natural resources and the level of economic development are highly skewed across states. Poverty distribution coincidently is linked to the distribution of ecosystems

[3]Dasgupta (2007) finds that economic development during 1970–2000 in the Indian subcontinent was either unsustainable or barely sustainable when the productive base of the countries is taken into account.

[4]Recently, the chief minister of Himachal Pradesh, Mr. P. K. Dhumal raised such kind of concerns. He said that Himachal preserves and conserves its vast natural potential for the benefit of the nation, but it has not been compensated. (Why can't Himachal get carbon credits for keeping India green? *The Indian Express*, 8 April 2008).

[5]The words "union government" and "central government" or "the center" are used interchangeably, and refer to the federal government.

[6]The Indian Constitution, a lengthy document, comprises 395 Articles and 12 Schedules. Since its inception, it is amended 104 times, thus, it is very much a living document.

[7]For details on Environmental Federalism in India, see Gupta (2001) and Mandal and Rao (2005).

and their health in India (ESPASSA, 2008). States that house most of the natural resources suffer damage from others' actions but remain uncompensated. This happens because natural resources are typically underpriced (Dasgupta, 2007b), and the revenues generated from the exploitation of natural resources are shared more generally.

This chapter is organized as follows. Section 4.2 makes an inquiry into the theory of fiscal federalism and analyzes the role of the theory in protecting and enhancing the quality of environment. An empirical investigation of environmental concerns in Indian intergovernmental fiscal relations is carried out in Section 4.3. Section 4.4 explores possible options for integrating ecological functions into intergovernmental fiscal transfers. An illustration of how ecological indicators could be included in the fiscal devolution mechanism is provided in Section 4.5. Some international practices of compensating the public sector actors for conservation activities and concluding remarks are discussed in Section 4.6.

4.2 Theory of Fiscal Federalism and the Environment

4.2.1 Decentralization and the Environment

Local public goods and services are provided more efficiently when resource allocation decisions are limited to the lowest governmental level (Oates, 1972). This makes it possible to respond more appropriately to the regionally heterogeneous preferences (Tiebout, 1956), introduce intergovernmental competition and checks and balances (Breton, 1996), and reduce coordination and transactions costs. The decentralization rule in resource allocation is applicable in the absence of economies of scale and externalities.

In the provision of local public goods, spatial externalities exist between jurisdictions. A match is not found among those who decide about a public good, those who pay for it, those who receive its benefits, and thus the good remains underprovided. The principle of fiscal equivalence helps in achieving the match (Buchanan, 1950; Olson, 1969) and intergovernmental fiscal transfers are supposed to ensure efficiency on this account (Buchanan, 1950; Rao, 2005).

Following the rule of decentralization, lower levels of governments are assigned the task of protecting the environment where appropriate. There are numerous studies demonstrating that decentralization works much better in environmental protection than does a top-down mechanism (e.g., Chopra et al., 1990). The presence of spatial externalities in the provision of environmental services calls for a differentiated approach in executing the decentralization rule. Appropriate solutions have to be sought according to the specific characteristics of the various environmental problems. This is reflected in the debate "regarding the competencies of the national or even supranational governmental level versus the state or local level in environmental standards setting" (Ring, 2008c; see also Oates, 1998, 1999).

Highly mobile environmental services and pollutants that easily cross administrative boundaries create far reaching spatial externalities and require more centralized solutions (Ring, 2002). For example, the problem of climate change requires centralized solutions if not global policies. Similarly, public goods such as basic and applied research, including that concerning the development of environmental policy instruments but also the dissemination of information on harmful environmental impacts or the development of pollution control techniques, tend to be underprovided at decentralized levels (Oates, 2001). In contrast, an environmental policy associated with less mobile environmental services/pollution is better suited for assignment to decentralized levels of government (Oates, 2001).[8]

Though land use related policies are suitable at local levels, spatial externalities may require different and more appropriate solutions. Organic farming, cultivation of high-agrobiodiversity crops, coexistence of genetically and nongenetically modified crops, and management of invasive species in shifting cultivation systems are examples where spatial externalities stemming from neighboring or proximate farms affect the returns and hence land use decisions (Lewis et al., forthcoming). Sigman (2005) estimates the environmental costs of water pollution generated downstream due to free-riding states when rivers cross state boundaries in the United States. These examples reveal that the misallocation of resources cannot be avoided simply by assigning the task at the appropriate level because these mechanisms do not provide adequate incentives for internalizing the externalities.

Whereas transboundary water pollution is associated with negative externalities, priority areas for water protection can involve positive externalities. Water protection zones are generally located in rural areas; villages bear the costs of protection but provide water services far beyond their boundaries. Conservation and sustainable use of biodiversity is another example of spatial externalities (Perrings and Gadgil, 2003). Spatial scale and mobility are very important determinants of species protection. In a study of federal and state spending under the Endangered Species Act in the United States, List et al. (2002) find the phenomenon of free riding on the part of the states. States tend to spend less relative to the federal government on those species that demand a large habitat area and whose preservation causes conflicts with economic development. Rewarding local communities for their conservation efforts is necessary to reconcile both local and global public benefits of conservation of natural resources (Perrings and Gadgil, 2003; Millennium Ecosystem Assessment, 2005; Ring, 2008c).

To summarize, intergovernmental fiscal transfers can help in achieving environmental sustainability because (1) economic and political incentives to states and local bodies often favor the destruction of natural resources since the benefits from conservation and costs of conservation are distributed asymmetrically across territorial units (Köllner et al., 2002); (2) the ecologically rich regions often have to

[8] According to the Principle of Subsidiarity the responsibility of providing public goods should be assigned to the smallest jurisdiction whose geographic scope encompasses the relevant benefits and costs associated with the provision of the good in question.

spend more on conservation measures than others and are in need of more resources given their fiscal capacity;[9] and (3) decentralization works better in the management and conservation of natural resources.

4.2.2 Designing Fiscal Transfers for Environmental Sustainability

Dahlby (1996) provides a formula for designing fiscal transfers in the provision of public goods provided by the local governments. Using the Atkinson and Stern (1974) condition of optimal provision of public good by a local government, Dahlby sets the matching grant rate equal to

$$m_{g_i} = \sigma_{g_i}\left(1 - \rho_{g_i}\right) + \left(\rho_{g_i} - \rho^{i}_{g_i}\right), \tag{4.1}$$

where m_{g_i} is the matching grant rate for public good g_i to its provider state i, σ_{g_i} is the fraction of direct benefits that go to people who reside outside i (positive externalities), ρ_i is the change in total revenue of state i per rupee spent on g_i, and $\rho^{i}_{g_i}$ is the additional revenue to state i from an additional rupee spent on g_i.

This formula has two components. The first term is the value of direct positive externalities that has to be financed through intergovernmental grants in the absence of net revenue spillovers. The second term represents net revenue spillovers. If the expenditure on public good is financed through nondistortionary taxes, then the matching rate would be $m_{g_i} = \sigma_{g_i}$. In the absence of spillover effects, but with distortionary taxes, the matching rate would be equal to the second term. In practice, public goods are financed through distortionary taxes and also create revenue spillovers; the matching rate would be equal to the sum of both of the terms.

The value of σ_{g_i} is not known by the central governments (here assuming that the provision of good i is revenue neutral), Dur and Staal (2008) show that two types of transfers – earmarked and lump-sum – help in mitigating the underprovision problem and in increasing the allocative efficiency of public good provision in the economy. The earmarked transfers are influenced by the level of public goods and are spent on these goods only.

Given the logic of public good provision, only the direct costs accruing for environmental services provide a basis for matching grants. The grantor finances the fraction of a recipient's expenditure that externally benefits other jurisdictions on the basis of (4.1). Clean-up activities such as air and water pollution abatement and their regulation, municipal solid waste management, reforestation, management of public parks, and development of renewable energy are good candidates for

[9] As mentioned, the ecologically rich states are economically poor in India and their own fiscal capacity to spend on conservation is limited.

earmarked grants. In the absence of the known value of spillover effects, appropriate procedures for matching contributions are required to be developed that approximately reflect the local and external benefits of relevant expenditures.

Note that some nature conservation activities affect the ability to develop productive activities and generate revenue in a number of ways, both for local governments and private land users. It is economically rational for local governments not to be interested in the protection of nature conservation and watershed protection activities given the mismatch between net benefits forgone by them and externally enjoyed benefits. Therefore, it would be rational to compensate local governments for the foregone opportunities. The compensations could be in the form of lump-sum fiscal transfers. Appropriate indicators of environmental performance need to be identified that might constitute a link between the environmental services and the corresponding costs needed for their provision (Ring, 2002), and are required to be used for modifying the existing formulas of lump-sum intergovernmental fiscal transfers.

4.3 Fiscal Federalism and the Environment in India

4.3.1 Fiscal Federalism in India

Similar to the assignment system, fiscal federalism in India is characterized by constitutional demarcation of revenue and expenditure powers among different levels of government. The finance commission facilitates the division of financial resources between different levels of governments. The federal finance commission recommends the distribution of the net proceeds of taxes and grants-in-aid from the center to states and also among the states, and the state finance commissions determine the revenues of local governments.

The main considerations before a federal finance commission are (1) how the proportion of central tax revenue to be shared is to be determined; (2) specifying criteria for deciding shares to be received by individual states; and (3) determining the weights attached to different allocation criteria (Government of India, 2004; Hazra et al., 2008). The tax devolution criterion involves three sets of considerations: (1) population, tax efforts, and fiscal discipline to correct vertical imbalance; (2) the income distance method[10] to correct horizontal imbalance; and (3) area to account for cost disabilities (Rangarajan and Srivastava, 2008). As of now, inter se sharing of taxes between the Union and states, according to the recommendation of 12th Finance Commission, is governed by the broad criteria given in Table 4.1.

[10]"Distance formula $= (Y_h - Y_i)P_i / \sum (Y_h - Y_i)P_i$, where Y_i and Y_h represent per capita state domestic product (SDP) of the *i*th and the richest state, P_i is the population of the *i*th state, $(Y_h - Y_i)$ for the *h* state is to be equivalent to that of the second highest per capita SDP state" (Rao, 2000).

Table 4.1 Criteria for intergovernmental transfers adopted by the 12th Finance Commissions in India

Criteria	Relative weights (%)
Population	25
Income (distance method)	50
Area	10
Tax effort (income weighted)	7.5
Fiscal discipline	7.5
Total	100

Source: The 12th Finance Commission Report (Government of India, 2004)

The formula-based disbursement constitutes the unconditional grants which the state governments use in the way they wish. In addition to these transfers, there are grants-in-aid, which are given for specific purposes; sometimes they are partly in the form of matching grants. The 12th Finance Commission recommended equalization grants for education and health with the aim of augmenting the equalization content of fiscal transfers (Rangarajan and Srivastava, 2008).

Besides the Finance Commission, the Planning Commission is also a major distributor of funds in India. It provides grants and loans to the states. The funds are distributed according to a formula evolved and modified by the National Development Council (NDC) from time to time. The criteria used by the Planning Commission consider population (60% weight) and fiscal management efforts (7.5% weight) for addressing vertical imbalance and the distance in per capita income from the national average (25% weight) for dealing with horizontal imbalance. The predominant weight given to population favors urban and populated areas.

In addition to these two agencies, states and local bodies also receive purpose-specific grants from various central ministries. Some of these grants are entirely funded by the central government, and some are shared cost programs. Moreover, state governments also get implicit transfers in the form of subsidized loans from the central government and priority sector borrowing from financial and banking system (Rao, 2000).

The main concerns before a state finance commission are (1) distribution of the revenue of the state between the state and local governments and determining the allocation of individual local governments; (2) assignment of tax and nontax powers to village panchayats and urban local bodies; and (3) determination of the grants-in-aid to the local governments from the consolidated fund of the state (Rao, 2000). The local governments also receive funds for implementation of the central government's schemes through the state government. Note that local governments have little flexibility in the use of these funds and rarely execute any development program (Rao, 2000).

In their recommendations, consecutive Finance Commissions tread cautiously in balancing concerns of equity versus efficiency, arriving at a scheme of fiscal transfers that is predictable and stable from socioeconomic aspect. However, the issues of fiscal spillovers in general and environment sustainability in particular had not found a place as it should, of the financial devolution mechanism in the country.

4.3.2 *Fiscal Transfers and Provision of Environmental Services in India*

Fiscal transfers can effectively address fiscal externalities, either vertical or horizontal, arising from expenditures on environmental activities by regional governments; these transfers help achieve national standards in environmental programs, like socioeconomic programs, and induce efficiency in the functioning of the economy (Dahlby, 1996). These transfers can be linked to the fiscal needs for environmental indicators where the respective revenues do not necessarily have to be used for environmental purposes. The transfers can be both conditional (earmarked) and nonconditional grants from the central government to state governments and from the state government to local governments.

Before discussing the role of fiscal transfers in the provision of environmental services, it is necessary to make some preliminary remarks. In addition to above mentioned fiscal transfers, state governments and local bodies are familiar with a number of other earmarked grants that include environmental functions – for example, an antipoverty program such as Bharat Nirman,[11] in which local bodies can take up projects that have a bearing on environmental conservation. Similarly, the Jawaharlal Nehru Urban Renewal Mission,[12] a program for urban areas, includes projects related to provision of water services, solid waste management, etc. A comprehensive analysis of fiscal grants should also consider additional policies and regulations in conjunction with fiscal equalization where the local jurisdiction can serve as applicant (Ring, 2002).[13]

4.3.2.1 Geographic Area as an Indirect Criterion for Fiscal Transfers

The geographic area of a state can be a starting point for considering environmental services in intergovernmental fiscal transfers (Ring, 2002). The provision of environmental services is directly linked to land uses. A state having less population density is supposed to delegate more land to agriculture and forestry and house valuable habitat for rare species. Similarly, within a state the areas that are remote from the district centers have low population density and provide more environmental services given the land use practices in these areas.

[11] Bharat Nirman is a rural development project initiated by Government of India in partnership with State Governments and Panchayat Raj Institutions for the period 2005–2009. Under Bharat Nirman, actions are proposed in the areas of irrigation, road, rural housing, rural water supply, rural electrification, and rural telecommunication connectivity.

[12] Jawaharlal Nehru National Urban Renewal Mission is a massive city modernization scheme launched by Government of India. It envisages a total investment of over $20 billion over a period of 5–6 years.

[13] Analysis of additional policies and regulations is beyond the scope of present chapter.

In India, federal finance commissions consider the "area" of a state as one of the criteria to account for cost disabilities in providing public goods in fiscal transfers. They define cost disabilities as circumstances such as excess rainfall, hilly terrain, and large and remote areas with low population density that are beyond the control of a state and lead to higher than average per capita costs for delivering the same level of services at an average level of efficiency. The use of area of a state as a criterion for determining its share stems from the additional administrative and other costs that a state with a larger area has to incur in order to deliver a comparable standard of service to its citizens (Government of India, 2004). The Finance Commissions recognize that the costs of providing services increase with the size of a state, but at a decreasing rate. Similarly, the State Finance Commissions use area and remoteness as criteria for financial devolution from the state governments to rural and urban local bodies.[14] The consideration of area as an indicator in fiscal transfers from the cost disability point of view recognizes economic functions; this criterion has also relevance for internalizing the environmental externalities.

Though one can argue that indirectly the ecological functions are already considered in the fiscal devolution in India, the existing regulations concerning area as an indicator in fiscal transfers at all levels predominately concentrate on socioeconomic functions. To account for ecological functions the weight attached to the area criterion should be adequately strengthened until an appropriate direct criterion of environmental sustainability is considered.

4.3.2.2 Provision of Environmental Services and Grants-in-Aid

The Finance Commissions recognize that the formula used for allocation of tax proceeds among states cannot take care of all dimensions of the fiscal needs of a state. Therefore, the lower levels of governments receive certain grants-in-aid. Some of the grants-in-aid are common for all the states and some are specific to a particular state given its needs. These grants-in-aid are purpose-specific earmarked grants.

In India, there is a total ban on green felling. State governments consider forests a net liability rather than a source of revenue. This made the 12th Finance Commission aware that maintenance of the forest area as required by the working plans had become a problem due to financial constraints. They emphasized the need for separate grants for the maintenance of forests. The 12th Finance Commission recognized the problem and recommended a grant of Rs. 10 billion spread over the award period of 2005–2010 for maintenance and preservation of forests that

[14]For example, the 2nd Uttaranchal Finance Commission used area and remoteness in addition to population, tax efforts, and deprivation index for devolution of finances to the local bodies (Second Uttaranchal Finance Commission Report as accessed on 24 June 2008, http://gov.ua.nic.in/sfc/Second.htm).

Table 4.2 Environment-related grants-in-aid recommended by the 12th Finance Commission

State	Environment-related activity	Amount (rupees billions)
Haryana	Water logging/salinity and declining water tables	1.00
Kerala	Inland waterways and canals	2.25
	Coastal zone management	1.75
Maharashtra	Coastal and ecotourism	2.50
Manipur	Loktak lake	0.115
Meghalaya	Zoological park	0.30
	Botanical garden	0.05
Mizoram	Bamboo flowering	0.40
Orissa	Consolidation and strengthening ecorestoration work in the Chilika lake	0.30
	Sewerage system for Bhubaneswar	1.40
Punjab	Stagnant agriculture	0.96
Rajasthan	Indira Gandhi Nahar Pariyojana	3.00
	Meeting drinking water scarcity in border and desert districts	1.50
Tamil Nadu	Sea erosion and coastal area protection works	0.50
West Bengal	Arsenic contamination of groundwater	6.00
	Problems relating to erosion by Ganga-Padma river in Malda and Murshidabad districts	1.90
	Development of Sundarbans Regions	1.00
Total		24.925

Source: The 12th Finance Commission Report (Government of India, 2004)

would be distributed among the states according to their forested area (Government of India, 2004).

The 12th Finance Commission also recommended total state specific grants-in-aid of the amount of Rs. 71 billion over its award period. Table 4.2 shows that about 35% of these grants-in-aid are allocated for environment-related activities.[15] Table 4.2 reveals the following points. First, except for the common pool of the earmarked grants-in-aid for the maintenance of forest areas, ecological functions are not included directly in the fiscal transfers. However, beyond area, the states get some earmarked grants for the provision of environmental services. Second, the fiscal equalization rules consider ecological functions by means of conditional grants. Most of the fiscal transfers are explicitly related to sewage disposal, water supply, waste disposal, urban and agricultural development, etc. Third, there is a widespread tendency to support the end-of-pipe infrastructure such as drinking water supply provision. Though a proportion of total grants-in-aid is *implicitly* kept aside for environmental management, it is by no means enough to internalize the externalities that cause environmental degradation.

[15] For the definition of activities or what these activities involve see the 12th Finance Commission Report, Chap. 10 (Government of India, 2004).

4.4 Fiscal Options for Integration of Environmental Services into Fiscal Transfers

A suitable way to mitigate the problem of underprovision of environmental services could be, among others, integration of these services into intergovernmental fiscal transfers. Emphasis must be given to precautionary environmental services such as nature conservation, landscape preservation, and soil and water protection. Both earmarked and lump-sum transfers are required for internalizing the spatial externalities, and a distinction has to be drawn between the two.

Simple assignment of functions does not lead to optimal provision of environmental services. For example, the Supreme Court's decision to ban green felling is restricting the ability of both governments and inhabitants in the forest-rich states to develop productive activities and generate revenue in a variety of ways. An investigation of the implications of the implementation of Joint Forest Management practices at local levels reveals that inadequate funds and arbitrary allocation of available funds are issues of serious concern. Similarly, the realization of full benefits of a centrally sponsored program requires some additional investment in related activities. Lump-sum transfers help in developing alternative productive activities that compensate for forgone revenues and opportunities. The existing formula of resource allocation needs to be modified in such as way that it takes into account the conservation activities performed by a state or local body and the stock of natural resources in the concerned state or local body.

Both the Finance and Planning Commission should reduce the weight of population assigned in their formulas for the disbursement of funds. The Planning Commission is also required to introduce environmental services in the disbursement formula. The environmental performance indicator should account for both efforts of concerned governments and the stock of natural resources in their territories. The inclusion of the indicator would not only compensate the states for their environmental efforts, it would also help reverse the "race to the bottom" approach followed by state governments in the implementation of environmental regulations.[16]

Earmarked grants to states for environmental management allow a greater degree of targeting and can be allocated by the Finance Commission as a separate grant – similar to the grants given for education or health. There is an advantage of this kind of grant. That is, they ensure that funds are directly channeled into the provision of the targeted good. The Finance Commission recommends the principles and the amount of grants-in-aid of revenues for the states and local bodies which are in need of assistance. Generally grants-in-aid are allocated on the basis of

[16] Per se, states in India cannot compete by lowering environmental standards, but it is possible that the states may get to "race to the bottom approach" by using lax enforcement of standards (Gupta, 2001). In the absence of systematic studies no conclusive statement can be made, however the possibility of the presence of race to the bottom cannot be completely ruled out.

socioeconomic considerations; the 13th Finance Commission should *explicitly* consider environmental services also while recommending these grants.

Another way to approach this is to base the transfers on the lines of the Fiscal Reform Facility,[17] wherein a particular environmental performance standard could be made conditional for the states and local bodies to apply for grants to address specific environmental objectives. An ecological fund can be designed to cofinance actions to help improve the environmental performance of states. The resources can also be used to finance ways in which human resources and built infrastructure can be improved to build resilience to environmental degradation. Similar funds can be designed at state levels to finance environment-related activities of local bodies.

To bring accountability in the use of funds, recipient states and local bodies would be required to meet certain conditions. These conditions include a clear objective as what would be achieved along with a statement of how the funds would be spent to achieve the goal with specified benchmarks to attain targets. Moreover, no state and local body would be allowed to get the funds unless they were spent on a *verifiable* project with measurable benefits towards environmental sustainability. That is, grants must be linked to physical outcomes measured by independent auditing and evaluation system assigned to outside agencies.[18]

Note that the options explored above do not in any way undermine the role of existing regulatory functions performed by the respective institutions, environment-related fiscal policies, and other earmarked grants received from different agencies by the states and local bodies for realizing their environmental goals and can continue without any conflict.

4.5 Integrating Ecological Indicators into Fiscal Transfers: An Illustration

To account for ecological services, the combination of forest, tree, and mangrove cover could be considered as an additional criterion for allocation of fiscal transfers. Forests help in maintaining biological variability and protect against natural risks such as landslides, soil erosion, and climate changes. Forests play also an important role in the hydrological cycle through the water flows originating in them. The *State of Forest Report 2005* (Forest Survey of India, 2005) provides figures not only for recorded forest area in 2005, but also of actual forest, tree, and mangrove covers for all states and union territories. *The Report* also provides area figures at the district level.

---

[17]The Fiscal Reform Facility was envisaged by the 11th Finance Commission to address the problem of revenue deficit and to help states converge to a stable, sustainable debt path. However, the scheme was later discontinued primarily because it did not reward past fiscal performance and did not provide enough incentives for a prudent fiscal behavior in the future.

[18]Grafton et al. (2004) developed such conditions for the use of funds available from agencies such as Global Environmental Facility (GEF) that funds projects meeting environmental sustainability criteria.

Table 4.3 A suggested formula for intergovernmental transfers

Criteria	Relative weights (%)
Income (distance method)	50
Tax effort (income weighted)	7.5
Fiscal discipline	7.5
Geographic area	12.5
Forest cover	7.5
Population	15
Total	100

Source: Author's own calculations

For illustrative purposes, it is assumed that the total amount to be allocated among state and union territories is Rs. 1,000 billion. The 13th Finance Commission, following its predecessor, recommends allocation of 65% of the total amount on the basis of income distance and fiscal efforts and discipline (Table 4.1), and the remaining 35% on the basis of population, geographic area and forest cover. It assigns 15% weight to population,[19] 12.5% to geographic area, and a weight of 7.5% to forest cover. The suggested formula is given in Table 4.3.

Table 4.4 demonstrates the illustrated allocation of the remaining amount of Rs. 350 billion. This table also provides the distribution of population, geographic area, and forest cover per thousand persons among states and union territories. Following the proposed formula, state and union territories with less forest cover area receive fewer lump-sum transfers, while the others make gains according to their forest cover area. Beneficiary states are the states containing a relatively good amount of forest cover – either hilly states like Tripura, Sikkim, Mizoram, Meghalaya, Manipur, Arunachal Pradesh or states in which the composition of the population below the poverty line is relatively higher (e.g., Madhya Pradesh, Jharkhand, Chhattisgarh, Orissa, Nagaland, and Andaman and Nicobar). The hilly states with high forest cover also contribute substantially towards the population below the poverty line (ESPASSA, 2008).

The states that would be getting lesser amounts due to lowering the weight of population are mostly the states having per capita state domestic product (SDP) exceeding Rs. 20,000. However, Bihar and Uttar Pradesh would also get lesser amounts following the proposed formula, and per capita SDP in these states is less than Rs. 10,000. Similarly Rajasthan, which is also a poor state, would be a loser. The losing states can be compensated through more grants-in-aid for clean-up activities.

This illustration demonstrated that inclusion of forest cover in the allocation formula not only helps internalize spatial environmental externalities but also makes disbursements more progressive.

[19]The *11th Commission* assigned 10% weights to population in devising the formula for tax devolutions.

Table 4.4 Illustrated allocations of 350 billion rupees according to the *12th Finance Commission* (TFC) formula and suggested formula

State	Population (%)	Geographic area (%)	Forest, tree, and mangrove cover (%)	Per capita SDP (2002–2003) (rupees)	Allocations following TFC formula (billion rupees)	Allocations following suggested formula (billion rupees)	Difference (billion rupees)	Forest cover per thousand population (km^2)
General category states								
Goa	0.13	0.11	0.32	60,787	0.44	0.57	0.13	1.82
Maharashtra	9.42	9.36	7.32	26,858	32.90	31.32	−1.59	0.58
Haryana	2.06	1.34	0.41	26,818	6.48	5.07	−1.41	0.15
Punjab	2.37	1.53	0.44	26,395	7.45	5.79	−1.66	0.14
Kerala	3.10	1.18	2.36	22,776	8.92	7.89	−1.03	0.57
Gujarat	4.93	5.96	3.01	22,624	18.28	17.10	−1.18	0.46
Tamil Nadu	6.07	3.96	3.71	21,740	19.12	16.83	−2.29	0.46
Karnataka	5.14	5.83	5.27	19,576	18.68	18.95	0.27	0.77
Andhra Pradesh	7.41	8.37	6.77	19,087	26.89	26.65	−0.24	0.69
West Bengal	7.79	2.70	2.17	18,494	22.18	16.69	−5.49	0.21
Uttarakhand	0.83	1.63	3.25	14,947	3.69	5.71	2.02	2.96
Rajasthan	5.49	10.41	3.13	12,641	24.14	23.60	−0.54	0.43
Chhattisgarh	2.03	4.11	7.81	12,369	9.18	14.03	4.86	2.90
Madhya Pradesh	5.87	9.38	10.64	11,500	24.04	28.50	4.46	1.36
Jharkhand	2.62	2.42	3.32	11,139	8.97	9.45	0.48	0.95
Orissa	3.58	4.74	6.88	10,164	13.68	16.44	2.76	1.44
Uttar Pradesh	16.16	7.33	2.89	9,963	47.72	35.56	−12.16	0.13
Bihar	8.07	2.86	1.05	5,606	23.03	16.47	−6.57	0.10
Special category states								
Himachal Pradesh	0.59	1.69	1.95	22,902	3.17	4.47	1.30	2.48
Mizoram	0.09	0.64	2.43	22,207	0.86	2.76	1.90	21.15
Nagaland	0.19	0.50	1.81	20,746	0.99	2.27	1.29	7.01
Sikkim	0.05	0.22	0.43	20,013	0.35	0.67	0.32	6.08

Tripura	0.31	0.32	1.07	18,550	1.10	1.67	0.57	2.59
Arunachal Pradesh	0.11	2.55	8.82	16,916	2.81	9.96	7.15	62.13
Meghalaya	0.23	0.68	2.25	16,803	1.25	2.88	1.63	7.50
Jammu & Kashmir	0.99	6.76	3.48	14,507	9.23	12.54	3.31	2.65
Manipur	0.22	0.68	2.23	12,878	1.24	2.85	1.62	7.51
Assam	2.59	2.39	3.77	12,247	8.86	9.69	0.83	1.09
Union territories								
Chandigarh	0.09	0.00	0.00	53,886	0.22	0.14	−0.08	0.03
Delhi	1.35	0.05	0.04	45,579	3.41	2.10	−1.31	0.02
Pondicherry	0.09	0.01	0.01	45,431	0.25	0.17	−0.08	0.09
A & N Islands	0.03	0.25	0.95	28,340	0.34	1.08	0.74	20.56
Dadra	0.02	0.01	0.03	NA	0.07	0.07	0.01	1.13
Daman	0.02	0.00	0.00	NA	0.04	0.03	−0.01	0.11
Lakshadweep	0.01	0.00	0.00	NA	0.02	0.01	0.00	0.48

Source: For population, per capita state domestic product: Economic Survey 2007–2008, and for geographic area and forest cover (forest cover + mangrove cover + tree cover): State of Forest Report, 2005

4.6 Conclusions

This chapter analyzed the role of intergovernmental fiscal transfers in achieving environmental sustainability. Simply assigning the functions at appropriate levels does not ensure optimal provision of environmental services. Optimality in resource allocation could be achieved by combining the assignment system with an appropriate incentive mechanism. Intergovernmental fiscal transfers help in internalizing spatial environmental externalities as they are used for internalizing the fiscal externalities. These transfers could be both, lump-sum and matching (earmarked) grants.

The significance of socioeconomic functions has a comparably long tradition in federal systems, including India. The respective consideration of environmental services, however, is yet to be recognized. It is found that in India, though the assignment of responsibility for protecting the environment is clear, the genesis of environmental degradation could be found in the incentive structure of governance. Though environmental functions are not directly considered in intergovernmental transfers, they find a place through the grants-in-aid route. About 35% of total grants-in-aid recommended by the 12th Finance Commission is allocated for the provision of environmental services. These grants are predominately for developing the end-of-pipe infrastructure. Consideration of "area" as an indicator of ecological functions to a certain limited extent might be considered the inclusion of environmental services in the disbursement of fiscal transfers, but it is considered only for socioeconomic considerations.

This study highlights the need for both lump-sum and earmarked grants for internalizing spatial externalities. Earmarked grants are better suited for environmental clean-up activities and for financing ways in which human resources and built infrastructure can be improved to build resilience to environmental degradation. Lump-sum transfers are better suited for precautionary activities such as nature preservation and soil and water protection. The study also underscores the need to find an appropriate biotic and abiotic indicator of environmental performance that constitutes a link between environmental services and corresponding costs for their provision. This indicator would be used to modify existing formulas of resource allocations for acknowledging the environmental services provided by the states and local bodies.

To understand the significance of intergovernmental fiscal transfers in internalizing environmental positive externalities, the study provided an illustration. The illustration demonstrated that inclusion of forest cover in the formula for lump-sum transfers benefits poor states that contain ecological resources. The states that are poor and have degraded environment can be compensated through grants-in-aid for clean-up activities.

It would be useful to discuss some international practices of compensating local jurisdictions for ecological functions. It is a common practice in the United States and Europe to compensate farmers or private land users for nature conservation activities. The same argument, however, applies to compensating local governments for environmental services provided within their boundaries. The potential benefits of introducing fiscal equalization principles into regional environmental funding have been recognized in some countries. Ring (2002) notes that in Germany, ecological functions are incorporated into intergovernmental fiscal relations at the local level through conditional grants. In Switzerland, Köllner et al. (2002) report that fiscal transfers for nature conservation are from the federal government to the cantonal level and these grants are project oriented. They develop an index to base intergovernmental fiscal transfers on biodiversity. Similarly, Hajkowicz (2007), using multiple criteria analysis (MCA), defines a needs index that forms the basis for fiscal equalization across regions for environmental management in Queensland, Australia. Portugal has set up a fiscal transfer scheme for explicitly rewarding local governments (municipalities) for conservation activities (Ring, 2008b).

During the early 1990s, the Brazilian state Paraná introduced ecological indicators alongside other indicators commonly used for lump-sum fiscal transfers (May et al., 2002; Ring, 2008b). To base the fund allocation on environmental indicators, an instrument known as ICMS-Ecólogico (ICMS-E) was introduced. Other states followed Paraná and now 12 Brazilian states consider explicitly ecological indicators in intergovernmental fiscal transfers. According to the Brazilian Constitution, 75% of the total amount of ICMS (a value-added tax on goods and services) revenue has to be passed on to municipalities according to their contribution to state ICMS. The state governments decide on further indicators for allocating the remaining 25%. Note that each state is independent about taking a decision on the inclusion of indicators and assigning corresponding weights for allocating the revenues to municipalities. The share of ICMS to be distributed for environmental indicators varies from 0.5 in Minas Gerais and São Paulo to 5% in Rondonia and Mato Grosso do Sul. As a result the total area in conservation units across all governmental levels increased by 165% during the period of 1992–2000 in Paraná (Ring, 2008b; May et al., 2002). The Brazilian case is an innovative example to learn from.[20]

Financial acknowledgment of the environmental services provided by the states and local bodies would raise environmental awareness and provide incentives for the protection and enhancement of ecosystem services. The inclusion of environmental services in the disbursement of fiscal transfers could also help in reducing poverty and regional disparities because distribution of poverty and ecosystems and their health is overlapping in the country.

[20]For details on the Brazilian case see Ring (2008b).

Part II
Industrial Development and Benefits and Costs of the Environmental Regulations

Chapter 5
Total Factor Productivity of Indian Industry

5.1 Introduction

The New Industrial Policy introduced in 1991 is considered a watershed event for the Indian economy that shattered the old order. Trade liberalization and deregulation became the central elements. Here it should be noted that the pickup in India's industrial growth precedes the 1991 liberalization by a full decade. Even a cursory glance at the industrial growth record shows that India's rate more than doubled during the 1980s, with very little discernible change in trend after 1991. During the first half of the 1980s the government's attitude towards business went from being outright hostile to supportive, which was further reinforced, in a more explicit manner, in the second half of 1980s. Rodrik and Subramanian (2004) have characterized the policy changes of the 1980s and 1991 as probusiness and promarket reforms, respectively. The former focuses on raising the profitability of the established industrial and commercial establishments. It tends to favor the incumbents by erasing restrictions on capacity expansion, removing price controls, and reducing corporate taxes. A promarket orientation, in contrast, removes the bottlenecks to markets and aims to achieve this through economic liberalization by favoring new entrants and consumers.

Looking into the underlying forces responsible for the changed growth process, the recent works by Burgess and Venables (2003) and Foster and Rozenzweig (2003) show that it is nonagricultural productivity that appears to be the driver of aggregate outcomes at state levels. A number of studies also have argued that manufacturing experienced a surge in productivity in the 1980s (Ahluwalia, 1995; Unel, 2003; RBI, 2004). For example, Unel shows that under the assumption of perfect competition, the average annual growth rate of total factor productivity (TFP) is 1.8% and under the assumption of a constant labor elasticity of 0.6, it is 3.1% over the period of 1979–1980 to 1997–1998. However, there is another set of studies, which contains evidence on the declining TFP growth in the post-reform years (see, for example, Das, 2003; Goldar, 2004). The role of TFP, estimated from the manufacturing sector in the spurt of growth of the Indian economy therefore remains an unresolved problem.

S. Kumar and S. Managi, *The Economics of Sustainable Development*, Natural Resource Management and Policy 32,
DOI: 10.1007/978-0-387-98176-5_5, © Springer Science+Business Media, LLC 2009

In the last two decades, the productivity growth measurement literature has been extended from the standard calculations of TFP employing a production function framework towards more refined decomposition methods. To overcome the shortcomings of the growth accounting approach and to identify the components of productivity change, techniques have been developed that are based on the decomposition of TFP index. A method of measuring productivity with growing popularity is the use of Malmquist index. After its use from a nonparametric perspective by Caves et al. (1982), who developed it as a way of measuring output produced per unit of input, Färe et al. (1994a) went further and employed Shepherd output distance functions and a nonparametric linear programming (LP) approach to measure productivity change for OECD countries.

The Malmquist index has several features that make it an attractive approach. First, it is a TFP index (Färe and Primont, 1995). Second, it can be constructed using distance functions, which are primal measures based only on input and output quantities rather than price. Third, the index can be decomposed into technical efficiency change, technical change, and scale effect components. Efficiency change can be further decomposed into pure efficiency change and scale components. The technical change component can also be decomposed into pure technical change, input-biased as well as output-biased technical change components. As efficiency and technical changes are analogous to the notions of technological innovation and adoption, respectively, the dynamics of the recent growth observed in the manufacturing sector of the Indian economy can be better appreciated. Finally, assumptions do not need to be made with regards to objectives of firms or regions in terms of, say, cost minimization or profit maximization objectives, which could be inappropriate in certain situations.

In contrast to the approach adopted by growth accounting and econometric studies, Ray (2002) uses nonparametric linear programming techniques to construct the Malmquist productivity index. In measuring the annual rates of change in productivity and technical efficiency in manufacturing for individual states in India, he uses the data for the period 1986–1987 to 1995–1996. Results of this study show that, on average, the annual rate of productivity growth has been higher in the 1990s in comparison to the 1980s. It has also been pointed out that some states have actually experienced a slowdown or even productivity decline in the 1990s. However, Ray's decomposition of the Malmquist productivity index contains no index reflecting the contribution of productivity change of biased technical change.

We extend the work of Ray (2002) not only by including the more number of years but also by further decomposition of the technical progress into pure technical progress, input-biased as well as output-biased technical progress. In the process it succeeds in determining whether during the reform period technical progress was labor or capital deepening.

The remainder of the chapter is structured as follows. Section 5.2 outlines the methodological issues related to the measurement of TFP. Empirical results derived from these models and discussions are presented in Section 5.3. The present analysis, therefore, allows us to present the efficiency and productivity scores and factors explaining the productivity. The final section summarizes the findings of the study.

5.2 Measurement of Total Factor Productivity

We use linear programming techniques to construct the Malmquist productivity index for the major states of India. Our analysis is confined to the measurement of TFP growth in the manufacturing sector, which is decomposed into efficiency and technological changes with an isoquant serving as the reference technology. Such a method also allows the determination of the nature of technological change, either capital or labor augmenting, in the Hicksian sense.

As noted above, to measure TFP in state manufacturing, we use nonparametric LP. The LP approach has two advantages over the econometric one in measuring productivity change (Grosskopf, 1986). First, it compares the states to the "best" practice technology rather than "average" practice technology as is done by econometric studies. Second, it does not require the specification of an ad hoc functional form or error structure. In the process, the LP approach allows the recovery of various efficiency and productivity measures in an easily calculable manner. Specifically, it is able to answer questions related to technical efficiency, scale efficiency, and productivity change.

We employ input distance function to construct the various measures of efficiency and productivity, which allows estimation of a multiple output, multiple input production technology. It gives the maximum proportional contraction of all inputs that still allows a state to produce a given level of manufacturing output. It is the reciprocal of input based Farrell measure of technical efficiency and provides the theoretical basis for the Malmquist productivity index.

Let $\mathbf{x}^t = (x_1^t, x_2^t, \ldots, x_N^t)$ denote an input vector at period t with $i = 1, 2, \ldots, N$ inputs and $\mathbf{y}^t = (y_1^t, y_2^t, \ldots, y_M^t)$ an output vector at period t with $j = 1, 2, \ldots, M$ where $\mathbf{x}^t \in \Re_+^N$ and $\mathbf{y}^t \in \Re_+^M$. The technology can be represented by the input requirement set as follows:

$$L^t(\mathbf{y}^t) = \{\mathbf{x}^t : (\mathbf{x}^t, \mathbf{y}^t) \in S^t\}, \quad t = 1, \ldots, T, \tag{5.1}$$

where $S^t = \{(\mathbf{x}^t, \mathbf{y}^t) : \mathbf{x}^t \text{ can produce } \mathbf{y}^t\}$ is the technology set at period t. The input requirement set provides all the feasible input vectors that can produce the output vector. The input distance function requires information on input and output quantity and is independent of input prices as well as behavioral assumptions on producers. Figure 5.1 illustrates the input distance function for a two input case. The frontier technology is given by the piecewise linear isoquant, $L^t(\mathbf{y}^t)$. Efficient production activities occur at the extreme points of the convex hull of the frontier (B and C). The vertical and horizontal segments of the frontier lines indicate the strong (free) disposability of inputs. Production activities inside the input requirement set indicate the presence of inefficiency in those activities. For example, production activity c is inside the input requirement set and therefore inefficient. Ob/Oc gives the technical efficiency of production activity c in terms of input distance function at period t. When the observation falls on the efficient range, the value of input distance function is equal to 1.

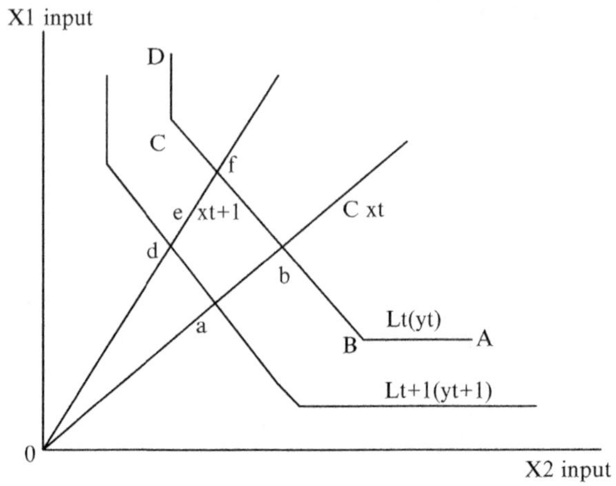

Fig. 5.1 Input oriented distance function and the Malmquist productivity index

Let there be $k = 1, 2, \ldots, K^t$ firms that produce M outputs $y_m^{k,t}$, $m = 1, \ldots, M$ using N inputs $x_n^{k,t}$, $n = 1, \ldots, N$, at each time period $t = 1, \ldots, T$. A piecewise linear requirement set at period t is defined as

$$
\begin{aligned}
L^t(y^t) = \{x^t : \quad & \sum_{k=1}^{K} z_k^t y_{km}^t \geq y_m^t \qquad && m = 1, \ldots, M, \\
& \sum_{k=1}^{K} z_k^t x_{kn}^t \leq x_n^t \qquad && n = 1, \ldots, N, \\
& z_k^t \geq 0 \qquad && k = 1, \ldots, K\}.
\end{aligned}
\tag{5.2}
$$

where z_k^t indicates intensity level, which makes the activity of each observation expand or contract to construct a piecewise linear technology (Färe et al., 1994b). The constraint $z_k^t > 0$ implies constant returns to scale (CRS). By controlling the intensity variable with additional constraints, i.e., $\sum_{k=1}^{K} z_k^t = 1$ and $\sum_{k=1}^{K} z_k^t \leq 1$ in the linear program, variable returns to scale (VRS) and nonincreasing returns (NRS) to scale can be imposed (Afriat, 1972). Let us define $D_i^t(\mathbf{x}^t, \mathbf{y}^t)$ as Shepherd's input distance function at period t with strong disposability of inputs assumption as

$$D_i^t(\mathbf{x}^t, \mathbf{y}^t) = \max\{\lambda : (\mathbf{x}^t/\lambda) \in L^t(\mathbf{y}^t)\}, \tag{5.3}$$

where $D_i^t(\mathbf{x}^t, \mathbf{y}^t)$ estimates the maximum possible contraction of $\mathbf{x}^t$ and can be termed as a measure of overall technical efficiency (OTE). OTE can be further decomposed into a product of pure technical efficiency (PTE) and input scale efficiency (ISE). That is, OTE = PTE × ISE. Pure technical inefficiency is due to over employment of inputs, while scale inefficiency is due to the states not operating in the range of CRS.

The value of input distance function under VRS provides the measure of PTE. ISE is then equal to ISE = OTE/PTE (Färe et al., 1994a).

The Malmquist productivity index (MALM) yields a convenient way of decomposing TFP change into technical change (TECH) and overall technical efficiency change (OTEC). In order to estimate the Malmquist productivity index from period t to $t+1$, additional distance functions required are

$$D_i^t(\mathbf{x}^{t+1}, \mathbf{y}^{t+1}) = \max\{\lambda : (\mathbf{x}^{t+1}/\lambda) \in L^t(y^{t+1})\}, \tag{5.4}$$

$$D_i^{t+1}(\mathbf{x}^t, \mathbf{y}^t) = \max\{\lambda : (\mathbf{x}^t/\lambda) \in L^{t+1}(\mathbf{y}^t)\}, \tag{5.5}$$

and

$$D_i^{t+1}(\mathbf{x}^{t+1}, \mathbf{y}^{t+1}) = \max\{\lambda : (\mathbf{x}^{t+1}/\lambda) \in L^{t+1}(\mathbf{y}^{t+1})\}. \tag{5.6}$$

The cross-period distance function, $D_i^t(\mathbf{x}^{t+1}, \mathbf{y}^{t+t})$, indicates the efficiency measure using the observation at period $t+1$ relative to the frontier technology at period t, and $D_i^{t+1}(\mathbf{x}^t, \mathbf{y}^t)$ shows the efficiency measure employing the observation at period t relative to the frontier technology at period $t+1$. In Figure 5.1, the input requirement set for period $t+1$ is given by $L^{t+1}(y^{t+1})$, and $D_i^{t+1}(\mathbf{x}^{t+1}, \mathbf{y}^{t+1})$ and $D_i^{t+1}(\mathbf{x}^t, \mathbf{y}^t)$ are given by Oe/Of and Oc/Oa, respectively. Cross-period distance functions take values of less than, equal to, or more than one. Similarly, $D_i^{t+1}(\mathbf{x}^{t+1}, \mathbf{y}^{t+1})$ is given by Oe/Od.

The MALM consists of four input distance functions to avoid choosing arbitrary base period and the geometric mean of two input based technical efficiency indices is taken to form

$$\text{MALM} = \left[\frac{D_i^{t+1}(\mathbf{x}^{t+1}, \mathbf{y}^{t+1})}{D_i^{t+1}(\mathbf{x}^t, \mathbf{y}^t)} \times \frac{D_i^t(\mathbf{x}^{t+1}, \mathbf{y}^{t+1})}{D_i^t(\mathbf{x}^t, \mathbf{y}^t)}\right]^{0.5}. \tag{5.7}$$

The MALM can be decomposed into OTEC and TECH as

$$\text{MALM} = \underbrace{\frac{D_i^{t+1}(\mathbf{x}^{t+1}, \mathbf{y}^{t+1})}{D_i^t(\mathbf{x}^t, \mathbf{y}^t)}}_{\text{OTEC}} \underbrace{\left[\frac{D_i^t(\mathbf{x}^t, \mathbf{y}^t)}{D_i^{t+1}(\mathbf{x}^t, \mathbf{y}^t)} \times \frac{D_i^t(\mathbf{x}^{t+1}, \mathbf{y}^{t+1})}{D_i^{t+1}(\mathbf{x}^{t+1}, \mathbf{y}^{t+1})}\right]^{0.5}}_{\text{TECH}}. \tag{5.8}$$

where the first term defines the changes in OTE from period t to $t+1$, i.e., moving closer to the isoquant or "catching up." The second term, i.e., the geometric mean (GM) in parentheses, represents changes in technology, i.e., a shift in the frontier from period t to period $t+1$. Recall that OTE = PTE×ISE. Therefore, OTEC can be further decomposed into pure technical efficiency change (PTEC) and input scale efficiency change (ISEC), where PTEC = PTE^{t+1} / PTE^t and ISEC = $\text{ISE}^{t+1}/\text{ISE}^t$. The MALM can be written as

$$\text{MALM} = \text{PTEC} \times \text{ISEC} \times \text{TECH}. \tag{5.9}$$

In the input oriented case all the indices can be interpreted as progress, no change, and regress, when their values are less than one, equal to one, and greater than one, respectively. Following Färe et al. (1997), the TECH can be decomposed into product of output-biased technological change (OBTECH), input-biased technological change (IBTECH), and the magnitude of technological change (MATECH). Thus,

$$\text{TECH} = \text{OBTECH} \times \text{IBTECH} \times \text{MATECH}, \tag{5.10}$$

where

$$\text{OBTECH} = \left[\frac{D_i^t(\mathbf{x}^{t+1}, \mathbf{y}^{t+1})}{D_i^{t+1}(\mathbf{x}^{t+1}, \mathbf{y}^{t+1})} \times \frac{D_i^{t+1}(\mathbf{x}^{t+1}, \mathbf{y}^t)}{D_i^t(\mathbf{x}^{t+1}, \mathbf{y}^t)}\right]^{0.5},$$

$$\text{IBTECH} = \left[\frac{D_i^{t+1}(\mathbf{x}^t, \mathbf{y}^t)}{D_i^t(\mathbf{x}^t, \mathbf{y}^t)} \times \frac{D_i^t(\mathbf{x}^{t+1}, \mathbf{y}^t)}{D_i^{t+1}(\mathbf{x}^{t+1}, \mathbf{y}^t)}\right]^{0.5},$$

and

$$\text{MATECH} = \frac{D_i^t(\mathbf{x}^t, \mathbf{y}^t)}{D_i^{t+1}(\mathbf{x}^t, \mathbf{y}^t)}.$$

Since we are considering only one output in the present study, there will be no output-biased technological change, i.e., OBTECH = 1, and (5.10) reduces to

$$\text{TECH} = \text{IBTECH} \times \text{MATECH}. \tag{5.11}$$

IBTECH measures the shift in the isoquant from period t to $t + 1$ due to changes in technology holding the level of output constant at $\mathbf{y}^t$. The definition of Hicks' neutral, capital- or labor-deepening technological change depends on, under constant capital-labor ratio, the marginal rate of substitution of labor for capital (MRS_{LK}) remaining constant, decreasing, or increasing (see Binswanger, 1974). Following Färe et al. (1995) and Weber and Domazlicky (1999) IBTECH is independent of outputs under CRS when states produce a single output. Figure 5.2 describes how the value of IBTECH and change in the capital–labor (K/L) ratio can be used to identify the capital- or labor-deepening character of technological change. Assume $y = 1$, $x_1 =$ labor (L), and $x_2 =$ capital (K). Let $L^t(1)$ represent the period t isoquant and $L_n^{t+1}(1)$, $L_1^{t+1}(1)$, and $L_2^{t+1}(1)$ Hicks' neutral, Hicks' labor-deepening (or capital-saving), and capital-deepening (or labor-saving) from period t to $t + 1$. A state is observed to use the input vector $\mathbf{x}^t = (L^t, K^t)$ in period t and $\mathbf{x}^{t+1} = (L^{t+1}, K^{t+1})$ in period $t + 1$ so that $(K/L)^{t+1} < (K/L)^t$. If IBTECH = 1, then $D_i^{t+1}(\mathbf{x}^t, 1)/D_i^t(\mathbf{x}^t, 1) = D_i^{t+1}(\mathbf{x}^{t+1}, 1)/D_i^t(\mathbf{x}^{t+1}, 1)$. In this case Oa = Of/Od, indicating Hicks' neutrality, since MRS_{LK} does not change. If the technology shifts instead to $L_1^{t+1}(1)$, then (Ob/Oa) < (Of/Oc) and IBTECH < 1. In this case, IBTECH < 1 coupled with the increase in the K/L ratio, indicates a capital-deepening

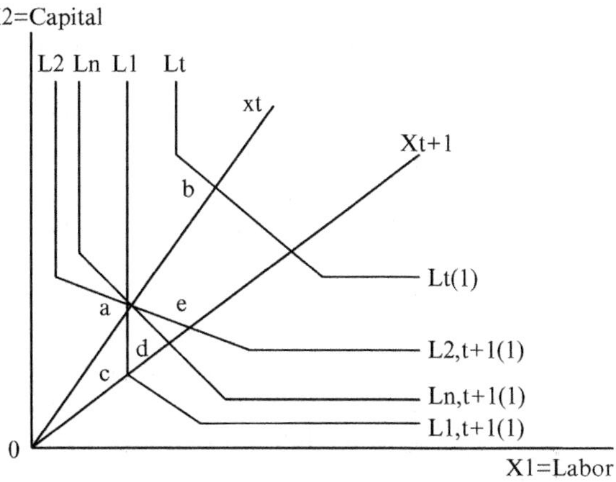

Fig. 5.2 Input-biased technological progress

Table 5.1 Input-biased technical change direction

	IBTECH > 1	IBTECH = 1	IBTECH < 1
$(K/L)^{t+1}(K/L)^{t}$	Labor-deepening	Neutral	Capital-deepening
$(K/L)^{t+1}(K/L)^{t}$	Capital-deepening	Neutral	Labor-deepening

(or labor-saving) technological bias and decrease in the *K/L* ratio indicates a labor-deepening technological bias. Finally, if the technology shifts to $L_2^{t+1}(1)$, then (Ob/Oa) > (Of/Oe) and IBTECH > 1. Therefore, IBTECH > 1 coupled with the increase in the *K/L* ratio indicates a labor-deepening (or capital-saving) technological bias. In other words, when the *K/L* ratio increases from period t to period $t + 1$, IBTECH < 1 indicates a capital-deepening technological bias and IBTECH > 1 indicates a labor-deepening technological bias. Table 5.1 summarizes the various kinds of input-biased technological change that may occur.

5.3 Total Factor Productivity of Indian Industry

We calculate productivity and its components for 15 major Indian states[1] over the period of 1982–1983 to 2000–2001. The period up to 1990–1991 is considered as pre-reform and the subsequent period is regarded as post-reform. The data used in

[1]The 15 major states are Andhra Pradesh (AP), Assam (ASS), Bihar (BIH), Gujarat (GUJ), Haryana (HAR), Karnataka (KAR), Kerala (KER), Madhya Pradesh (MP), Maharashtra (MAH), Orissa (ORI), Punjab (PUN), Rajasthan (RAJ), Tamil Nadu (TN), Uttar Pradesh (UP), and West Bengal (WB). These 15 major states account for approximately 95% of population and industrial output in the country and are therefore representative.

this study for calculating productivity and its various components come from the Annual Survey of Industries (ASI) for the relevant years. The manufacturing sector is modeled as an industry producing a scalar output measured by the gross value added at constant prices by employing the factor inputs, labor and capital. Using gross value added at constant prices is a common practice in the Indian empirical literature (e.g., Unel, 2003; Ahluwalia, 1991; Balakrishnan and Pushpagandan, 1994; Goldar, 1986). One advantage of using the gross value added rather than gross output is that it allows comparison between the firms that are using heterogeneous raw materials (Griliches and Ringstad, 1971). The use of gross output in place of gross value added necessitates the use of raw materials, which may obscure the role of labor and capital in the productivity growth (Hossain and Karunaratne, 2004). Another advantage is that use of gross value added accounts for differences and changes in the quality of inputs (Salim and Kalirajan, 1999).

The input–output data covered by the ASI for individual states are the aggregates of all establishments in the state. The number of establishments covered by the census varies widely across the states. Therefore, following Ray (1997a, 2002), state-level input–output quantity data for the "representative establishment" are constructed by dividing the state-level aggregate values of the variables by the number of establishments covered in the state. The advantage of using the state-level average data is that it imposes fewer restrictions on the production technology.[2] Moreover, such kind of averaging reduces the effects of random noise due to measurement errors in inputs and outputs.

Except for the labor input, which is measured by the total number of persons engaged in an average establishment, ASI reports fixed capital stock and gross value added data in value terms. Nominal values of gross value added were deflated by the wholesale price index for manufactured goods. Fixed capital stock was deflated by the price index for *new* machinery and transport equipment. Both of these variables are measured at 1981–1982 prices at all-India level.[3] Measuring the capital stock input is problematic. In many studies capital stock is measured by the book value of fixed assets while in others its flow is measured by summing rent, repairs, and depreciation expenses or perpetual inventory created from annual investment data. Needless to point out that each of these measures has its own shortcomings. For example, the book value and perpetual inventory methods do not address the question of capacity utilization, whereas the flow measure may be questioned on the ground that the depreciation charges in the financial accounts may be unrelated to actual depreciation of hardware. Thus following Ray (2002) in the present study. capital is measured by the book value of fixed assets. But to the

[2]The firm-level input–output pairs are feasible, although not individually reported. Therefore, by the assumption of convexity, the average input–output bundle will always be feasible. The aggregate input–output bundle will be feasible only under the condition of additivity of technology (Ray, 2002).

[3]To the extent that price indices at the state levels deviate from the all-India indices, the nonlabor variables for individual states will be distorted. But nonavailability of price indices at the individual state level precluded a more refined construction of data.

extent that the true capital input is distorted, it is distorted uniformly in all the states. Therefore, the relative performance of states should not be affected seriously by this shortcoming.

Contemporaneous CRS, VRS, and NRS technology sets were constructed from the state-level input–output data for each year. Own period input distance functions were computed for each year under the CRS, VRS, and NRS assumptions. Similarly, cross-period input distance functions were also computed for every pair of adjacent years. Yearly MALM and its components were computed for all the states in adjacent years.

5.3.1 Technical Efficiency Estimates

Since the basic components of the Malmquist index is related to measures of technical efficiency, we first report these results. Values of unity imply that the state is on the isoquant in the associated year while those exceeding unity imply that it is above the isoquant or technically inefficient. Table 5.2 provides the geometric means of the components of OTE for the 15 states. On average, inputs employed in state manufacturing could have been contracted by 26.6% = $(1 - 1/1.362) \times 100$, 28 and 25% in the overall, pre-reform and post-reform[4] periods, respectively. The average output loss due to pure technical inefficiency was 13%, 16%, and 11%, and the output loss due to scale inefficiency was 33%, 35%, and 32.5%, respectively, for all the three periods. It implies that the promarket reform has helped in increasing the technical efficiency of Indian states.

The state-wise results of technical efficiency are presented in Table 5.3. Maharashtra, which is an industrially developed state, is the most efficient among the states under consideration. It was on the isoquant during the pre-reform era and experienced only 1.2% overall technical inefficiency during the post-reform era, and all the inefficiencies were due to input scale inefficiency. The table also reveals that the most inefficient states in terms of overall technical efficiency were Punjab in the pre-reform period, West Bengal in the post-reform period, and Andhra Pradesh over the entire period of study. Except for six states (Assam, Kerala, Madhya Pradesh, Maharashtra, Tamil Nadu, and West Bengal), all others experienced gains in OTE in the post-reform period in comparison to the pre-reform years. Here it should be noted that the inefficiency in majority of the states is due to scale.

Table 5.4 reports the states operating in the range of CRS, decreasing returns to scale (DRS) and increasing returns to scale (IRS) year-wise. To determine the scale of returns a state operates in, following Grosskopf (1986), we estimate technical

[4]The terms pre-reform and probusiness reform are used synonymously as they refer to same period in the present study. Like that the terms post-reform and promarket are used synonymously in the present study.

Table 5.2 Efficiency results, geometric means (year-wise)

Year	Overall technical efficiency	Pure technical efficiency	Input scale efficiency
1982–1983	1.448	1.257	1.317
1983–1984	1.273	1.160	1.235
1984–1985	1.310	1.130	1.289
1985–1986	1.507	1.176	1.498
1986–1987	1.348	1.150	1.340
1987–1988	1.338	1.125	1.310
1988–1989	1.437	1.179	1.417
1989–1990	1.422	1.131	1.407
1990–1991	1.414	1.105	1.374
1991–1992	1.296	1.079	1.281
1992–1993	1.391	1.120	1.346
1993–1994	1.505	1.149	1.505
1994–1995	1.369	1.132	1.346
1995–1996	1.282	1.102	1.274
1996–1997	1.282	1.103	1.281
1997–1998	1.343	1.071	1.332
1998–1999	1.316	1.154	1.310
1999–2000	1.361	1.145	1.343
2000–2001	1.275	1.089	1.249
Probusiness reforms	1.387	1.156	1.352
Promarket reforms	1.340	1.114	1.325
Overall	1.362	1.134	1.338

Source: Authors' own calculations

efficiency under CRS (T^{CRS}), VRS (T^{VRS}), and NRS (T^{NRS}).[5] In our study most of the states were operating in the range of IRS. Maharashtra operates in the range of CRS in 15 out of 19 years, while Assam operates in the same range in the pre- and post-reform years. Thus the operation of most of the states in the range of IRS helps to explain the cause of inefficiency observed.

5.3.2 Total Factor Productivity Estimates

Next we calculate the Malmquist productivity index along with its components for each state. Instead of presenting the year-wise disaggregated results, we turn to a summary description of the average performance of all states.[6] Recall that if the value of the Malmquist index or any of its components is greater than unity, then it denotes regression or deterioration in performance between any two adjacent years. Also it may be necessary to note that these measures capture the performance relative to the best practice one.

[5] If $T^{CRS} = T^{VRS}$ the state operates in the range of constant returns to scale (CRS). If $T^{CRS} \neq T^{VRS} = T^{NRS}$ the state operates in the range of decreasing returns to scale (DRS). Finally, if $T^{CRS} = T^{NRS} < T^{VRS}$ the state operates in the range of increasing returns to scale (IRS).

[6] The disaggregated results for each state and year can be had from the authors on request.

Table 5.3 Efficiency results, geometric means (state-wise)

States	Overall technical efficiency			Pure technical efficiency			Input scale efficiency		
	Overall	Pre-reforms	Post-reforms	Overall	Pre-reforms	Post-reforms	Overall	Pre-reforms	Post-reforms
Andhra Pradesh	1.840	1.917	1.773	1.032	1.011	1.052	1.824	1.917	1.744
Assam	1.037	1.002	1.070	1.017	1.000	1.033	1.035	1.002	1.066
Bihar	1.120	1.126	1.114	1.074	1.029	1.117	1.068	1.048	1.086
Gujarat	1.218	1.264	1.173	1.042	1.069	1.017	1.217	1.262	1.173
Haryana	1.371	1.445	1.301	1.209	1.252	1.168	1.361	1.435	1.292
Karnataka	1.196	1.286	1.112	1.097	1.134	1.062	1.194	1.282	1.112
Kerala	1.376	1.364	1.429	1.127	1.204	1.061	1.365	1.352	1.429
Madhya Pradesh	1.303	1.278	1.329	1.190	1.164	1.216	1.203	1.169	1.237
Maharashtra	1.019	1.013	1.025	1.000	1.000	1.001	1.012	1.000	1.025
Orissa	1.482	1.548	1.419	1.352	1.388	1.318	1.413	1.480	1.350
Punjab	1.775	1.935	1.629	1.105	1.119	1.090	1.775	1.935	1.629
Rajasthan	1.501	1.568	1.438	1.158	1.305	1.027	1.490	1.544	1.438
Tamil Nadu	1.354	1.281	1.432	1.052	1.078	1.027	1.354	1.281	1.432
Uttar Pradesh	1.527	1.690	1.379	1.265	1.406	1.138	1.520	1.676	1.379
West Bengal	1.642	1.482	1.819	1.476	1.302	1.674	1.541	1.315	1.805

Source: Authors' own calculations

Table 5.5 reports the annual average values of Malmquist index along with those obtained from its decomposition. It can be seen from the table that the Malmquist index does not show a steady upward trend. On the contrary, it indicates productivity decline in 1983–1984, 1987–1988, 1989–1990, 1991–1992, and again in 2000–2001. In the midst of such variations, however, the average annual rate of productivity growth is higher during the post-reforms period than in its preceding regime. The TFP has increased by 1.7% and 3.0% per annum during pre- and post-reform years, respectively. On average, the improvement can be ascribed to technical progress (TECH) (0.4% and 2.8%, respectively) and efficiency improvement (OTECH) (1.2% and 0.2%, respectively). Further decomposition of technical progress indicates that during these two periods the magnitude of pure technical progress (MATECH) was −0.2 and 1.6% whereas that of IBTECH was 0.6% and 1.2%. A decomposition of efficiency improvement reveals that in the pre-reform years, the efficiency improvement was governed by the gain in pure technical efficiency (PTEC) (1.7%), while in the subsequent period, the improvement in scale efficiency (ISEC) and pure technical efficiency change equally influenced the gain in the overall efficiency change. In nutshell, it can be said that in the pre-reform period, three-fourths of improvement in the TFP was governed by the technical efficiency improvement, whereas in the post-reform years it was the technical progress that governed the growth in TFP.

Results of the present study confirm those of Ray (2002). Ray found that TFP increased from 0.17% per year during the pre-reform era (up to 1990–1991) to 1.45%

Table 5.4 Returns to scale in states

Year	Constant returns to scale	Decreasing returns to scale	Increasing returns to scale
1982–1983	TN	BIH, GUJ, HAR, KAR, KER, MP, MAH, ORI, RAJ, UP, WB	AP, ASS, PUN
1983–1984	ASS, BIH, MAH, TN	MP, WB	AP, GUJ, HAR, KAR, KER, ORI, PUN, RAJ, UP
1984–1985	ASS, MAH	WB	AP, BIH, GUJ, HAR, KAR, KER, MP, ORI, PUN, RAJ, TN, UP
1985–1986	ASS, MAH	-	AP, BIH, GUJ, HAR, KAR, KER, MP, ORI, PUN, RAJ, TN, UP, WB
1986–1987	ASS, MAH	-	AP, BIH, GUJ, HAR, KAR, KER, MP, ORI, PUN, RAJ, TN, UP, WB
1987–1988	ASS, MAH	BIH, MP, WB	AP, GUJ, HAR, KAR, KER, ORI, PUN, RAJ, TN, UP
1988–1989	ASS, GUJ	BIH, MP, MAH, ORI	AP, HAR, KAR, KER, PUN, RAJ, TN, UP, WB
1989–1990	ASS, MAH	BIH, ORI	AP, GUJ, HAR, KAR, KER, MP, PUN, RAJ, TN, UP, WB
1990–1991	ASS, MAH	MP, ORI	AP, BIH, GUJ, HAR, KAR, KER, PUN, RAJ, TN, UP, WB
1991–1992	BIH, KAR, MAH	ORI	AP, ASS, GUJ, HAR, KER, MP, PUN, RAJ, TN, UP, WB
1992–1993	KAR, MAH	MP, ORI	AP, ASS, BIH, GUJ, HAR, KER, PUN, RAJ, TN, UP, WB
1993–1994	ASS, BIH	–	AP, GUJ, HAR, KAR, KER, MP, MAH, ORI, PUN, RAJ, TN, UP, WB

(*continued*)

Table 5.4 (continued)

Year	Constant returns to scale	Decreasing returns to scale	Increasing returns to scale
1994–1995	KAR, MAH	MP	AP, ASS, BIH, GUJ, HAR, KER, ORI, PUN, RAJ, TN, UP, WB
1995–1996	ASS, MAH	MP	AP, BIH, GUJ, HAR, KAR, KER, ORI, PUN, RAJ, TN, UP, WB
1996–1997	BIH, GUJ, KAR	KER, MAH, ORI	AP, ASS, HAR, MP, PUN, RAJ, TN, UP, WB
1997–1998	BIH	ORI	AP, ASS, GUJ, HAR, KAR, KER, MP, MAH, PUN, RAJ, TN, UP, WB
1998–1999	ASS, BIH, GUJ, MAH	ORI	AP, HAR, KAR, KER, MP, PUN, RAJ, TN, UP, WB
1999–2000	ASS, MAH	BIH	AP, GUJ, HAR, KAR, KER, MP, ORI, PUN, RAJ, TN, UP, WB
2000–2001	MAH	BIH, MP	AP, ASS, GUJ, HAR, KAR, KER, ORI, PUN, RAJ, TN, UP, WB

AP Andhra Pradesh, *ASS* Assam, *BIH* Bihar, *GUJ* Gujarat, *HAR* Haryana, *KAR* Karnataka, *KER* Kerala, *MP* Madhya Pradesh, *MAH* Maharashtra, *ORI* Orissa, *PUN* Punjab, *RAJ* Rajasthan, *TN* Tamil Nadu, *UP* Uttar Pradesh, *WB* West Bengal
Source: Authors' own calculations

per year during the post-reform years. Although the rates of growth in TFP obtained by Ray (2002) are different from the ones in the present study, direction of change in both is found to be same, that is, positive growth in the decades of 1980s and 1990s. Another feature common to both the studies is the higher growth rate of TFP in the post-reform period compared to its preceding period. The difference in magnitude of estimated growth rates in TFP might be due to difference in orientation of the methodology. While Ray used the output orientation in the measurement of Malmquist index, the present study employed input distance functions for that purpose.

The performance of TFP in each state is given in Table 5.6 as average annual rates of growth over the period 1982–1983 to 2000–2001. The table also contains the TFP growth rates for the pre- and post-reform periods. As it is difficult to summarize the disaggregated results, we include some of their general features. The

Table 5.5 Malmquist productivity index and its decomposition, geometric means (year-wise)

Year	OTECH	PTEC	ISEC	IBTECH	MATECH	TECH	MALM	$(K/L)^{t+1}/(K/L)^t$
1983–1984	0.879	0.923	0.938	1.005	1.155	1.161	1.021	1.131
1984–1985	1.029	0.974	1.044	0.987	0.980	0.967	0.995	1.105
1985–1986	1.151	1.040	1.163	0.996	0.851	0.847	0.975	1.065
1986–1987	0.894	0.978	0.895	0.989	1.076	1.064	0.952	1.092
1987–1988	0.992	0.978	0.978	1.001	1.014	1.016	1.008	1.059
1988–1989	1.074	1.048	1.082	0.989	0.841	0.831	0.893	1.015
1989–1990	0.990	0.959	0.993	0.993	1.024	1.017	1.006	1.012
1990–1991	0.994	0.977	0.977	0.988	0.978	0.967	0.961	1.116
1991–1992	0.917	0.976	0.932	0.995	1.149	1.144	1.048	0.972
1992–1993	1.073	1.038	1.050	0.979	0.945	0.925	0.993	1.079
1993–1994	1.082	1.026	1.119	0.983	0.900	0.884	0.957	1.101
1994–1995	0.910	0.985	0.894	1.001	1.096	1.097	0.998	1.048
1995–1996	0.936	0.974	0.946	0.993	1.054	1.047	0.980	1.161
1996–1997	1.000	1.001	1.005	0.999	1.001	1.000	0.999	1.074
1997–1998	1.048	0.971	1.040	0.996	0.931	0.927	0.971	1.121
1998–1999	0.980	1.077	0.983	0.988	1.001	0.989	0.969	0.973
1999–2000	1.034	0.992	1.025	0.958	0.799	0.766	0.792	1.435
2000–2001	0.937	0.951	0.929	0.991	1.183	1.173	1.098	0.927
Probusiness reforms	0.988	0.983	0.997	0.994	1.002	0.996	0.983	1.062
Promarket reforms	0.998	1.001	0.997	0.988	0.984	0.972	0.970	1.094
Overall	0.993	0.992	0.997	0.991	0.993	0.984	0.977	1.078

OTECH overall technical efficiency change index, *PTEC* pure technical efficiency change index, *ISEC* input scale efficiency change index, *IBTECH* input-biased technological change index, *MATECH* magnitude of pure technological change index, *TECH* technological change index, *MALM* Malmquist productivity index, $(K/L)^{t+1}/(K/L)^t$ change in capital-labor ratio over previous year

Source: Authors' own calculations

Table 5.6 Decomposition of Malmquist index: average annual percentage changes (state-wise)

States	Overall			Pre-reforms			Post-reforms		
	OTECH	TECH	MALM	OTECH	TECH	MALM	OTECH	TECH	MALM
Andhra Pradesh	0.662	−0.809	−0.142	−4.666	−0.214	−4.890	5.719	−1.409	4.391
Assam	−1.360	−3.511	−4.918	−0.204	−2.210	−2.418	−2.529	−4.828	−7.479
Bihar	0.663	3.599	4.238	2.838	1.686	4.477	−1.561	5.475	3.999
Gujarat	1.224	4.504	5.673	−0.618	1.392	0.783	3.031	7.518	10.321
Haryana	0.071	1.063	1.133	−0.251	0.656	0.407	0.391	1.469	1.854
Karnataka	−0.043	−1.096	−1.139	2.376	−0.310	2.073	−2.522	−1.887	−4.457
Kerala	0.635	−2.137	−1.488	−0.003	0.089	0.086	1.270	−4.412	−3.086
Madhya Pradesh	1.602	5.371	6.887	2.413	1.653	4.026	0.785	8.949	9.663
Maharashtra	0.472	2.757	3.217	0.943	0.425	1.363	0.000	5.035	5.035
Orissa	2.584	5.693	8.130	8.188	1.734	9.780	−3.362	9.492	6.449
Punjab	2.475	−0.080	2.397	3.218	−0.072	3.148	1.727	−0.088	1.640
Rajasthan	3.440	5.522	8.772	6.201	1.664	7.763	0.598	9.228	9.771
Tamil Nadu	−1.217	−1.667	−2.905	−3.363	−1.987	−5.417	0.884	−1.349	−0.453
Uttar Pradesh	0.954	5.610	6.511	4.741	2.477	7.101	−2.983	8.642	5.917
West Bengal	−1.726	−1.072	−2.817	−4.488	−0.480	−4.990	0.963	−1.667	−0.689

OTECH overall technical efficiency change index, *TECH* technological change index, *MALM* Malmquist productivity index

Source: Authors' own calculations

disaggregated results reveal widespread regional variation in productivity changes. In the study period, 9 out of 15 states experienced productivity improvement. While in the pre-reform period 11 states witnessed growth in TFP, the corresponding number was 10 in the post-reform years. In the pre-reform period four states (Orissa, 9.8%; Rajasthan, 7.8%; and Uttar Pradesh, 7.1%) witnessed the growth in TFP more than 5% per year, whereas in the post-reform years six states (Gujarat, 10.3%; Rajasthan, 9.8%; Madhya Pradesh, 9.7%; Orissa, 6.5%; Uttar Pradesh, 5.9%; and Maharashtra, 5.04%) registered more than 5% annual change in TFP. The table reveals that the variation in TFP has decreased in the post-reform period in comparison to its preceding years. The coefficient of variation in its growth rate among the states was 301.7% and 187.5% during the pre- and post-reform periods.

The most significant factor behind the improvement in TFP during the period of study could be found in technical progress, as evident from the positive rates of technical change in eight states. Here it should be noted (see Table 5.6) that in the pre-reform era, nine states exhibit technical regress, whereas in the post-reform period only the states of Andhra Pradesh (−1.4%), Assam (−4.8%), Karnataka (−1.9%), Kerala (−4.4%), Punjab (−0.09%), Tamil Nadu (−1.35%), and West Bengal (−1.7%) exhibited technological regression. Also during the decade of 1980s the contribution of OTE improvement was substantial. But in the 1990s, it was technical progress that contributed significantly to the TFP progress. During both the decades, the progress in TFP in Punjab was only due to the presence of "catch-up" effect while it was due to innovation in Maharashtra.

Table 5.7 shows the decomposition of OTEC (catch-up effect). During the entire period, out of 15 states, 11 exhibit the presence of the catch-up effect (positive change in OTECH). In four states the contribution of change in PTE was zero, while in another two this effect was negative. The remaining nine states witnessed a

Table 5.7 Decomposition of efficiency change index, geometric means (state-wise)

States	Overall			Pre-reforms			Post-reforms		
	OTECH	PTEC	ISEC	OTECH	PTEC	ISEC	OTECH	PTEC	ISEC
Andhra Pradesh	0.993	0.995	0.993	1.047	0.990	1.047	0.943	1.000	0.943
Assam	1.014	1.008	1.014	1.002	1.000	1.002	1.025	1.016	1.025
Bihar	0.993	1.000	1.000	0.972	1.000	1.000	1.016	1.000	1.000
Gujarat	0.988	0.983	0.989	1.006	0.969	1.008	0.970	0.998	0.970
Haryana	0.999	0.998	1.003	1.003	0.999	1.009	0.996	0.996	0.996
Karnataka	1.000	0.997	1.001	0.976	0.978	0.978	1.025	1.016	1.025
Kerala	0.994	0.983	0.994	1.000	0.976	1.001	0.987	0.991	0.987
Madhya Pradesh	0.984	1.000	1.000	0.976	1.023	1.027	0.992	0.977	0.974
Maharashtra	0.995	1.000	1.000	0.991	1.000	1.000	1.000	1.000	1.000
Orissa	0.974	0.980	0.983	0.918	0.923	0.923	1.034	1.042	1.046
Punjab	0.975	0.978	0.975	0.968	0.965	0.968	0.983	0.990	0.983
Rajasthan	0.966	0.961	0.973	0.938	0.928	0.952	0.994	0.996	0.994
Tamil Nadu	1.012	1.000	1.012	1.034	1.000	1.034	0.991	1.001	0.991
Uttar Pradesh	0.990	0.978	0.995	0.953	0.949	0.961	1.030	1.009	1.030
West Bengal	1.017	1.019	1.026	1.045	1.054	1.054	0.990	0.986	0.998

OTECH overall technical efficiency change index, *PTEC* pure technical efficiency change index, *ISEC* input scale efficiency change index
Source: Authors' own calculations

positive change. In the pre-reform period, the highest catch-up effect was in Orissa, whereas in Andhra Pradesh it was noticed during the post-reform years. In Orissa, the change in scale of production and improvement in PTE equally contributed to the positive effect, while in Andhra Pradesh the positive changes were due to improvement in scale effects only.

Table 5.8 provides the decomposition of technical change into pure and input-biased changes. The table also provides the annual average estimates of change in capital-labor ratio. During the pre-reform period, Uttar Pradesh exhibits the highest growth in the pure technical change (3.2%) followed by Orissa (1.7%) and Rajasthan (1.7%). It was Assam which records the highest negative change in the magnitude of pure technical change during the decade of 1980s. In the decade of 1990s, Orissa (9.3%), Rajasthan (8.7%), Madhya Pradesh (8.2%), Uttar Pradesh (8%), Gujarat (6.1%), and Bihar (4.6%) had the highest growth rates in pure technical progress. During this decade, seven states witnessed a negative change in pure technical progress and in the two states, Maharashtra and Punjab, there was stagnation.

Recall that if capital–labor ratio increases and IBTECH < 1, then it implies capital-using technical bias. On the other hand, IBTECH > 1 implies labor-using technical bias. If the capital–labor ratio decreases, then IBTECH < 1 indicates labor-using bias and IBTECH > 1 shows capital-using technical bias. In the present analysis except for 1991–1992, 1997–1998, and 2000–2001, the capital–labor ratio has increased over its previous year (Table 5.5). During the pre-reform era, the average annual change in the capital ratio was 6.2%, whereas it was 9.4% during the post-reform period. Moreover, during both of the periods, the value of IBTECH was less than unity, implying the presence of capital-using technical bias in Indian manufacturing. This finding concurs with the finding of Pradhan and Barik (1999). Pradhan and Barik also find the absence of labor-using technical progress in Indian manufacturing. Moreover, the manufacturing sector exhibits neutral technical bias for 2 years (1987–1988 and 1994–1995) and labor-using technical bias for 4 years. But we do not observe any consistent trend in input-biased technical change either in favor of capital or labor (Table 5.5).

The state-wise picture of the change in technical bias can be judged from Table 5.8. The table reveals that all the states witnessed an increase in average capital–labor ratio. In the post-reform era, all except Kerala exhibit capital–using technical bias. In Kerala the technical bias was almost neutral. The finding on capital–using technical bias of the 1990s is a significant departure from the preceding decade, when 7 out of 15 states (Karnataka, Kerala, Madhya Pradesh, Orissa, Punjab, Rajasthan, and West Bengal) exhibited almost neutral technical progress. In one of the states (Uttar Pradesh) technical progress was slightly in favor of labor.

5.3.3 *Innovative States and Convergence*

It should be noted that the technical progress change index for any particular state between two adjacent years merely depicts the shift in the isoquant at the output

Table 5.8 Decomposition of technological change, geometric means (state-wise)

States	Overall				Pre-reforms				Post-reforms			
	IBTECH	MATECH	TECH	$(K/L)^{t+1}/(K/L)^t$	IBTECH	MATECH	TECH	$(K/L)^{t+1}/(K/L)^t$	IBTECH	MATECH	TECH	$(K/L)^{t+1}/(K/L)^t$
Andhra Pradesh	0.994	1.015	1.008	1.092	0.994	1.008	1.002	1.165	0.993	1.021	1.014	1.023
Assam	0.961	1.077	1.035	1.128	0.965	1.060	1.022	1.150	0.958	1.094	1.048	1.106
Bihar	0.994	0.970	0.964	1.048	0.998	0.985	0.983	1.050	0.989	0.956	0.945	1.046
Gujarat	0.986	0.969	0.955	1.120	0.987	1.000	0.986	1.121	0.985	0.939	0.925	1.119
Haryana	0.985	1.004	0.989	1.094	0.993	1.001	0.993	1.118	0.978	1.008	0.985	1.070
Karnataka	0.998	1.013	1.011	1.103	1.002	1.001	1.003	1.092	0.994	1.025	1.019	1.115
Kerala	1.002	1.020	1.021	1.067	0.997	1.002	0.999	1.093	1.006	1.038	1.044	1.042
Madhya Pradesh	0.995	0.951	0.946	1.041	0.998	0.985	0.983	1.071	0.992	0.918	0.911	1.012
Maharashtra	0.963	1.010	0.972	1.094	0.977	1.020	0.996	1.126	0.949	1.001	0.950	1.063
Orissa	0.999	0.944	0.943	1.099	1.000	0.983	0.983	1.167	0.997	0.907	0.905	1.035
Punjab	0.998	1.003	1.001	1.064	0.997	1.003	1.001	1.122	0.998	1.002	1.001	1.010
Rajasthan	0.997	0.947	0.945	1.066	1.000	0.983	0.983	1.084	0.995	0.913	0.908	1.049
Tamil Nadu	0.994	1.023	1.017	1.097	0.989	1.031	1.020	1.136	0.998	1.016	1.013	1.060
Uttar Pradesh	1.000	0.944	0.944	1.077	1.008	0.968	0.975	1.116	0.993	0.920	0.914	1.039
West Bengal	0.996	1.014	1.011	1.077	1.003	1.002	1.005	1.125	0.990	1.027	1.017	1.030

IBTECH input-biased technological change index, *MATECH* magnitude of pure technological change index, *TECH* technological change index, *MALM* Malmquist productivity index, $(K/L)^{t+1}/(K/L)^t$ change in capital-labor ratio over previous year

Source: Authors' own calculations

level observed for that state. A value of technical change index less than unity does not necessarily imply that the state under consideration did actually push the overall isoquant inward. Thus in order to determine the states that were shifting the frontier or were "innovators" (see Färe et al., 1994b), the following three conditions are required of various input distance functions for a given state k':

(a) $\text{TECH}_t^{t+1} < 1$
(b) $D_i^t(\mathbf{x}^{t+1}, \mathbf{y}^{t+1}) < 1$
(c) $D_i^{t+1}(\mathbf{x}^{t+1}, \mathbf{y}^{t+1}) = 1$

The condition (a) indicates that the isoquant shifts in case of fewer inputs for the given level of output. With a given output vector, in period $t + 1$ it is possible to decrease the input bundle relative to period t. This measures the shift in the relevant portions of the isoquant between periods t and $t + 1$ for a given state. The condition (b) indicates the production in period $t + 1$ that occurs outside the isoquant of period t (i.e., technical change has occurred). It implies that the technology of period t is incapable of producing the output vector of period $t + 1$ with the input vector of period $t + 1$. Hence the value of input distance function$(\mathbf{x}^{t+1}, \mathbf{y}^{t+1})$ relative to the reference technology of period t is less than one. The condition (c) specifies that the state must be on the isoquant in period $t + 1$. Table 5.9 shows the states that were innovators. Out of 18 two-year periods, Maharashtra and Assam shifted the isoquant five times each while Bihar achieved the feat three times and Gujarat two times.

In a recent study Aghion et al. (2003) find that promarket reform gives rise to a larger increase in productivity in the states that were closer to the frontier when the

Table 5.9 States causing inward shift in isoquant over the previous year

Year	States
1983–1984	–
1984–1985	Maharashtra
1985–1986	Assam, Maharashtra
1986–1987	–
1987–1988	Assam
1988–1989	Assam, Gujarat
1989–1990	–
1990–1991	Maharashtra
1991–1992	–
1992–1993	Maharashtra
1993–1994	Bihar, Assam
1994–1995	–
1995–1996	–
1996–1997	Gujarat, Bihar
1997–1998	Bihar
1998–1999	Assam
1999–2000	Maharashtra
2000–2001	–

Source: Authors' own calculations

reforms were initiated. So the growth enhancing effect should be smaller for the representative firm in the state that is farther from the frontier. On the other hand, the convergence theory could be restated in terms of the relationship between productivity and technical inefficiency. Such a relationship would state that the states that were near the production frontier would record a lower level of productivity growth than those farther away. Therefore, the positive relationship between productivity level and lagged technical inefficiency would indicate the presence of convergence hypothesis (Lall et al., 2002).

In the present exercise we find that the states that were closer to the frontier in the efficiency estimation at the beginning of post-reform are not having the higher growth rate in TFP index. The correlation coefficient between the technical efficiency scores in 1991–1992 and cumulative Malmquist index in 2000–2001 (assuming that the value of the Malmquist index is unity in 1991–1992) is 0.22, which is statistically significant at the 95% level of confidence interval. Moreover, we find that the states that were farther from the frontier in 1991–1992 have gained not only due to increase in technical efficiency but also have experienced the higher growth rate of technical progress. This indicates that there is a tendency towards convergence in the productivity growth rates across the states. This finding concurs with Ray (2002) and does not conform to Aghion et al. Here, it should be noted that if a state is technically efficient and is on the production frontier, then it is maximizing its productive potential and there is little to be gained from adopting technology or knowledge from elsewhere. But only the states that were technically efficient were innovative in the sense that they were able to shift the isoquant inwards (see Table 5.9). It implies that although there is a tendency of convergence in manufacturing productivity growth among Indian states during the post-reform period, only those that are efficient at the beginning of the reform remain innovative.

5.4 Conclusions

In this chapter we use state-level data on manufacturing from the Annual Survey of Industries for the years 1982–1993 through 2000–2001 to measure the Malmquist index of productivity growth. The index is also decomposed into technical change and efficiency change. The efficiency change is further decomposed into pure technical efficiency and input scale efficiency changes. The technical change is decomposed into magnitude of pure technical change and input-biased technical change. Such a decomposition of technical change helps in identifying the direction biases in favor of labor or capital.

We found that in the pre-reform period TFP had grown at the rate of 1.7% per year while in the post-reform era the corresponding growth rate was 3%. While in pre-reform periods the growth rate in TFP was due to gains in technical efficiency, in the post-reform era it was influenced by the technical progress. Another interesting result of the present exercise is the nature of technical progress in Indian manufacturing. It was seen that the capital intensity of Indian firms is increasing in the recent years.

Although regional difference in TFP persists, it appears that the variation has declined in the post-reform period. The majority of the states tried to be nearer the isoquant in the post-reform era in comparison to the pre-reform years. Most of the states are also operating under increasing returns to scale, and the gain in TFP in the post-reform era was due to gain in technical progress. In contrast, in the pre-reform period it was due to efficiency improvement. During the 1990s, capital intensity of the manufacturing sector seemed to have increased as the technical progress was in favor of capital. The states which were exhibiting either neutral or labor-using technical bias in the pre-reform period also show capital-using technical change during the post-reform era. It is also found that although there is a tendency of convergence in terms of TFP growth rate among Indian states during the post-reform era, only those that were technically efficient at the beginning of the reform remained innovative.

Beyond measuring of state TFP growth rates, the present analysis demonstrates the richness of a linear programming technique that allows for an investigation of important research questions on the underlying processes that influence TFP growth. Notwithstanding the striking feature of the techniques used here, data limitations involved in estimation remains an important factor. It is therefore necessary to be cautious while applying these results to policy formulation.

Chapter 6
Valuing the Benefits of Air Pollution Abatement

6.1 Introduction

In the course of past decades, India has undergone economic development. Real gross national product has grown at an average rate of more than 6% in the last decade. This economic growth was fuelled by processes of industrialization, urbanization, and population growth and was not achieved without sacrifices. Air pollution must be counted among those sacrifices, and the levels of air pollution in urban areas often exceed national air quality standards for several pollutants. An increase in the air pollution level raises public mortality and morbidity (Krupnick et al., 1990; Cropper et al., 1997a; Chhabra et al., 2001). Cropper et al. report the results of a study relating levels of particulate matter to daily deaths in Delhi between 1991 and 1994. This study finds a positive, significant relationship between particulate pollution and daily nontraumatic deaths, as well as deaths from certain causes (respiratory and cardiovascular problems) and for certain age groups. Chhabra et al. find evidence of elevated rates of respiratory morbidity among those dwelling in highly polluted areas of Delhi after adjusting for several confounders. Daily counts of emergency room visits for acute asthma, acute exacerbation of chronic obstructive airway disease (COAD), and acute coronary events are related to daily levels of pollutants, particularly total suspended particulate (TSP) recorded a day earlier using time series approach. Therefore, governments increasingly need information about the costs and benefits associated with reduced levels of pollution to assist them in pollution control measures.

Air quality affects the utility of individuals and an economic value exists. There are several ways to capture this economic value, viz., dose-response, revealed preferences, and contingent valuation methods (CVMs). The dose-response method assumes a relationship between air quality and morbidity (and/or mortality). It puts a price tag on air quality without retrieving people's preferences for the good. The revealed preference methods assume that the consumers are aware of the costs/benefits of air quality, and are able to adjust their locations to reveal their preferences. Markets should be functioning perfectly and consumers should be well

S. Kumar and S. Managi, *The Economics of Sustainable Development*, Natural Resource Management and Policy 32,
DOI: 10.1007/978-0-387-98176-5_6,

informed (Freeman, 1993). In developing countries like India, markets neither are functioning perfectly nor are consumers well informed. Moreover, dose-response and revealed preferences methods do not consider the nonuse values that form a substantial portion of the total economic value of environmental resources. Therefore, in conducting demand assessment studies in a developing country context (including India), the CVM continues to be extensively used by researchers (Whittington and Swarna, 1994; Griffin et al., 1995; Choe et al., 1996; Bateman and Willis, 1999; Ready et al., 2002; Ahmad et al., 2002).

This study applies both, revealed preference and contingent valuation methods to estimate the economic value that people in an urban area in India, viz., Panipat Thermal Power Station (PTPS) colony in Panipat Haryana, place on improving the air quality.

We employ the dose–response method, based on the Gerking and Stanley (1986) model, to estimate the economic benefits of air quality improvement. This model establishes an association between air pollution and health based on consumer choice, i.e., a health-oriented consumer choice model. The estimates of willingness to pay (WTP) for reduction in air pollution, obtained from the analysis of the consumer choice based dose–response of the residents of PTPS colony, reveal that income and health status were significant determinants. These estimates range from 1% to 2% of monthly income.

Similarly, the estimates from the analysis of the responses to the CVM questions reveal that bid value and health status were significant determinants of one's WTP for air quality improvement initiatives. These estimates were about 2% of their monthly income. Thus, findings of the present exercise support the wisdom of acting now to protect the environment before it is too late. However, we do not contend that these results are generalizable to all the environmental problems or locations; we doubt that the situation of PTPS colony with respect to air quality preservation benefits was in some way unique since the respondents were working for a polluting company.

We compare the WTP elicited through CVM and mitigating behavior. Such a comparison acts as a validity check for the WTP figures reported by the respondents in the CVM survey. Economic theory posits that, ceteris paribus, the estimates of WTP elicited through CVM should be greater than WTP through indirect methods. It is of independent interest to see if WTP to avoid illness is, in India, as large a fraction of the total damage as it is in the developed world, given the differences in cost and availability of medical care and perception of illness between western countries and developing countries.

This chapter is organized as follows. Section 6.2 provides an overview of PTPS, in brief. Information on the questionnaire and survey format is contained in Section 6.3. Applications of the revealed preference method and CVM are discussed in Sections 6.4 and 6.5, respectively. Section 6.6 compares the results of these two applications on the same set of data. The chapter closes with some concluding remarks.

6.2 Panipat Thermal Power Station

This thermal power station is located at Panipat on the Assand-Panipat road at a distance of 12 km. from Panipat city. The Haryana Power Generation Corporation (formally Haryana State Electricity Board) owns it. It came into existence in 1972. It has five units. These units generate approximately 2.4 GWh. units of electricity per year. In this plant, to meet the environmental standards electrostatic precipitators (ESPs) are provided to extract ash from the furnace outlet. However, due to an old design, unit numbers 1 and 2 have not been armed with ESPs. To these units mechanical precipitators (MPs) are provided which are not up to standards. To meet the environmental standards in these units, the ESPs construction work was in full swing when the survey was conducted. Units 3–5 have already been armed with the ESPs.

The circular area of about 5 km surrounding PTPS is very thinly populated; it is comprised mostly of agricultural fields and is without inhabited villages. However, the power station has its own residential colony adjoining the plant for its employees. In this locality, there is a community guesthouse, a health center, and a school for the residents. The health center provides medical facilities only to the residents of this locality. It had three doctors and 13 other employees of class II and III. The total number of employees working in the plant was 2,613, including the employees of the health center. Most of employees reside in the colony and use the services of the health center.

Ambient air pollution monitoring was done by the Shri Ram Institute of Industrial Relations, Delhi, on weekly intervals at three points: the plant, the guesthouse, and a nearby village, Khakharana. The descriptive statistics of the ambient air pollutants, monitored during 2 January 1995–1 December 1995, are given in Table 6.1. The descriptive statistics reveal that the oxides of sulfur and nitrogen were within the permissible limits and not violating the National Ambient Air Quality Standards fixed by the Central Pollution Control Board for the residential areas. Here it should be noted that the situation of PTPS colony with respect to air quality is in some way unique. Since Indian coal has high content of ash, the level of PM_{10} was violating the ambient standards both on an annual average basis and on a 24 h basis.[1] Thus, the widespread criticality of PM_{10} was the *main* cause of

Table 6.1 Descriptive statistics of ambient air quality at PTPS colony Panipat (average annual $\mu g/m^3$)

	NO_2	SO_2	PM_{10}
Mean	31	18	245
Standard deviation	8	7	109
Median	33	16	204
Maximum	50	38	583
Minimum	10	6	83

Source: Panipat Thermal Power Station

[1] The Central Pollution Control Board has put the following: National Ambient Air Quality Standards for the residential areas: for SO_2, 60 $\mu g/m^3$ annual average and 80 $\mu g/m^3$ 24 h; for NO_x, 60 $\mu g/m^3$ annual average and 80 $\mu g/m^3$ 24 h; and for PM_{10}, 60 $\mu g/m^3$ annual average and 100 $\mu g/m^3$ 24 h.

deteriorating air quality, and about 2,400 families living in the PTPS colony were exposed to the polluted air.

6.3 Questionnaire and Survey Format

The PTPS colony survey was conducted in April–May 1996. It provides information on the health, activity patterns, and lifestyles of members of 155 households, which is about 6% of the total households of the colony. Every respondent was asked 17 questions. The survey included attitudinal, behavioral, and demographic questions. The questions used in empirical analysis can be divided into six categories:

1. Sociodemographic characteristics
2. Monthly salary and asset income (if any)
3. Health stock measures
4. Consumption of medical services
5. Questions about environmental awareness
6. Contingent valuation question

With regard to sociodemographic characteristics, six questions were put to the respondents, i.e., age, sex, family background (rural/urban), years of schooling, marital status, and designation. The sample consists mostly of married males (147 out of 155), and two-thirds have rural backgrounds (104). The age of the respondents ranges from 25 to 57 years. The education standard of the respondent can be judged from the number of years of schooling. In our survey, these years range from 10 to 18 years for most of the sample units (138). With respect to designation, the survey attempted to include all categories of employees, and their positions vary from class IV employee to class I officer.

The second category of questions relates to the respondents' earnings. Monthly gross salary and asset income, if any, are considered. As all but a few of the respondents are government employees, all have only monthly salary as a sole source of income. The average monthly salary for the sample is Indian rupees (Rs.) 5,037.

Health is treated as a multidimensional rather than a one-dimensional variable. More specifically, the PTPS colony health survey contains three types of variables measuring health for adults. These variables are defined as (1) subjectively reported health status (whether health is considered excellent (16), good (89), fair (47), or poor (31)); (2) existence of any illness and if yes, then name it; and (3) length of suffering from the disease. Taken separately, each of these variables may measure a different dimension of the health stock.

The measure of medical care consumption comes from a yes/no (95/60) question asking if a doctor usually was seen during a month; if yes, then the months of the visits, since the data on air pollution are available after weekly intervals for the same year. These kinds of measure indicate whether the respondent has received regular medical attention over time.

To determine the environmental understanding of the respondent, three more questions were posed. These questions are defined as (1) subjectively reported environmental knowledge (whether the respondent is fully aware [15], highly aware [37], aware to some extent [90], or ignorant [13] about environmental problems); (2) cause of above stated disease, if any (polluted surroundings [86], lack of proper facilities [35], excessive workload [17], and any other [41]); and (3) the measure of the use of any environmental safety measure (a yes/no [7/148] question).

The last question that was posed to the respondents was a referendum question in the form of a payment card revealing different bid values (as a percentage of monthly salary) for improving the ambient air quality, which is deteriorated due to widespread criticality of PM_{10}. Each respondent was confronted with a series of money amounts ranging from 1% to 5% of their monthly salaries in 1% increments and asked to circle their maximum WTP for the program. The respondents were also given the option of marking "zero" for WTP.

While stating the WTP for air quality, the respondent would implicitly be valuing several benefits from reduced damage costs. These include the avoidance of air-borne illness along with their associated costs (lost work, lost leisure time, expenditures on defensive behavior, and treatment costs). However, paying for safe air quality means that households have to give up some other uses of its present income. The respondent was made aware of these aspects in creating the appropriate hypothetical market. The respondents were also made aware of free riding, as the good to be valued was a public good.

The questionnaire was administered to the respondent in face-to-face interviews. The questionnaire was pretested through a series of focus group interviews conducted in the field. This pretesting helped us to design the questionnaire in such a way that respondents generally accepted our questions. Table 6.2 provides the descriptive statistics and definitions to major variables.

Table 6.2 Descriptive statistics of the major variables used in the study

Variable name	Description of variable	Mean	Standard deviation
X_1	Sex, male = 1, female = 2	1.065	0.246
X_2	Age	37.568	6.517
X_3	Rural = 1, urban = 2	1.329	0.471
X_4	Years of schooling	12.929	3.224
X_5	Married = 1, unmarried = 2	1.026	0.159
X_6	Income (Rs. per month)	5,036.774	1,788.392
X_7	Environmental awareness: fully aware = 1, highly aware = 2, some extent = 3, ignorant = 4	2.652	0.769
X_8	Health status: excellent = 1, good = 2, fair = 3, poor = 4	2.239	0.655
X_9	Use of safety measure: yes = 1, no = 2	1.955	0.208
X_{10}	Bid value (percentage of monthly salary)	1.677	1.248
Observations	155		

Source: Primary Survey

6.4 Application of Revealed Preference Method

The indirect valuation approach is usually applied to environmental problems. That is, if there is some damage and it is linked to a cause, the relationship between that cause and its effect is a dose–response linkage. Once a dose–response relationship is established, the indirect approach then utilizes valuations that are applied to the "responses." For example, consider the linkage between air pollution and health. If health effects are established, values for life and for illness are applied, but such a mechanical relationship in the dose–response function does not take into account consumer behavior. In the absence of stated preferences, it becomes necessary to have estimates of WTP or willingness to accept (WTA) on the basis of a consumer choice model aimed at measuring the strength of association between health effects and septic air pollutants. Therefore, the monetary valuation of morbidity due to emission is calculated on the basis of the Gerking and Stanley (1986) model.

In this model individuals produce health capital in a utility-maximizing framework and are able to adjust their behavior in order to defend against reduction in air quality. These adjustments, which involve substituting medical care or other health-producing activities for reduced air quality, form the basis for the method used in making the benefits, or WTP, calculations. This method is empirically implemented using survey data on the residents of PTPS, Panipat. From a policy standpoint, the empirical results are of interest because they support the notion that individuals are willing to pay for better health resulting from air quality improvements.

The remainder of this section is divided into two sections. Section 6.4.1 describes the health model and the method derived for estimating WTP for improved air quality. Section 6.4.2 outlines the empirical estimation strategy used and presents the empirical results.

6.4.1 The Model

Individuals derive utility (*U*) from the consumption of two classes of goods:

1. Their own stock of health capital (*H*)
2. Representative consumption good (*X*) that yields direct satisfaction, but does not affect health.

Hence, we write

$$U = U(X, H). \tag{6.1}$$

The stock of health capital is determined by the production function:

$$H = (M; \alpha, \delta), \tag{6.2}$$

where

$$H_{\mathrm{M}} > 0, \quad H_{\alpha} > 0, \quad H_{\delta} >=< 0,$$

where M denotes medical care (from which the individuals derive no direct utility), α denotes air quality, δ denotes a set of other exogenous variables, such as education, that affect H, and subscripts denote derivatives.

Utility is maximized subject to (6.2) and the full income budget constraint shown in (6.3):

$$Xq_x + Mq_{\mathrm{M}} + WT_{\mathrm{L}} = WT + A, \tag{6.3}$$

where $q_i = (P_i + WT_i)$, $i = X, M$; P_i is the money price of commodity i, W is the wage rate, T_i is the time required to consume one unit of commodity i, T_{L} is the time lost from market and non-market activities due to illness, T is the total time available to the consumer, and A is an exogenously determined amount of asset income. T_{L} is related to health stock according to

$$T_{\mathrm{L}} = G(H), \tag{6.4}$$

where $G_{\mathrm{H}} < 0$.

This model can be manipulated in order to derive a compensating variation (CV) type expression for the marginal WTP for improved air quality. Totally differentiating the utility function and setting $\mathrm{d}U = 0$,

$$\mathrm{d}U = 0 = U_x \mathrm{d}X + U_{\mathrm{H}} H_{\mathrm{M}} \mathrm{d}M + U_{\mathrm{H}} H_{\alpha} \mathrm{d}\alpha + U_{\mathrm{H}} H_{\delta} \mathrm{d}\delta. \tag{6.5}$$

Then, totally differentiate the full income budget constraint, holding $\mathrm{d}q_i = \mathrm{d}W = \mathrm{d}T = 0$ for $i = X, M$:

$$d\,(WT) = 0 = q_x dX + (q_{\mathrm{M}} + WG_{\mathrm{H}} H_{\mathrm{M}})\mathrm{d}M - \mathrm{d}A + WG_{\mathrm{H}} H_{\alpha} \mathrm{d}\alpha + WG_{\mathrm{H}} H_{\delta} \mathrm{d}\delta. \tag{6.6}$$

The Lagrangean of the objective function is

$$L = U[X, H(M; \delta, \alpha)] + \lambda[WT + A - Xq_x - Mq_{\mathrm{M}} - WG(H)]. \tag{6.7}$$

and the first-order conditions for the model are

$$\begin{aligned} &\partial L/\partial X = U_X - \lambda q_x = 0 \quad \text{or} \quad U_x/q_x = \lambda, \\ &\partial L/\partial H = U_{\mathrm{H}} H_{\mathrm{M}} - \lambda(q_{\mathrm{M}} + WG_{\mathrm{H}} H_{\mathrm{M}}) = 0, \\ &\text{or,} \quad (U_{\mathrm{H}} H_{\mathrm{M}})/(q_{\mathrm{M}} + WG_{\mathrm{H}} H_{\mathrm{M}}) = \lambda. \end{aligned} \tag{6.8}$$

From (6.8)

$$U_x/q_x = U_H H_M/(q_M + WG_H H_M) = \lambda. \quad (6.9)$$

Equations (6.6), (6.8), and (6.9) yield

$$\partial A/\partial \alpha = -H_\alpha q_M/H_M. \quad (6.10)$$

This equation indicates that the individual is willing to pay more for a given air quality improvement, the greater the associated improvement in health. Also, that bid is higher, the lower the productivity of medical services and higher their costs. Therefore, if medical services are as expensive, but are an ineffective means of improving health, the individual is willing to pay more for improved air quality. Moreover, this equation is relatively straightforward to implement empirically, since utility terms have been eliminated.

6.4.2 Empirical Estimates of WTP

This section presents empirical estimates of WTP for improved air quality. As shown in (6.10), the magnitude of the WTP term hinges critically on the estimation of the health production function. The approach taken to estimate this function is considered in Part 1. Part 2 presents the empirical estimates.

Part 1: Estimation Approach

In estimating the health production function, *H* is treated as a multidimensional, rather than a unidimensional variable. More specifically, the survey contains three types of variables for measuring *H*. These variables are defined as:

1. Subjectively reported health status whether health is considered excellent, good, fair, or poor
2. Existence of chronic illness
3. Days of suffering from these chronic conditions

Taken separately, each of these variables may measure a different dimension of the health stock. In any case, the perspective that the health stock is better treated as a multidimensional, rather than a unidimensional, variable underlines the approach taken for estimating $\partial A/\partial \alpha$.

This approach can be illustrated by expressing (6.2) as

$$F(H, M; \alpha, \delta) = 0. \quad (6.11)$$

Assuming that the conditions of the implicit function hold, i.e., that (1) the function *F* has continuous partial derivatives F_H, F_M, F_α, and F_δ and (2) $F_M \neq 0$.

Equation (6.11) can be rewritten as

$$M = M(H; \alpha, \delta). \tag{6.12}$$

This alternative specification of the production function has three features that are worth elaborating:

1. In the empirical work described below, it allows for the possibility that H may best be measured as a set of health indicators rather than as a single variable, since H now appears on the right-hand side.
2. $M_\alpha = -(F_\alpha/F_\mathrm{M}) = -(F_\alpha/F_\mathrm{H})(F_\mathrm{H}/F_\mathrm{M}) = -H_\alpha/H_\mathrm{M}$. Therefore, in order to obtain the marginal WTP, from (6.10) M_α need only be multiplied by q_M, the price of medical care.
3. In the subsequent empirical analysis, (6.12) is overidentified by exclusion restrictions, since one jointly dependent variable appears as a regressor and three predetermined variables, q_M, W, and A have been excluded.

Part 2: Empirical Estimates

The basic equation to be estimated is

$$\mathrm{MED} = \mathrm{MED}(\mathrm{PM}_{10}\mathrm{Chro}, \mathrm{Length}, \mathrm{Age}, \mathrm{School}, \mathrm{F/B}, \mathrm{Income}, \mathrm{Envknow}), \tag{6.13}$$

where MED is a discrete variable, whether a doctor has been consulted during the last 1 month; PM_{10}, the ambient concentration of particulates, i.e., particles measuring less than 10 μm in diameter, measured in $\mu g/m^3$; Chro, dummy variable indicating whether the illness is chronic (1) or not (0); Length, length of illness in days; Age, age of the respondent in years; School, years of schooling of the respondent; and F/B, family background of the respondent whether rural or urban; Income, monthly income of the respondent in Rupees; and Envknow, awareness about environmental problems, a dummy variable 1–4.

In (6.13), the aerometric variable PM_{10} is pollution rather than air quality. Hence, the expected sign on the coefficient of this variable is positive, implying that it must be multiplied by minus one in computing $\partial A/\partial\alpha$. Moreover, the expected signs of Chro and length should be positive, since increase in the magnitude of these variables is associated with increase in the use of medical care.

The expected signs of the five socioeconomic and demographic variables should be as follows:

- The coefficient of age should be positive if the aging process reduces the efficiency with which the health stock is produced.
- The coefficient of school should be negative if years of schooling increase the efficiency with which health is produced.
- The coefficient of family background (F/B) should be positive if rural areas tend to have lower health stock.
- The coefficient of income and environmental knowledge (subjectively reported) should be negative since these variables increase the health stock.

Equation (6.13) was specified as a restricted Cobb-Douglas function. Additionally, because of the discrete nature of the dependent variable MED, Logit and Probit models were used for estimation purposes. As is commonly done in such analysis, for each observation ten observations are taken, one for each PM_{10} level, since the data are available for 10 months on PM_{10}. The other explanatory variables have the same value in all ten observations. The discrete dependent variable takes value one if the MED is yes, and zero otherwise. Estimates the two models are presented in Table 6.3.

With respect to the air pollution variable PM_{10}, the coefficient is positive, but not significantly different from zero at the 10% level. The coefficient of length and chro are, as expected, at a significance level of 1% in both the models. For socioeconomic demographic variables, the signs are according to expectations and are significantly different from zero, at either the 5% or 10% level.

With caution, the results from Table 6.3 can be used to illustrate WTP estimates for a reduction in PM_{10} levels. These benefit estimates are offered advisedly because of the caveats enumerated concerning the model, as well as the above outlined data problems. Since the sign of the PM_{10} is expected but insignificant, this is used in making the benefit calculation.

Because PTPS colony experiences a large number of days each year when the average PM_{10} level substantially exceeds both national and World Health Organization (WHO) standards, large reductions in ambient concentrations are necessary in order to meet the standard. Therefore, a reduction in PM_{10} level of 67% of the mean has been used to calculate benefits (mean level for PTPS colony is 248.6 $\mu g/m^3$ and WHO standard is 75 $\mu g/m^3$). Illustrative WTP estimates are calculated based on Logit and Probit models, respectively. For a 67% reduction in ambient mean PM_{10} level concentrations, the monthly WTP ranges from Rs. 21 to Rs. 52.5 for the Logit model and Rs. 12.15 to Rs. 30.45 for the Probit model in 1996. The WTP estimates

Table 6.3 Estimates of the medical care function

Variable	Models			
	Logit		Probit	
	Coefficient	*t*-Value	Coefficient	*t*-Value
Constant	0.514	0.178	0.356	0.23
Y_1/(length of disease)	0.0002	3.20	0.0001	3.261
Y_2/(chronic)	0.830	4.614	0.458	4.546
Y_3/(schooling)	0.039	1.231	0.026	1.486
Y_4 (age)	0.783	1.698	0.458	1.814
Y_5 (family B/G)	0.295	1.846	0.162	1.844
Y_6 (PM_{10})	0.121	0.545	0.070	0.567
Y_7 (income)	−0.758	−2.497	−0.45	−2.756
Y_8 (knowledge)	−0.261	−2.497	−0.45	−2.756
Log-likelihood	−624.206		−622.454	
RMSE	0.456		0.456	
M_α	0.097		0.056	

Source: Authors' own calculations

reported are computed using the means of the independent variable. These estimates appear to be small. This may be because the production function method is able to capture only the short-term illness effects of air quality improvements. The reduction in mean levels of PM_{10} allows us to calculate the WTP for each $\mu g/m^3$. That is a WTP of Rs. 21/173.6 per month per microgram (lower bound for the Logit model), which amounts to Rs. 1.45 per person per year. These estimates account only for the effects of the improvements in air quality on illness. A total estimate might also account for reduced materials damage, minor symptomatic discomforts, and improved visibility.

Brandon and Hommann (1995) estimate the annual per capita hospital cost and emergency treatment cost caused by pollution to be in the range $86.49–$216.13 for India. Using the exchange rate $1 = Rs. 30, their estimate works out to be in the range of Rs. 216–540 per month. They estimate the health impacts of air pollutants through the use of mechanical dose–response functions drawn from epidemiological studies done in other developed countries. A World Bank review (Ostro, 1990) of such studies is used as a basis for estimating the health impacts for India. There are uncertainties in applying the results directly to cities in developing countries. That said, dose–response functions are not necessarily applicable to all populations within a certain geographic area: more people breathe similarly polluted air than people consume similarly polluted water. Moreover, for valuation they used the U.S. values in the Indian context by employing the ratio of national per capita incomes. More careful epidemiological modeling attempts to capture the exposure of individuals to pollutants, not of total populations to ambient conditions. Such work has not been widely done in India, except the more recent study by Cropper et al. (1997b) for Delhi. Cropper et al. quantify an exposure response function which describes the relationship between rise in air pollution level and increase in mortality rates in Delhi. Their study does not make any monetary valuation. However, neither of these two studies is based upon models of consumer choice. The mechanical dose-response function generally does not take into account the defensive or averting expenditures that individuals can and do make to protect their well-being (Mishan, 1974; Fisher and Zeckhauser, 1976; Courant and Porter, 1981; Shibata and Winrich, 1983; Harford, 1984).

6.5 Application of Contingent Valuation Method

In the case of public goods like air quality, individuals face a quantity rather than a price constraint. Public goods have much higher income elasticity than marketed goods. This may be particularly true in a developing country where air quality is considered a luxury good which is afforded only when adequate food, clothing, and shelter have been acquired. Hence, the income effect due to a change in air quality provision undermines the consumer surplus of welfare change. Hicksian compensating surplus (CS) (i.e., WTP to ensure that the change occurs) and equivalent surplus (ES) (i.e., WTA if gain does not occur), could be used to measure change in the level of welfare in quantity constrained utility functions.

Measurement of consumers' preferences for air quality improvement initiatives allows one to quantify the individuals' WTP. Consumers' preferences can be elicited either using revealed or stated preference data. The main differences between the two methods lay in the data origin and collection procedures. Revealed preference data are obtained from the past behavior of consumers. In CVM, the economic value placed by an individual for improved air quality is contingent upon a hypothetical scenario that is presented to the respondent for valuing. By means of an appropriately designed questionnaire, a hypothetical market is described where the good or service in question can be traded. The contingent market defines the good itself, the context in which it is provided, and the way it is financed. Respondents then express their maximum WTP for the good/service.

The choice of elicitation format for WTP questions in CVM surveys has already passed through a number of distinct stages (Hanley et al., 2001). For assessing WTP, CVM studies use either an open-ended elicitation format (which asks the respondent to state the sum he/she is willing to pay) or a closed-ended referendum type elicitation format (where the respondent is asked whether or not he/she would be willing to pay a particular amount for the good being valued). The open-ended CVM method is now rarely used because it has been found to be vulnerable to a range of biases; for example, respondents find open-ended questions too difficult to answer because they are not accustomed to paying for non-market goods and services. Ordinary Least Squares regression is used for the estimation under the open-ended CVM version.

The advantage of a closed-ended elicitation format is that it is convenient for the respondent to consider the suggested price options, particularly where the good is non-marketed. A more compelling reason for using the closed-ended format is that strategic biases in the responses can be better controlled. The preference data generated using this method are encoded in binary forms, as respondents are only given the option of answering yes or no, i.e., a dichotomous choice, which implies the adoption of a random utility function. In this case, the coefficient values are obtained through the estimation of a binary Logit or Probit model using the maximum likelihood procedure.

After receiving the endorsement of the NOAA experts panel in 1993 (Arrow et al., 1993), the use of dichotomous choice questions has substantially increased. Dichotomous choice response is defined as a simple referendum where a simple offer is made or in a double referendum where a second offer is made conditional to the response given to the first offer. In this situation, the respondents respond whether they are willing to pay a certain amount of money. The question is simple and easy to answer; however, it supplies a limited quantity of information.

The other option is to use a payment card method or sequential referendum method. In the payment card method, the respondent is asked to circle (indicate) the highest WTP from an ordered set of values ranging, say, from zero to "rupees *X* or more" per month. In the referendum method, the respondent gets the opportunity to vote on a single offer amount. A sequential referendum would increase the valuation efficiency but at the expense of a greater difficulty in the estimation of the distribution, as well as endogeneity in the follow-up question. The present exercise used the payment card method to quantify the economic value of air quality improvement at PTPS colony.

Literature on CVM frequently emphasizes the conditional analysis of the WTP, where one of the main interests is to analyze the conditional relation of WTP and its determinants (*X*), which describes consumer characteristics. McFadden (1994) shows that when necessary auxiliary information is available to reweigh the sample, there is a gain in statistical efficiency of a conditional approximation. This result does not require the *X*s to be free of measurement errors, because errors simply contribute to the nonconditional variance. However, the conditional approximation is not robust to inconsistencies between sample data and auxiliary distributions of *X*s, such as mean differences due to changes in the phraseology of the question, codification, or the period of data collection (Torero et al., 2003).

The estimation of the distribution of WTP function is carried out in the literature with parametric or nonparametric approximations. The parametric approximation, widely used in the CVM analyses, specifies a parametric model in which a relation between the WTP and the consumer's intrinsic characteristics is stated. The non-parametric approximation estimates the distribution of the WTP without assuming any parametric specification of the preferences distribution. Both approximations have advantages and disadvantages. The parametric approximation is susceptible to misspecification errors. On the other side, the nonparametric approximation does not permit a conditional analysis (An, 2000). The main advantages of the parametric method are that it makes it relatively easy to impose preferences axioms, combines experiments and, primarily, allows one to extrapolate the calculation to different populations without constraining exclusively to the sampled population (Torero et al., 2003). In the present study, we use Logit model to estimate the parameters of the WTP function.

Most of the published studies on CVM belong to the developed countries and there is little treatment at all of the use of CVM in developing countries. The initial applications of the CVM in developing countries were primarily in two areas: water supply and sanitation (Whittington et al., 1988, 1990, 1993; McConnell and Ducci, 1988; Briscoe et al., 1990; Singh et al., 1993; Altaf et al., 1993, 1994; Anand and Perman, 1999); and recreation, tourism, and national parks (Grandstaff and Dixon, 1986; Shyamsundar and Kramer, 1993; Menkhaus, 1994; Hadkar et al., 1995). The areas of application are growing rapidly, however, and now include surface water quality (Choe et al., 1996); health (Whittington et al., 1996; Alberini et al., 1997; Alberini and Krupnick, 2000; Yoo and Chae, 2001); social forestry (Kohlin, 2001); and telecom services (Torero et al., 2003).

Our approach[2] for eliciting WTP to avoid illness is different from that used in the above-cited studies, in that the commodity to be valued is defined by the respondent, as opposed to being defined for the respondent by the researcher (imagine the

[2] Alberini and Krupnick (2000) also adopted a similar approach in their study of Taiwan.

health damage they were experiencing). The advantage is that the respondent is familiar with the "good" to be valued. It is also unique in the sense that the bid values are not absolute figures, but they are in terms of the percentage of their monthly income. This reduces the hypothetical bias, although it requires individuals to compute WTP as percentage of their income. Moreover, in CVM surveys, the values elicited are not based on real income decisions and a no budget constraint choice is meaningless.

6.5.1 The Model

Let us consider the household utility function for the affected household in the PTPS colony as

$$U^h = U^h(X^h, Q), \quad h = 1, 2, \ldots, H, \tag{6.14}$$

where U^h is the utility of the hth household; X^h, a Nx1 vector of private goods consumed by the hth household; and Q, quality of air, the public good.

The dual of the utility maximizing problem can be stated as expenditure (E) minimization problem:

$$\text{Minimize}\, E = E(p, Q, U) \tag{6.15}$$

and

$$\text{WTP} = W_{\text{Q}} = -\partial E(p, Q, U)/\partial Q, \tag{6.16}$$

where W_{Q} is interpreted as the reduction in income that is just sufficient to maintain utility at its level corresponding to no improvement in the quality of air. The CS for measuring the welfare change using expenditure function can be defined as follows:

$$\text{WTP} = E(p, Q', U, S^h) - E(p, Q^0, U, S^h), \tag{6.17}$$

where $Q' \geq Q^0$, p is the price vector, and S^h, a vector of socioeconomic characteristics of the hth household. If the reference utility is the utility that the household gets with polluted air, then it is the difference in the minimum expenditure required for the household to be as well off with the unpolluted air as it was with polluted air.

Given (6.17) the hth household will respond "yes" to a particular bid "B" if

$$\text{WTP}(p, Q', Q^0, U, S^h) \geq B \tag{6.18}$$

and "no" otherwise. The probability P of accepting the bid related to the quality of air Q' will be

Table 6.4 Central tendency measures formulas

Description	Symbol	Formula
Mean, $E(\text{WTP})$, $-\infty < \text{WTP} < \infty$	C^{+}	β_0/β
Median WTP	C^{*}	β_0/β
Truncated mean, $E(\text{WTP})$, $0 < \text{WTP} < \infty$	C'	$\ln(1+\exp(\beta_0))/\beta$
Truncated mean, $E(\text{WTP})$, $0 < \text{WTP} < B_{\max}$	C''	$1/\beta \ln[(1+\exp(\beta_0))/(1+\exp(\beta_0-\beta B_{\max}))]$

$B_{\max}$ is the maximum bid

$$P(\text{yes}) = P(B - \text{WTP} < \eta), \tag{6.19}$$

where η is the unobserved random component of WTP function. It could be logistically or standard normally distributed. If η is assumed to be distributed logistically, it becomes a Logit model. The probability of an affirmative response to the bid B when the household has a vector X of explanatory variables is

$$P(\text{yes}/X) = \frac{e^{\beta X}}{1+e^{\beta X}}. \tag{6.20}$$

Maximum likelihood routine can be used to estimate the Logit model. Once the Logit model is estimated the estimate of parameters allows identification of the cumulative distribution function of WTP (Hanemann, 1984). For the Logit model, Hanemann (1984, 1989) and Vaughan et al. (1999) provide the WTP formulas for the unrestricted expected value, the median and the truncated expected value that restricts WTP to be positive (Table 6.4).

Table 6.4 shows the different central tendency measures for the probability model. Following the notation used by Hanemann (1984, 1989), letter C in Table 6.4 is an abbreviation to identify the measure of the central tendency of WTP. β_0 is called augmented intercept and is equivalent to the intercept coefficient of the model plus other estimated parameters ($i = 1,\ldots,n$) (except the bid parameter (β)), multiplied by the sample mean of the explanatory variables (X).

6.5.2 *Results*

The data used are obtained from the PTPS colony survey. The survey yielded 155 useable interviews, and the findings are based on these interviews. The WTP elicitation procedures were well within the respondents' ability. To obtain consistent parameter estimates using nonlinear estimation procedure, the Logit model used here, the estimates take into account the nonrandom sampling (Maddala, 1983).

Each respondent gives answers to five bid values. The bid values range from 1% to 5% of monthly salary of the respondent. Therefore, we enter five observations for each respondent, one for each bid level. The other explanatory variables have the same values in all five observations. The discrete dependent variable takes value one if the bid is accepted and zero, otherwise. This data entering procedure introduces correlation among the errors of each household. Although it provides unbiased parameter estimates, it gives imprecise estimates of z-statistics, since they are deflated. Following Briscoe et al. (1990), this problem is dealt with through a straightforward procedure, which is consistent with the "bootstrapping" literature. Two types of runs were made. First, the point estimates of the parameters were estimated from a model that used all the data. However, to estimate the true z-statistics of these estimates, one observation was picked at random from each group of observations for each household, thus reducing the sample to the number of respondents. This procedure was followed ten times; the reported values for the z-statistics are based on the average values of the z-statistics from these ten runs. The estimated parameters of Logit model are presented in Table 6.5.

Table 6.5 shows that only two parameters, i.e., health status and bid values, are statistically significant. The coefficient of health status is positive and statistically different from zero at 0.05 level. Recall that health status is a self-reported index, whose value ranges from 1 to 4. The respondent indicates value one for excellent health status, and the value increases as the quality of self-reported health status declines. This implies that the probability of acceptance of higher bid values is positively associated with the decline in self-reported health status. The coefficient of bid value is negative and statistically significant at the 0.01 level. This implies that the probability of acceptance of a bid is negatively associated with its value.

Table 6.6 reports the estimates of household's WTP. Applying the expected value and median formulas produces the WTP estimates for the untruncated mean, median, the mean truncated at zero but untruncated from above, and the truncated mean confined between zero and the maximum bid. The expected WTP varies between Rs. 89.39 per month (untruncated mean/median) to Rs. 114.07 per month (truncated at zero but untruncated from above mean). The expected value of WTP is

Table 6.5 Multivariate models of the determinants of household's WTP for air quality improvement

Variable	Coefficient	z-Value
Constant	0.774	0.110
X_1 (sex)	0.617	1.012
Ln X_2 (age)	−0.266	−0.266
X_3 (family background)	−0.122	−0.322
Ln X_4 (schooling)	−0.226	−0.475
X_5 (marital status)	−0.804	−0.886
X_7 (environmental knowledge)	−0.142	−0.637
X_8 (health status)	0.477*	1.942
X_9 (use of safety measure)	−0.666	−1.116
X_{10} (bid value)	−0.602**	−6.427
Log-likelihood	−171.764	
LR Stat. (9 d.f.)	48.316	
RMSE	0.248	

*Significant at 5% critical level, **significant at 1% critical level
Source: Authors' calculations

Table 6.6 Estimates of expected willingness to pay (WTP)

Central tendency measures		WTP (Rs.)
Mean, E(WTP), $-\infty <$ WTP $< \infty$	C^+	89.36
Median WTP	C^*	89.36
Truncated mean, E(WTP), $0 <$ WTP $< \infty$	C'	114.07
Truncated mean, E(WTP), $0 <$ WTP $< B_{max}$	C''	102.86

Source: Authors' calculations

Rs. 102.86, when the mean is truncated between zero and maximum value of bid. The variation of benefits shows that it is not possible to find a single number that correctly represents households' WTP for improved air quality. However, the net benefits of improved air quality remain positive irrespective of the measure of central tendency and the households are willing to pay about 2% of their monthly income for air quality improvement initiatives.

6.6 Comparison Between WTP Obtained from CVM and Mitigation Behavior

The health production function was used to help arrive at an estimate of the opportunity cost of illness for the population under consideration. Illustrative WTP estimates is calculated for a 67% reduction in ambient mean PM_{10} level concentrations. The monthly WTP ranges from Rs. 21 to 52.5 for Logit model.

The comparison between the results of two methods, however, should be interpreted carefully because the estimates of WTP from the two valuation approaches are not measuring precisely the same thing. The ratio of mean WTP from CVM to mitigation behavior approaches ranges between 1.7 and 2.17 for untruncated mean/median to a mean that is truncated at zero but untruncated from above. This ratio is 1.96 when the mean WTP from CVM is confined between zero and maximum bid. The ratio is both reassuring and surprising. It is reassuring in the sense that it is consistent with the economic theory, which predicts that WTP from CVM to avoid the ill health effects of pollution is greater than the mitigating behavior approach. It provides support for the validity of the WTP amounts announced by the respondents in the contingent valuation survey.

The mitigating behavior method measures only the use values lost by households due to deterioration in air quality. Smith (1993) states that all indirect methods measure what might be available as the privately capturable aspects of the environmental services being valued. Each method must link the non-market service to a private choice. To that extent that environmental services have public good aspects and this "publicness" has value in addition to the private aspects. The CVM captures both the use and nonuse values associated with improvements in their quality. Based on the CV data, it is not possible to distinguish between the use and nonuse values of the respondents. The problem comes in interpreting the CV estimates of WTP of nonusers. It is not necessary for a nonuser who indicates a positive WTP to think only of use values. Such a respondent may receive several use-related economic benefits other than enhanced recreated opportunities (Choe et al., 1996).

Despite differences in the economic, cultural, and institutional conditions between the United States and India, the ratio of WTP from CVM to mitigating behavior is similar in both countries. Our finding concurs with the finding of Alberini and Krupnick (2000). They computed the ratio of WTP from CVM to cost of illness for the Taiwanese between 1.48 at very low levels of particulate matter and 2.26 at the highest levels. For the United States, Rowe and Chestnut (1985) report that ratio of WTP from CVM to cost of illness is equal to about 1.61 when the symptoms being valued are asthma symptoms and the respondents are asthmatic residents of Los Angeles. Alternative calculations for the same group of Los Angeles asthmatics results in a ratio of 3.7. All this makes our results surprising.

6.7 Conclusions

In this study an attempt is made to verify the economic rationale of air quality preservation by the dose–response technique (indirect valuation approach) and CVM to derive a CV type expression for the marginal WTP for improved air quality. The marginal WTP expression obtained using the health production function is quite simple in that it involves only one price (that of medical care) and two partial derivatives from the health production function or medical care function (therefore air pollution and medical care). Moreover, this expression does not involve any utility terms, so that empirical estimation is relatively straightforward. The WTP expression was estimated using health and air pollution (PM_{10}) data and responses to a questionnaire administered to a set of sample individuals from the affected area. These estimates range from Rs. 21 to 52.2 for the Logit model and Rs. 12.15 to 30.45 for the Probit model per month for a 67% reduction in ambient mean PM_{10} concentrations.

Other estimates are obtained using the CVM technique. This method attempts to assess the benefits of a hypothetical (proposed) project of improved air quality at PTPS colony, Panipat, India. The CV technique has been carefully adapted to relate it to the local situation. There are clear economic linkages between preservation efforts and the benefits derived from these efforts. It is shown that in a developing country like India, people's preferences are well formed to place values on air quality preservation. The benefits to the local residents are estimated, and estimates ranges from Rs. 89 to 114 for untruncated mean/median to truncated mean.

The study makes two contributions to the literature on non-market valuation of goods and services in developing countries. *First*, it compares the estimated value to households of air quality improvements in a developing country like India with two methods applied on same set of data. *Second*, it suggests that the use of non-market valuation methods in developing countries can be both practical and feasible. However, estimates of economic benefits such as presented in this study should not be used as the sole basis for evaluating air quality improvement projects since the respondents are working for a polluting company.

Nevertheless, we believe this study does provide important, policy-relevant information for evaluating air pollution abatement investments in thermal power sector in India. Moreover, the analysis of the CV survey responses confirms that people are aware of air pollution problems. They do feel that they have lost valuable health benefits because of air pollution, and are willing to pay for improved air quality.

Chapter 7
Environmental Regulation and Production Efficiency

7.1 Introduction

The development of the power sector in India has proceeded so far with little attention paid to its environmental implications. Such a course of development, however, seems difficult to continue in the face of growing degradation of environmental quality and increasing public awareness of environmental problems in the country. The share of the thermal-power sector is about two-thirds of India's total electricity production. In the thermal-power sector, coal contributes the largest share of fuel consumption. Shrestha and Acharya (1992) have noted the fairly substantial contribution of thermal power to air pollution in India. Being a negative externality, this pollution adversely affects the welfare of society. Thermal-power plants in India have been asked to make compliance decisions to meet environmental standards that can involve the investment of millions of rupees. How could these pollution control efforts affect the production efficiency of this sector? The objective of this chapter is to study the impact of compliance decisions on the production efficiency of India's thermal-power sector.

In recent years, an active debate has emerged on the impact of environmental regulations and pollution abatement on the production efficiency and the competitiveness of the firm. The discussion was initiated by Porter (1990, 1991) and Porter and van der Linde (1995), who hypothesized that environmental regulation improves a firm's overall production efficiency relative to unregulated firms. Moreover, the hypothesis rests heavily on the inference that stiffer environmental regulations result in greater production efficiency. The conventional textbook economic theory contradicts this hypothesis, and economists have tended to be critical of its theoretical and empirical foundations (see, e.g., Jaffe et al., 1995; Jaffe and Palmer, 1997; Simpson and Bradford, 1996; Hetemaki, 1996).

The empirical literature on the relationship between environmental regulations and production efficiency remains in its infancy. Most of the studies that have focused on efficiency and made use of micro data have been based on a nonparametric linear programming approach (Bernstein et al., 1990; Färe, 1988; Yaisawarng and

S. Kumar and S. Managi, *The Economics of Sustainable Development*,
Natural Resource Management and Policy 32,
DOI: 10.1007/978-0-387-98176-5_7,

Klien, 1994). Further, the studies on stochastic frontier functions include little analysis that explicitly takes into account pollution or environmental regulations in the estimation of production efficiency. This may be a significant shortcoming, because most of the studies are concerned with industries generating substantial amounts of pollution and/or environmental regulations that do not have any relation to production efficiency. Nevertheless, the nonparametric linear programming studies, which explicitly incorporate pollution and environmental regulations, indicate that these factors have a great impact on production efficiency. The present study extends this literature by examining the effect of environmental regulation on the production efficiency of the thermal-power sector in India.

This study uses the output distance function (which is the reciprocal of the output-based Farrell measure of technical efficiency) as an analytical tool to examine the impact of environmental regulation and pollution abatement on the production efficiency of the Indian thermal-power sector. Here, the stochastic output distance function is estimated simultaneously with a model that explains what causes the plants to be inefficient. The factors responsible for inefficiency are estimated, employing the framework introduced by Battese and Coelli (1995) in the production function context. This approach enables us to analyze explicitly the impact of environmental regulations and pollution abatement on production efficiency.

The first section that follows sets out the theoretical and econometric models; the second contains the data, estimation procedure, and results; and the last offers concluding remarks.

7.2 Output Distance Function and Its Econometric Estimation

Suppose that a coal-burning electric utility plant employs a vector of inputs $x \in \Re^N_+$ to produce a vector of outputs $y \in \Re^M_+$, where $\Re^N_+$ and $\Re^M_+$ are nonnegative N- and M-dimensional Euclidean spaces, respectively. The relationship between inputs and outputs is captured by the plant's technology, which can be expressed as a mapping $P(x) \subseteq \Re^M_+$ from an input vector x into the set of feasible output vector. It is assumed that the output set $P(x)$ satisfies the maintained axioms of Färe (1988). The output distance function is defined on the output set as

$$D_0(x,y) = \min_\theta\{\theta : y/\theta \in P(x)\}. \tag{7.1}$$

Equation (7.1) gives the largest radial expansion of the output vector for a given input vector, which is consistent with that output vector belonging to $P(x)$. The assumptions for the output set imply a set of properties for the output distance function (for properties of output distance function, see Färe and Primont, 1995). In particular, a well-defined output distance function will always be homogenous of degree one in outputs. Given the properties of the output distance function, the following pair of relationships holds between the efficient frontier and the output distance function:

$$D_O(x,y) \leq 1 \Leftrightarrow y \in P(x), \tag{7.2a}$$

$$D_O(x,y) = 1 \Leftrightarrow y \in \text{Isoq}\,P(x), \tag{7.2b}$$

where $\text{Isoq}\,P(x)$ is the frontier of the output set. Thus, the value of the output distance function must be less than or equal to one ($D_O \leq 1$) for feasible output.

7.2.1 Econometric Output Distance Function

The econometric formulation of the output distance function if (7.1) can be expressed as

$$D_O = f(x,y)\exp\varepsilon, \tag{7.3}$$

where ε is the random disturbance term and is assumed to be independently and identically distributed (iid) as $N(0,\sigma_\varepsilon^2)$. In econometric estimation, the basic problem with the output distance function is the inability to observe the dependent variable. Further, if the function is assumed to be efficient (i.e., $D_O = 1$), the left-hand side of the equation is invariant, an intercept cannot be estimated, and ordinary least-squares (OLS) parameter estimates will be biased.

To have the solution of the problem, let us utilize the property that the output distance function is homogenous of degree +1 in outputs (see Grosskopf et al., 1995a; Lovell et al., 1994):

$$\lambda D_O(x,y) = D_O(x,\lambda y). \tag{7.4}$$

Now suppose $\lambda = 1/y_m$, then

$$1/y_m D_O(x,y) = D_O(x,y/y_m). \tag{7.5}$$

From (7.2a) and (7.2b),

$$1/y_m D_O(x,y) \geq D_O(x,y/y_m). \tag{7.6}$$

Equation (7.6) can be converted into a stochastic frontier model for D_O and then introducing the composed error term

$$\ln(1/y_{mk}) = \ln D_{Ok}(x_k, y_k/y_{mk}) + u_k + v_k, \tag{7.7}$$

where $k = 1, 2, \ldots, K$ denotes for plants, v refers to the random shocks and noise, and u represents the production inefficiency. It is assumed that v_k is iid as $N(0, \sigma_v^2)$, and u is assumed to be distributed independently of v and to satisfy $v_k \leq 0$. After having estimated (7.7), $E[u_k|v_k + u_k]$ is calculated for each plant from which plant-specific measures are computed as

$$D_{0k}(x, y) = \exp\{-E[u_k|v_k + u_k]\}. \tag{7.8}$$

The composed error structure was originally formulated in a production function setting by Aigner et al. (1977), and in the context of output distance function it was first used by Grosskopf and Hayes (1993) and later by Hetemaki (1996). This framework is extended to incorporate a model for u_k, which is employed to estimate simultaneously the technical inefficiency and its determinants.

7.2.2 A Model for Determinants of Technical Inefficiency

Battese and Coelli (1995) proposed a framework, in a production setting, to estimate simultaneously the magnitude of inefficiency and its determinants. This framework is applied here in distance function setting. Assume the u_k in (7.8) be defined as

$$\exp(-u_k) = \exp(-Z_k\delta - w_k). \tag{7.9}$$

where the u_ks are assumed to be independently distributed such that u_k is obtained by truncation (at zero) of the normal distribution with mean $Z_k\delta$ and variance σ^2; Z_k is an $(1 \times h)$ vector of plant-specific variables; δ is an $(1 \times h)$ vector of unknown coefficients of the plant-specific inefficiency variables; and w_k accounts for the residual efficiency and is defined by truncation of the normal distribution with zero mean and variance σ^2, so that the point of truncation is $-Z_k\delta$, i.e., $w_k \geq -Z_k\delta$. In other words, the Z_k variables shift the mean of the technical inefficiency error term.

In this model, the explanatory variables of technical inefficiency may not enter into the distance function directly but affect technical inefficiency. The appropriate content and term of the Z vector are not obvious. The Z vector should reflect the reason why inefficiency may arise, i.e., why the plants are not operating on the output distance frontier. Here we examine two factors that may contribute to inefficiency. The factors reflect the intensity of environmental regulation (whether a plant is meeting environmental standards or not) and capacity utilization rate (plant load factor - PLF).

The parameters of (7.7) and (7.9) may be estimated simultaneously by the maximum-likelihood method following the approach of Battese and Coelli (1995). The likelihood function is expressed in terms of the variance parameters, $\sigma_s^2 = \sigma_v^2 + \sigma^2$ and $\gamma \equiv \sigma^2/\sigma_s^2$ (for detail see Battese and Coelli, 1995).

7.3 Production Efficiency of Thermal-Power Sector in India

7.3.1 Data

For the present study, the data requirement is of only quantities of different inputs and outputs. Since the study deals with the thermal-power sector in India, the required data are collected from a single source, *Performance Review of Thermal Power Stations 1991–92* (CEA, 1993). The process of fossil-fuel electricity generation typically uses the conventional inputs, namely, fuel, labor, and capital, to produce the desired output. This study requires environmental variables as well. Plants in our sample use coal as their primary fuel; because coal is bundled with ash, sulfur, and carbon, these plants produce sulfur dioxide (SO_2), nitrogen oxides (NO_x), carbon dioxide (CO_2), and particulate matters as a byproduct of electricity generation through the coal-burning process. Meanwhile, the plants have to comply with regulated limits on these emissions. Plants have to invest in pollution abatement equipment. As a result, these byproducts are classified as undesirable outputs. It is unfortunate that statistics on these variables are not published in India. There are some engineering restrictions between the consumption of coal and the production of these emissions, and data on these variables can be constructed. Nonetheless, these data cannot be employed in regression analysis.

Capital input in a power station has been calculated almost in the same manner as adopted by Dhryms and Kurz (1964):

$$K = SFT/10^3, \tag{7.10}$$

where K is the capital input, million kilowatt hours (10^6 kWh); T, number of hours in a year; S, station size in megawatts (MW); and F, availability factor ratio of the station.

Keeping in view the fact that a power plant may consist of sets of different sizes, the Central Electricity Authority (CEA), defines the availability factor of a plant in the following way:

$$F = \sum_j Z_j E / T \sum_j Z_j \quad j = 1, 2, \ldots, w, \tag{7.11}$$

where Z_j is the size (in MW) of jth set in the station; E_j, number of hours jth set was available for generating electricity during a year; and W, number of sets in the station.

Moreover, three types of statistics also are used in the analysis: plants meeting the environmental standards, plant load factor, and the size of the plant. All these statistics are available in the *Performance Review of Thermal Power Stations, 1991–92*. The PLF is defined as

$$\text{PLF}(\%) = \text{Energy generated during the period} \times 100/C \times h, \tag{7.12}$$

where C is the total capacity in megawatts (MW) and h is the total hours in the period under review, i.e., 1 year.

Therefore, the PLF can be considered as a capacity utilization rate of a plant. In the study, 33 plants are considered for analysis, and all the above-mentioned statistics are available for them.

7.3.2 Estimation Procedures and Results

The specification and estimation of the empirical model consist of the following stages: specification of the functional form for the output distance function; specification of the preferable model for technical inefficiency; and computation of technical efficiency scores for that model. The specification of the functional form for the output distance function is somewhat problematic. In principle, a flexible functional form (e.g., translog) would be preferred to simple Cobb-Douglas functions. However, the tradeoff in using a flexible functional form in the large number of included parameters usually causes serious multicollinearity problems. Singularity of the Hessian matrix due to multicollinearity restricts the estimation of the stochastic function. Although the deterministic approach allows for the computation of this model specification, the fitted values would be all near or equal to one, and the parameter values would be extremely sensitive even to minor charges in model specification or data (Hetemaki, 1996). However, the translog functional form is used for further analysis (Kumar, 1999).

In the stochastic estimation, in accordance with (7.7) and the linear homogeneity of the input distance function of degree +1 in outputs, for each observation, the LHS and RHS variables are multiplied by λ (λ is the reciprocal of one of the output), which is electricity in our model. It should be noted that in the literature two important methodological issues have been raised concerning the estimation of a function like (7.7), namely, the problem of endogeneity and plant-specific effects, which are discussed next.

On the endogeneity issue, a pragmatic approach is taken and tested for potential endogeneity bias using the Hausman specification test. Since it is unclear how to test simultaneity in the stochastic frontier maximum-likelihood model used in the present study, the test was computed on the OLS estimation of (7.7).

Equation (7.7) is estimated simultaneously with the model for determinants of inefficiency, i.e.,

$$Z_i = \delta_0 + \delta_1 \mathrm{ENVS}_i + \delta_2 \mathrm{PLF}_i + w_i. \quad (7.13)$$

In the determinants of inefficiency, two variables are taken: whether the plants are meeting environmental standards (ENVS) or not, i.e., by using dummy variables, one for the plants that are meeting these standards and zero otherwise. PLF is the capacity utilization rate of the thermal-power station and w_i is as defined earlier.

In the stochastic function, the maximization of the log-likelihood function, stated above, yield consistent and efficient estimates of the parameters of the model equation (7.7) and the determinants of inefficiency. The model is estimated using the Frontier 4.1 program (Coelli, 1994).

7.3.3 Results

We start by considering the results for the production structure and then turn to an analysis of the results for the inefficiency portion of the model. Table 7.1 gives the results from computation of the model: the translog specification. The results of the model show that some of the input and output parameters were not significant at the 5% level. In particular, the coefficient for the labor input variable has an expected positive sign and is significant. It deserves some explanation and is presented later. Also, the value of the coefficient of the capital is, as expected, low and insignificant. Table 7.2 offers production efficiency and the determinants of inefficiency. In this table, the null hypothesis – that there is no technical inefficiency in thermal-power plants – is considered for the various specifications. This test is equivalent to testing the null hypothesis that the parameters are zero, i.e., $\delta_0 = \delta_1 = \delta_2 = 0$ or $\delta_1 = \delta_2 = 0$. If the null hypothesis is accepted, the δ_i term could

Table 7.1 Maximum-likelihood estimates of stochastic output distance function

Parameter	Value[a]
β_0	−9.39 (7.8)
β_1	−11.81 (−2.83)
β_2	0.034 (0.014)
β_3	12.89 (2.28)
β_{11} (K2)	−7.5 (−1.99)
β_{22} (C2)	−0.13 (−0.16)
β_{33} (L2)	−12.48 (−1.22)
β_{12}(CK)	−1.47 (0.24)
β_{13} (CL)	20.36 (2.76)
β_{23} (LK)	0.77 (0.13)
δ_0	2.08 (2.84)
δ_1 (PLF)	−0.03 (−2.92)
δ_2 (ENVS)	0.61 (2.64)
$\gamma = \sigma_u^2/(\sigma_v^2 + \sigma_u^2)$	0.002 (0.08)
Log-likelihood	−21.38
Mean efficiency	0.58
Number of observations	33

Source: Authors' calculations

[a]Value in parentheses is the *t*-statistic; *C* consumption of coal, *K* capital, *L* labor; *PLF* plant load factor (capacity utilization); *ENVS* dummy variable (1, if the plant is meeting environmental standards, 0 otherwise)

Table 7.2 Tests of hypotheses for functional form and parameters of the inefficiency model

Null hypothesis	LR test statistic	Critical (λ)[a]	Decision value at 5%
Ho: $\delta_1 = \delta_2 = 0$	9.382	3.84	Reject Ho
Ho: $\delta_0 = \delta_1 = \delta_2 = 0$	9.382	5.99	Reject Ho

Source: Authors' calculations

[a] $\lambda = -2\{\log[\text{likelihood}(H_0)] - \log[\text{likelihood}(H_1)]\}$

be omitted and the model estimated using OLS. The test results reject the null hypothesis. The rejection of null hypothesis is significant at either 1% or 5%. In summary, these tests indicate that both the PLF and ENVS variables should be included in the model.

Coelli (1995) indicated that the interpretation of the γ parameter is not as clear in the above specifications as it is in the conventional half-normal stochastic frontier model. According to him, the γ parameter may be interpreted loosely in the present context as an indication of the amount of unexplained variation in the technical inefficiency effects, relative to the sum of this value and the variance of v_i. The value of γ lies between zero and one. If it is zero, then the variance of the inefficiency effects is zero and the model reduces to a traditional mean response function. On the other hand, a high value of the γ indicates that the model of determinants of inefficiency accounts for the bulk of the variation in the technical inefficiency. In all the model specifications shown in Table 7.1, the absolute value of γ is very low and statistically insignificant.

The sign of the δ_i coefficients are of particular interest. The coefficient of the capacity utilization rate (PLF) is negative, as one would expect. Thus, the higher the capacity utilization, the more efficient is the plant.

The efficiency scores computed from the model and capacity utilization rate indicate that there is a strong link between these variables. In our study, the least-efficient plant (Muzaffarpur) has only 20% efficiency and the most efficient (Korba STPS) has as high as 70.95% efficiency. The "all-India" plant availability of thermal units was at 72.8%; by comparison, the average PLF of these units was 55.3% in the study year (1991–1992). The nonavailability of coal of appropriate quality, shortfalls in the availability of gas, equipment deficiencies, and equipment failure are some of factors that have contributed to the low average PLF of the thermal-power stations and, in turn, to low efficiency (Planning Commission, 1994).

What should be the direction of the sign of the coefficient ENVS? There is substantial controversy in the literature. According to economics textbook versions, this sign should be positive, since introducing environmental regulations to areas where they were not previously applied or tightening existing regulations inevitably results in lower profits for the firm. Porter (1990, 1991) has hypothesized that environmental regulations can improve a firm's overall production efficiency and competitiveness relative to firms not regulated, and the sign of the coefficient of ENVS should be negative. Porter and van der Linde (1995) argue:

> that properly designed environmental standards can trigger innovation that may partially or more than fully offset the costs of complying with them. Such "innovation offsets," as we

> call them, cannot only lower the net costs of meeting environmental regulations, but even lead to absolute advantages. (p. 98)

The state of the discussion recently has been summarized in the *Journal of Economic Perspectives* (Palmer et al., 1995; Porter and van der Linde, 1995). It has been summed up thusly:

> Fundamentally, [pollution] is a manifestation of economic waste and involves unnecessary, inefficient or incomplete utilization of resources or resources not used to generate their highest value. (Porter and van der Linde, 1995)[1]

Thus, the sign of this coefficient has paramount significance. A negative sign for the estimated coefficient of ENVS indicates that meeting the environmental standards will result in a decrease in the value of the technical inefficiency effect. But if the sign is positive and statistically significant, either at the 5% or 1% level, the more restrictive the regulations are, the more inefficient the production process would be. This result is contrary to one of the central arguments in Porter's hypothesis. Instead, this suggests that plants not meeting environmental standards do much better than those meeting these standards. Nonetheless, one should keep in mind that it is precisely the intensity of stringency of the regulation, rather than regulation, per se. This is the key to Porter's argument:

> Stringent regulation can actually produce greater innovation and innovation offsets than lax regulation. Relatively lax regulation can be dealt with incrementally and without innovation, and often with 'end of pipe' or secondary treatment solution. (Porter and van der Linde, 1995)

In India, the environmental standards are met either through "end-of-pipe" or secondary treatment solutions.

The technical efficiency scores rely on the value of μ_i. The details of obtaining the values for the conditional expectation of $\exp(-\mu_i)$, given the value of $\varepsilon_i = v_i - \mu_i$, is described earlier. The results from the model indicate that an average plant produces 58% of the output that could be produced with the same bundle of inputs by a technically efficient plant; when the technical efficiency scores are compared across plants, the results reflect high variation. The coefficient of variation is 47.16%.

Table 7.3 reflects that only 21.21% fall in the range of 0.96–1.00 technical efficiency. The percentage of the plants in the range of 0.51–0.75 is also 21.21. Approximately 50% of plants are up to the range of 0.5; in other words, they are producing only half of the output of a theoretically efficient plant with the given bundle of inputs.

It should be noted that the above results are in line with the findings of Singh (1991) for Indian thermal-power plants. His observations also are based on plant-level cross-section data from the same source and indicate that there is large difference between the efficiency levels of the least and most efficient plants. But there is great difference between the average technical efficiency of the two studies.

---

[1] On the level of material flow, this statement is correct but it does not mean that it is optimal from an economic point of view.

Table 7.3 Frequency distribution of plant-specific technical efficiency (TE): econometric estimates

TE internal	Number of plants[a]
$0 \leq 0.25$	5 (15.15)
0.26–0.50	11(33.33)
0.51–0.75	7 (21.21)
0.76–0.85	2 (6.06)
0.85–0.95	1 (3.03)
0.96–1.00	7 (21.21)

Source: Authors' calculations
[a]Figures in parentheses are percentages

In Singh's study, it is 0.73; in this study, it is only 0.58. There is difference as well in the methodologies of these studies. Singh (1991) used the production function approach with linear programming estimation and the labor variable was dropped.

Further, we examine the relationship between relative technical efficiency and size of the plant. The correlation between technical efficiency and plant size is positive, as expected, i.e., 0.336, but the value of coefficients is relatively low. The positive correlation means that the technical efficiency of a larger plant is higher. This can be explained in the following manner. For two and half decades, installing sets of only large size has increased production capacities. Such sets exhibit one distinct feature. There are fuel economies across the sets, i.e., the fuel consumption per unit of generation falls as the set size rises (Singh, 1991).

In Table 7.1, the sign of the labor coefficient remains unexplained. The possible explanation could be as follows. CEA reported that manpower employed in thermal-power stations in 2.88 men/MW, considering all capacity groups of units (CEA, 1993). For the 200- to 500-MW capacity group versus the group of less than 200 MW of units, manpower employed was 2.5 and 7.00 men/MW, respectively. There is an inverse relationship between plant size and labor employed per MW, and there is a positive correlation between technical efficiency and plant size that explains the positive sign of the coefficient of labor input in the estimation of output distance function (Table 7.4).

7.4 Conclusion

This study uses the output distance function as an analytical tool to examine the relationship between environmental regulations and production efficiency, which is a more general representation of production technology. The stochastic output distance function was estimated simultaneously with a model that explained the causes of inefficiency. The results of this study contradict the M.E. Porter hypothesis, i.e., environmental regulations lead to production inefficiency. But, according to Porter, it is not regulations as such; it is the intensity or stringency of regulations that encourages firms to adopt "pollution prevention methods." These methods

Table 7.4 Plant-specific score for technical efficiency: econometric estimation

Sr. No.	Thermal-power station	Value	Rank[a]
1	Badarpur	0.520	19
2	I.P. Station	0.462	11
3	Panipat	0.487	15
4	Bhatinda	0.325	7
5	Kota	0.947	28
6	Panki Ext.	0.206	3
7	Singrauli	0.802	24
8	Gandhinagar	0.507	18
9	Ukai	0.417	9
10	Wanakbori	0.658	22
11	Korba STPS	0.998	33
12	Bhusawal	0.964	27
13	Chandapur	0.996	31
14	Koradi	0.507	17
15	Khaperkheda	0.995	30
16	Parli	0.483	13
17	Paras	0.843	25
18	Trombay	0.363	8
19	Ramagundam	0.746	23
20	Ramagundam (STPS)	0.869	26
21	Raichur	0.440	10
22	Ennore	0.306	6
23	Mettur	0.585	21
24	Tuticorin	0.997	32
25	Muzaffarpur	0.128	1
26	Patratu	0.167	2
27	Chandrapur	0.485	14
28	Santaldih	0.501	16
29	Titagarh	0.527	20
30	South Gen. Stat.	0.992	29
31	Durgapur DPL	0.219	4
32	Farakka	0.464	12
33	Bondigoan	0.226	5

Source: Authors' calculations
[a] *1* least efficient, *33* most efficient

restrict economic waste, and pollution is a manifestation of economic waste. If any pollution standards are met in India, they are met through the "end-of-pipe" treatment. It is found that the average level of efficiency is only 0.58 and the coefficient of variation is quite high. Moreover, here it is found that there is a positive association between plant size and production efficiency and between the rate of capacity utilization and production efficiency. This reveals that the energy crisis in India can be resolved, to some extent, through better utilization of existing capacity. We suggest that environmental standards be met strictly, so that the benefits of Porter's hypothesis can be generated and economic waste in the form of pollution can be reduced, that is, a stringent environmental policy should be demanded.

Chapter 8
Cost of Environmentally Sustainable Industrial Development

8.1 Introduction

It is now known that sustainable industrial development requires the preservation of the environment. Industries create a demand not only for waste-receptive services from the environmental media – air, forests, land, and water – but also for material inputs supplied by environmental resources (e.g., wood in the paper and pulp industry). Environmental resources can ensure a sustainable supply of these services if they are preserved at their natural regenerative level or if the demand for waste-receptive services is equal to the waste-assimilative capacity of environmental resources. Given that the demand for environmental services from various economic activities can exceed the natural sustainable levels of supply at a given time, and if measures are not taken to reduce this excess demand to zero, environmental resources can be degraded. The cost of reducing the demand for environmental services to the natural sustainable level of supply is regarded as the cost of sustainable use of environmental resources and, in the case of industrial demand for environmental services, it is the cost of sustainable industrial development. The measurement of this cost of sustainable industrial development is the main objective of this chapter.

As a part of environmental regulation, a firm faces a supply constraint on environmental services in the form of prescribed standards for effluent quality. The effluent standards are normally fixed such that the demand for the services of environmental media does not exceed the natural sustainable level of supply. The firm has to spend some of its resources to reduce the pollution loads to meet the effluent quality standards. The firm with a resource constraint will have fewer resources left for the production of its main product after meeting the standards. Therefore, the opportunity cost of meeting these standards is in the form of a reduced output of the firm. If all the firms in the industry meet the standards, the value of the reduced output of firms is the cost of sustainable industrial development. How can we estimate this cost for a competitive firm facing environmental regulation? It has to be estimated by studying the firm's behavior in making

S. Kumar and S. Managai, *The Economics of Sustainable Development*, Natural Resource Management and Policy 32,
DOI: 10.1007/978-0-387-98176-5_8,

decisions regarding pollution loads and the choice of pollution abatement technologies. In some recent studies, the technology of a polluting firm is modeled on one of the two basic approaches using conventional methods of the theory of production: (a) considering effluent as an additional input in the production or profit function and (b) by including abatement capital as an additional input in a cost function. In some studies, the pollution abatement technology is modeled with the assumption that it is nonseparable from the technology of main products, while in others it is modeled with the assumption it is separable. In response to environmental regulation, firms may adopt different types of technologies to reduce pollution. Jorgenson and Wilcoxen (1990) identify three different responses of firms. First, the firm may substitute less-polluting inputs for more-polluting ones. Second, the firm may change the production process to reduce emissions. Third, the firm may invest in pollution abatement devices. In practice, a firm may adopt a mix of these methods. The first two methods are nonseparable with the production processes of main products, while the third method is known as the end-of-the-pipe method.

Starting from the early 1980s, a number of empirical studies have examined the impact of environmental regulation on the economic performance of firms (see Myers and Nakamura, 1980; Pittman, 1981, 1983; Gollop and Roberts, 1983; Conrad and Morrison, 1989; Jorgenson and Wilcoxen, 1990; Barbara and McConnell, 1990; Gray and Shadbegian, 1995). The ultimate aim of these studies has been to measure the effect of pollution regulation on total factor productivity (TFP) growth. Most of these studies are based on production and cost profit functions, with the pollution variable modeled indirectly using abatement capital expenditure as one of the inputs. Ideally, the technology of water- or air-polluting firms must be described as one of joint production of good and bad outputs, the bad output being the pollution. The assumption of free disposal (a multiproduct firm can produce less of one output without reducing the outputs of other goods) that is normally made in the conventional production theory cannot be applied to describe the technologies of polluting firms. Shephard (1974: 205) noted that:

> ... for the future where unwanted outputs of technology are not likely to be freely disposable, it is inadvisable to enforce free disposal of inputs and outputs. Since the production function is a technological statement, all outputs, whether economic goods are wanted or not, should be spanned by the output vector *y*.

Also, conventional studies have implicitly assumed that the firms are operating on the production frontier and that pollution control does not have an impact on production efficiency. However, many recent studies have shown that these assumptions are unlikely to hold in many cases (see Färe et al., 1989, 1993; Hakuni, 1994; Yaisawarng and Klien, 1994; Porter and van der Linde, 1995; Coggins and Swinton, 1996; Kumar, 1999). Finally, the profit or cost functions used to represent production technology require firm-specific prices, especially input prices,[1] the reliable data of which are difficult to obtain. As will be shown in this chapter, the

[1] See recent studies on pollution abatement cost functions in India (e.g., Mehta et al., 1995; James and Murty, 1996; Pandey, 1999; Misra, 1999).

distance function approach for describing the production technology of a firm will potentially avoid all these problems.

The remainder of the chapter is planned as follows. Section 8.2 describes the methodology for measuring the cost of sustainable development. Section 8.3 provides information about the data and also highlights the methods of estimation of the output distance function. Section 8.4 presents estimates of shadow prices of bad outputs, scale economies, and technical efficiency for water-polluting industries in India. Finally, Section 8.5 provides concluding comments.

8.2 Measuring Cost of Sustainable Industrial Development

8.2.1 Output Distance Function

The conventional production function defines the maximum output that can be produced from an exogenously given input vector, while the cost function defines the minimum cost to produce the exogenously given output. The output and input distance functions generalize these notions to a multioutput case. The output distance function describes "how far" an output vector is from the boundary of the representative output set, given the fixed input vector. The input distance function shows how far the input vector is from the input vector corresponding to the least cost for producing a given vector of outputs.

Suppose that a firm employs a vector of inputs $x \in \Re_+^N$ to produce a vector of outputs $y \in \Re_+^M$; $\Re_+^N$, $\Re_+^M$ are nonnegative N- and M-dimensional Euclidean spaces, respectively. Let $P(x)$ be the feasible output set for the given input vector x and $L(y)$ is the input requirement set for a given output vector y. Now the technology set is defined as

$$T = \{(x, y) \in \Re^{M+N}, y \in P(x), x \in L(y)\}. \tag{8.1}$$

The output distance function is defined as

$$D_O(x, y) = \min\{\theta > 0 : (y/\theta) \in P(x)\} \quad \forall x \in \Re_+^N. \tag{8.2}$$

Equation (8.2) characterizes the output possibility set by the maximum equiproportional expansion of all outputs consistent with the technology set (8.1). We now turn to the properties of the output distance function. The output distance function can be used to measure the Debreu–Farrell technical efficiency (DF) (Debreu, 1951; Farrell, 1957). In terms of the above output set, the Debreu–Farrell measure can be defined as $\text{DF}(y, x) = \max\{\theta : \theta y \in P(x)\}$; and in terms of the output distance function $\text{DF}(y, x) = 1/D_O(x, y)$. Thus, the DF measure is the reciprocal of the value of the distance function, and it gives the factor by which all output could be expanded proportionately if the production units were operating on the frontier. It is clear that $D_O(x, y) \leq 1$. If $D_O(x, y) = 1$, the firm can be regarded as 100% efficient. For $D_O(x, y) \leq 1$, the firm produces in the interior and could be characterized as $100 \times D_O$ percent efficient.

The output distance function has, among others, the following properties (for a detailed description, see Färe, 1988):

1. $D_O(0, y) = +\infty$ for $y \geq 0$, that is, no free lunch.
2. $D_O(x, 0) = 0$ for all x in $\Re_+^N$, that is, inaction is possible.
3. $x' \geq x$ implies that $D_O(x, y) \geq D_O(x', y)$, that is, the more input the less efficient.
4. $D_O(x, \mu y) = \mu D_O(x, y)$ for $\mu > 0$, that is, positive linear homogeneity.
5. $D_O(x, y)$ is convex in y.

The assumptions about the disposability of outputs become very important in the context of a firm producing both good and bad outputs. The normal assumption of strong or free disposability about the technology implies

$$\text{if } (y_1, y_2) \in P(x) \text{ and } 0 \leq y_1^* \leq y_1,\ 0 \leq y_2^* \leq y_2 \Rightarrow (y_1^*, y_2^*) \in P(x).$$

That means we can reduce some outputs given the other outputs or without reducing them. This assumption may exclude important production processes, such as undesirable outputs. For example, in the case of water pollution, bio-oxygen demand (BOD), chemical oxygen demand (COD), and suspended solids (SS) are regulated, and the firm cannot freely dispose of waste. The assumption of weak disposability is relevant to describe such production processes. The assumption of weak disposability implies

$$\text{if } y \in P(x) \text{ and } 0 \leq \theta \leq 1 \Rightarrow \theta y \in P(x).$$

That means a firm can reduce the bad output only by decreasing simultaneously the output of desirable produce.

8.2.2 *Derivation of Shadow Prices of Bad Outputs*

The idea of deriving shadow prices using output and input distance functions and duality results is originally from Shephard (1970). A study by Färe et al. (1990) is the first to compute shadow prices using the (input) distance function and nonparametric linear programming methods. Färe et al. (1993) is the first study to derive the shadow prices of undesirable outputs using the output distance function. The derivation of absolute shadow prices for bad outputs using the distance function requires the assumption that one observed output price is the shadow price. Let y_1 denote the good output and assume that the observed good output price (r_1^0) equals its absolute shadow price (r_1^s) (i.e., for $m = 1$, $r_1^0 = r_1^s$). Färe et al. (1993) have shown that the absolute shadow prices for each observation of undesirable output ($m = 2, \ldots, M$) can be derived as (see Färe, 1988 for derivation)

$$(r_m^s) = (r_1^0)\frac{\partial D_0(x,y)/\partial y_m}{\partial D_0(x,y)/\partial y_1}. \tag{8.3}$$

The shadow prices reflect the trade off between desirable and undesirable outputs and the actual mix of outputs, which may or not be consistent with the maximum allowable under regulation (Färe et al., 1993: 376). Further, shadow prices do not require that the plants operate on the production frontier.

8.2.3 *Scale Economies*

Economies of scale for a multioutput production firm can be defined in terms of an output distance function[2] as

$$\frac{\mathrm{d}\theta/\theta}{\mathrm{d}\varepsilon/\varepsilon} = \frac{\sum_{n=1}^{N}(\partial D_0/\partial x_n)x_n}{y_1 + \sum_{m=1}^{M}(\partial D_0/\partial y_m)y_m}, \tag{8.4}$$

since $D_0(x,y) = y/F(x)$ (see Färe, 1988 for proof); where $n = 1, 2, \ldots, N$ inputs, $m = 1, 2, \ldots, M$ outputs, $\mathrm{d}\theta/\theta$ proportional increase in outputs, and $\mathrm{d}\varepsilon/\varepsilon$ proportional increase in inputs.

If the value of this function is equal to 1, it means the firm is operating under constant returns to scale, and if its value is greater than or less than one, then there are increasing or decreasing returns to scale, respectively. Having estimated the output distance function, the economies of scale for each firm can be computed by this formula.

8.3 Estimation of Output Distance Function

8.3.1 *Translog Output Distance Function and Data*

In order to estimate the shadow prices of pollutants (bad outputs) for Indian water-polluting industries using (8.3), the parameters of the output distance function have to be estimated. The translog functional form[3] is chosen for estimating the output distance function for Indian water-polluting industries, which is given as follows:

---

[2]See Pittman (1981) for the definition of scale economies in the production function setting for the firms producing multiple outputs.

[3]Many earlier studies for estimating shadow prices of pollutants have used the translog functional form for estimating the output distance function. These include Pittman (1981), Färe et al. (1990), and Coggins and Swinton (1996).

$$\ln D_0(x,y) = \alpha_0 + \sum_{n=1}^{N} \beta_n \ln x_n + \sum_{m=1}^{M} \alpha_m \ln y_m + \frac{1}{2}\sum_{n=1}^{N}\sum_{n=1}^{N} \beta_{nn'}(\ln x_n)(\ln x_{n'}) + \frac{1}{2}\sum_{m=1}^{M}\sum_{m=1}^{M} \alpha_{mm'}(\ln y_m)(\ln y_{m'}) + \sum_{n=1}^{N}\sum_{m=1}^{M} \gamma_{nm}(\ln x_n)(\ln y_m), \quad (8.5)$$

where x and y are, respectively, Nx1 and Mx1 vectors of inputs and outputs.

The data used in this chapter are from a recent survey of water-polluting industries in India.[4] These survey data provide information about characteristics of the main plant as well as the effluent treatment plant for the years 1994–1995. The data about the main plant are given for sales value, capital stock, wage bill, fuel cost, and other material input costs. The data about the effluent treatment plant are given for wastewater volume, influent and effluent quality for BOD, COD, and SS, capital stock, wage bill, fuel, and material input cost for a sample of 60 firms. The firms in the sample belong to chemicals, fertilizers, pharmaceuticals, drugs, iron and steel, thermal power, refining, and other industries. For estimating the output distance function, the technology of each plant is described by joint outputs – sales value (good output) and COD, BOD, and SS (bad outputs) – and inputs – capital, labor, fuel, and materials (see Table 8.1).

Water-polluting firms in Indian industry are supposed to meet the standards set for the pollutants (30 mg/L for BOD, 250 mg/L for COD, and 100 mg/L for SSP) by the Central Pollution Control Board. Command and control regulatory instruments are used to make the firms realize the standards. All 60 firms in the sample have effluent treatment plants; in addition, some firms are using process changes in

Table 8.1 Descriptive statistics of variables used in the estimation of the output distance function

Variable	Maximum	Minimum	Mean	Standard deviation
1. Sales	24,197.4	6.32	1,335.972	3,348.053
2. BOD	1,368,203.0	138.70	116,859.060	234,767.140
3. COD	10,005,560.0	335.80	934,810.750	1,954,634.800
4. SS	15,658,500.0	642.40	1,637,753.900	2,799,843.000
5. Capital cost	66,288.7	11.10	4,207.929	11,545,509
6. Wage bill	1,341.9	0.05	85.577	191.099
7. Power cost	16,150.0	2.58	779.090	2,505.045
8. Material cost	892.5	0.13	123.360	207.692

Note: Sales, wage bill, power cost, material cost, and capital cost are in Rs. million at 1994–1995 prices and BOD, COD, and SS are in kilograms
Source: Primary Survey

[4] A Survey of Water Polluting Industries in India, Research Project on "Fiscal Instruments for Water Pollution Abatement in India," Institute of Economic Growth, Delhi (1996).

production to achieve effluent standards. However, there is a large variation in the degree of compliance among the firms measured in terms of ratio of standard to effluent quality. The laxity of formal environmental regulation by the government, use of command and control instruments, and the absence of information regulation[5] by the communities in the neighborhood of the firms can be regarded as factors responsible for large variations in compliance to the pollution standards by the firms.

8.3.2 Estimation of Output Distance Function: Programming Model

In this section, a linear programming technique is used to estimate the parameters of a deterministic translog output distance function (Aigner and Chu, 1968). Let $k = 1, 2, \ldots, K$ index the observations in the data set. The following problem is solved to estimate the parameters

$$\max \sum_{k=1}^{K} [\ln D_0(x^k, y^k) - \ln 1] \tag{8.6}$$

subject to

(i) $\ln D_0(x^k, y^k) \leq 0$
(ii) $(\partial \ln D_0(x^k, y^k))/(\partial \ln y_1^k) \geq 0$
(iii) $\sum_{m=1}^{M} \alpha_m = 1,\ \sum_{m'=1}^{M} \alpha_{mm'} = \sum_{m=1}^{M} \gamma_{nm} = 0$
(iv) $\alpha_{mm'} = \alpha_{m'm},\ \beta_{nn'} = \beta_{n'n}$

Here the first output is desirable, and the rest of $(M - 1)$ outputs are undesirable. The objective function minimizes the sum of the deviations of individual observations from the frontier of technology. Since the distance function takes a value of less than or equal to 1, the natural logarithm of the distance function is less than or equal to 0, and the deviation from the frontier is less than or equal to 0. Hence the maximization of the objective function is done implying the minimization of the sum of deviations of individual observations from the frontier of technology. The constraints in (i) restrict the individual observations to be on or below the frontier of the technology. The constraints in (ii) ensure that the desirable output has a nonnegative shadow price. The constraints in (iii) impose homogeneity of degree +1 in outputs (which also ensures that technology satisfies weak disposability of outputs). Finally, constraints in (iv) impose symmetry. There is no constraint imposed to ensure nonnegative values to the shadow prices of undesirable outputs.

---

[5]For empirical evidence about informal regulation by the local communities, see Murty et al. (1999) and World Bank (1999).

Table 8.2 provides the linear programming estimates of the output distance function for the Indian water-polluting industries. For empirical evidence about informal regulation by the local communities, see Murty et al. (1999) and World Bank (1999).

8.3.3 Stochastic Output Distance Function

The stochastic output distance function for estimation is given as follows:

$$D_0 = f(x, y; \alpha, \beta) + \varepsilon, \tag{8.7}$$

where D_0 is the distance measure, $f(\cdot)$ is the production technology, x is a vector of inputs, y is a vector of outputs, α, β are vectors of parameters to be estimated, and ε is the additive error term. The error term may be generated for various reasons. Typically, it may include errors introduced by measurement, data collection, functional form specification, computational procedures, or factors known to the production units but not to the econometrician. Fuss et al. (1978), Brown and Walker (1995), and Griliches and Mairesse (1995) have provided a detailed analysis of the different factors that can generate random errors in production models.

The basic problem with distance functions that concerns econometric estimation is that one does not observe (have data on) the dependent variable. Further, if one sets the distance function equal to its efficient (frontier) value, $D_0 = 1$, the left-hand side of the distance function is invariant, an intercept cannot be estimated, and OLS parameter estimates will be biased. Further, if the distance function is expressed in logarithms, the left-hand side of the distance function will be zero for all observations (i.e., $D_0 = \ln(1) = 0$). In order to avoid these problems, Lovell et al. (1994), Grosskopf et al. (1995a), Grosskopf and Hayes (1993), Coelli and Perelman (1996), and Kumar (1999) utilize the property that the output distance function is homogeneous of degree one in outputs. Thus, for each observation to be used in estimating the distance function, a value that is unique to that observation can be used to multiply all output values on the right-hand side and the value of the distance function on the left-hand side. Thus, for an output distance function the following relationship (ignoring the error term) holds

$$\lambda D_0(x, y) = D_0(x, \lambda y) \quad \text{for any} \quad \lambda > 0. \tag{8.8}$$

In the literature, one of the outputs typically is chosen arbitrarily as a scaling variable. For example, if we chose the Mth output, and set $\lambda = 1/y_m$, (8.8) may be written as

$$1/y_m D_0(x, y) = D_0(x, y/y_m). \tag{8.9}$$

Table 8.2 Parametric estimate of the output distance function for water-polluting industries in India (linear programming)

Variables	Parameters	Values
y_1	α_1	0.173
y_2	α_2	−0.481
y_3	α_3	0.147
y_4	α_4	0.160
x_1	β_1	0.191
x_2	β_2	−0.493
x_3	β_3	−0.302
x_4	β_4	−0.560
y_1^2	α_{11}	−0.147
y_2^2	α_{22}	0.097
y_3^2	α_{33}	0.117
y_4^2	α_{44}	−0.013
y_1y_2	α_{12}	1.004
y_1y_3	α_{13}	−0.795
y_1y_4	α_{14}	−0.084
y_2y_3	α_{23}	−0.204
y_2y_4	α_{24}	0.021
y_3y_4	α_{34}	0.003
x_1^2	β_{11}	0.059
x_2^2	β_{22}	0.072
x_3^2	β_{33}	0.132
x_4^2	β_{44}	−0.131
x_1x_2	β_{12}	−0.005
x_1x_3	β_{13}	0.074
x_1x_4	β_{14}	0.051
x_2x_3	β_{23}	0.009
x_2x_4	β_{24}	−0.178
x_3x_4	β_{34}	−0.082
y_1x_1	γ_{11}	−0.125
y_1x_2	γ_{12}	0.045
y_1x_3	γ_{13}	−0.215
y_1x_4	γ_{14}	0.428
y_2x_1	γ_{21}	−0.055
y_2x_2	γ_{22}	−0.303
y_2x_3	γ_{23}	−0.580
y_2x_4	γ_{24}	−0.136
y_3x_1	γ_{31}	0.011
y_3x_2	γ_{32}	0.245
y_3x_3	γ_{33}	0.512
y_3x_4	γ_{34}	0.065
y_4x_1	γ_{41}	−0.044
y_4x_2	γ_{42}	0.083
y_4x_3	γ_{43}	0.014
y_4x_4	γ_{44}	0.054
Constant	α_0	−0.598

y_1 turnover (Rs. million), x_1 capital cost (Rs. million), y_2 BOD (tons), x_2 wage bill (Rs. million), y_3 COD (tons), x_3 power cost (Rs. million), y_4 SS (tons), x_4 material cost (Rs. million)
Source: Estimated

Now imposing some logarithmic functional form on the output distance function in accordance with most of the empirical literature, (8.9) becomes

$$\ln(D_0/y_m) = f(x, y/y_m, \alpha, \beta), \tag{8.10}$$

where f denotes some logarithmic functional form, such as the translog and the parameters. Alternatively, (8.10) may be expressed as

$$\ln D_0 - \ln y_m = f(x, y/y_m, \alpha, \beta) \tag{8.11}$$

or

$$-\ln y_m = f(x, y/y_m, \alpha, \beta) - \ln D_0. \tag{8.12}$$

Given the data, the parameters in (8.12) can be estimated in various ways, depending on the estimation criteria chosen. Basically, the objective of the estimation method is to generate parameter estimates that fit the data as closely as possible while maintaining the requirement that $0 \leq D_0 \leq 1$, which in the logarithmic case implies $-\infty \leq \ln D_0 \leq 0$.

Aigner et al. (1977) uses the stochastic frontier ML method in a production function context. This approach is based on the composed error term idea, in which a symmetric error term accounts for noise and an asymmetric error term accounts for production inefficiency. For the inefficiency component of the error term, one assumes a functional form and estimates simultaneously all the technology parameters and the parameter(s) of the distribution of the inefficiency term. Adding a symmetric error term, v, to (8.12), and denoting the distance to the frontier term, $-\ln(D_0)$, by μ, the stochastic frontier output distance function is obtained as

$$-\ln(y_m) = f(x, y/y_m, \alpha, \beta) + v + \mu. \tag{8.13}$$

In the literature it typically has been assumed that v is distributed $N(0, \sigma_v^2)$ and independently from μ, while μ is assumed to be either half-normal, truncated normal, exponential, or gamma distributed (see Green, 1993a, b). It appears that the most popular choice for application has been the half-normal distribution and maximum-likelihood estimation (Coelli, 1995). After having estimated (8.13), $E\langle\mu|v + \mu\rangle$ is computed for each plant from which plant-specific efficiency measures are calculated as

$$D_0(x, y) = \exp[-E\langle\mu|v + \mu\rangle]. \tag{8.14}$$

In order to estimate simultaneously the magnitude of inefficiency and the determinants of inefficiency, the framework proposed by Battese and Coelli (1995b) in a production function setting is applied to the distance function framework. Let (8.14) be defined as

$$\exp(-v) = \exp(-Z\delta - w), \tag{8.15}$$

Table 8.3 Maximum-likelihood estimate of the stochastic frontier output distance function for water-polluting industries in India

Variable	Coefficient	Parameter estimate	*T*-statistic
Constant	β_0	−1.458*	−3.892
y_1/y_4	β_1	0.661*	3.038
y_2/y_4	β_2	0.0096***	1.775
y_3/y_4	β_3	−0.052	−0.130
x_1	α_1	−0.079***	−1.847
x_2	α_2	−1.167*	−7.033
x_3	α_3	−0.333	−0.667
x_4	α_4	0.738***	−1.712
$(y_1/y_4)^2$	β_{11}	−0.017***	−1.423
$(y_2/y_4)^2$	β_{22}	−0.06	−0.352
$(y_3/y_4)^2$	β_{33}	−0.013	−0.111
x_1^2	α_{11}	−1.029	−0.572
x_2^2	α_{22}	−0.093***	−1.454
x_3^2	α_{33}	−0.0009	−0.013
x_4^2	α_{44}	−0.150**	−2.443
y_1y_2	β_{12}	−0.058	−1.186
y_1y_3	β_{13}	−0.045	−0.997
y_1x_1	β_{11}	−0.031	−0.738
y_1x_2	β_{12}	0.009	0.292
y_1x_3	β_{13}	0.005	0.0901
y_1x_4	β_{14}	0.013	0.379
y_2y_3	β_{23}	−0.023	−0.082
y_2x_1	γ_{21}	0.061	0.533
y_2x_2	γ_{22}	−0.138	−1.029
y_2x_3	γ_{23}	−0.142***	−1.321
y_2x_4	γ_{24}	0.069	0.760
y_3x_1	γ_{31}	0.073	0.779
y_3x_2	γ_{32}	0.0141***	1.323
y_3x_3	γ_{33}	0.169***	1.445
y_3x_4	γ_{34}	−0.1005***	−1.221
x_1x_2	α_{12}	−0.168*	−2.760
x_1x_3	α_{13}	0.209**	2.029
x_1x_4	α_{14}	−0.061	−0.889
x_2x_3	α_{23}	0.045	0.485
x_2x_4	α_{24}	0.008	0.105
x_3x_4	α_{34}	0.217**	2.171
Constant	δ_0	0.259*	2.623
BOD ratio	δ_1	−0.0057	−0.198
COD ratio	δ_2	−1.183*	−3.161
SS ratio	δ_3	0.0046***	1.747
	$\gamma = \sigma_\mu^2/\sigma_\mu^2 + \sigma_v^2$	0.0018**	2.366
	Log-likelihood	5.98	
	@	9.009***	

Notes: @ Likelihood ratio test of one-sided error with number of restrictions equal to 5
y_1 turnover (Rs. million), x_1 capital cost (Rs. million), y_2 BOD (tons), x_2 wage bill (Rs. million), y_3 COD (tons), x_3 power cost (Rs. million), y_4 SS (tons), x_4 material cost (Rs. million)
*Significant at 1% level, **significant at 5% level, ***significant at 10% level
Source: Estimated

where μ is assumed to be independently distributed, such that μ is obtained by truncation of the normal distribution with mean $Z\delta$ and variance σ^2; Z is a vector of plant-specific variables, and w stands for the unexplained part of the efficiency.

Here the model is estimated with the translog specification, and the determinants of inefficiency are taken as the ratios of effluent to influent of all the three pollutants, that is, BOD, COD, and SS. Estimation of the output distance function is done simultaneously with the model for determinants of inefficiency. The model was estimated using the Frontier 4.1 program (Coelli, 1994).

Table 8.3 gives the results from the estimation of the full translog specification. The results from the restricted translog and Cobb–Douglas specifications are not presented here, since the values of the log-likelihood ratio statistics are low for these specifications. The results for the translog model show that some of the parameters associated with the input and output variables are not significant even at the 10% level.

8.4 Estimates of Shadow Prices, Scale Economies, and Technical Efficiency

8.4.1 Shadow Prices

Table 8.4 provides estimates of industry-specific shadow prices for bad outputs, BOD and COD, based on the parameters of the translog output distance function estimated using the programming approach. These shadow prices are negative, reflecting desirable output and revenue foregone as a result of reducing the effluent by one unit (ton) per year. For instance, the average shadow price for water-polluting Indian industries is Rs. 0.246 million for BOD and Rs. 0.0775 million for COD per ton. That means reduction of BOD by one ton reduces production by Rs. 0.246 million worth of positive output. The average shadow price of total suspended solids (TSS) is zero. This zero shadow price implies that TSS can be disposed of at zero cost at the margin by the factories. Alternatively, the pollution abatement process may be such that reduction of BOD or COD may jointly reduce TSS such that the additional cost of reducing TSS is zero.

There is a wide variation of shadow prices of pollutants across firms and across industries as shown in Table 8.4 and Appendix. The range of shadow prices for BOD is Rs. 5,266–460,189 per ton while for COD, it is Rs. 528–77,462 per ton. This wide variation can be explained by the variation in the degree of compliance as measured by the ratio of pollutant effluent load and sales value and the different vintages of capital used by firms for the production of desirable output and pollution abatement.

The shadow prices of BOD and COD, which may be interpreted as the marginal costs of pollution abatement, are found to be increasing with the degree of compliance of firms. Taking the index of noncompliance by the firms as the ratio of

Table 8.4 Shadow prices of BOD and COD for water-polluting industries in India (Rs. per ton) (linear programming parameter estimates)

Industry	No. of firms	BOD shadow prices	COD shadow prices
All firms	60	−246,496	−77,462
Fertilizer	4	−41,343	−10,195
Sugar	11	−179,433	−66,486
Distillery	5	−91,606	−34,390
Chemical	11	−438,988	−127,164
Refinery	2	−460,189	−163,597
Tannery	4	−138,681	−72,671
Iron and steel	1	−6,785	−528
Paper and paper products	16	−5,266	−837
Drug	4	−737,638	−67,774
Others	2	−436,806	−68,407

Source: Estimated

effluent of BOD or COD to the sales value, it is found that the higher the index, the lower the shadow price. That means, the dirtier the industry, the lower the shadow price. Considering the logarithm of shadow price as a dependent variable and the logarithm of effluent to sales ratios as an independent variable, the estimated relationships between shadow prices and the index of noncompliance for BOD and COD are given as follows:

In (BOD shadow proice) = −0.226 −0.710 In(BOD effluent to sales ratio)
$R^2 = 0.277$ (−0.358) (−4.712)
In (COD shadow price) = −3.531 −0.270 In(COD effluent to sales ratio)
$R^2 = 0.004$ (−3.493) (−0.470)
Note. Figures in brackets are *t*-values

In the case of BOD, there is a statistically significant negative relationship between the shadow price and the noncompliance index. However, in the case of COD, the relationship is negative but not statistically significant.

Also, the estimates show that the shadow prices of undesirable outputs fall with the pollution load reductions obtained by the firms in the case of BOD and COD. That means, as found in the earlier studies of Indian water-polluting industries (Mehta et al., 1995; Murty et al., 1999; Pandey, 1999; Misra, 1999), that these results also show there are scale economies in water pollution abatement, implying that the higher the pollution load reduction, the lower the marginal abatement cost. The logarithms of shadow prices are regressed separately against the logarithms of BOD and COD loads reduced (the difference between the influent and effluent loads) by the firms, the results of which are given as follows:

In BOD shadow proice = −0.772 −0.353 In(BOD load reduced)
$R^2 = 0.111$ (−0.918) (−2.697)
In COD shadow price = −1.953 −0.448 In(COD load reduced)
$R^2 = 0.151$ (−1.042) (−3.215)
Note. Figures in brackets are *t*-values

Table 8.5 Scale economies and efficiency measures for water-polluting industries in India (econometric estimation)

Industry	No. of firms	Scale economies	Efficiency
All firms	60	0.686	0.899
Fertilizer	4	1.017	0.803
Sugar	11	0.999	0.909
Distillery	5	0.338	0.796
Chemical	11	0.421	0.887
Refinery	2	1.173	0.889
Tannery	4	0.66	0.875
Iron and steel	1	0.551	1.000
Paper and paper products	16	0.527	0.949
Drug	4	0.744	0.893
Others	2	1.236	0.994

Source: Estimated

8.4.2 Technical Efficiency

Given the estimate of the econometric model of the output distance function in Section 8.3, the firm-specific measures of technical efficiency can be estimated using (8.4). The technical efficiency scores rely on the value of the unobservable distance function predicted. The descriptive statistics for the technical efficiency scores are given in Table 8.5 (column 4). The mean level of efficiency for Indian water-polluting industries is 0.899 if all the outputs, that is, good as well as bad outputs, are taken simultaneously. It means that the Indian industries are operating below the frontier, and their production of desirable output can be increased.

What do the results of the econometric model estimated in Section 8.3 say about the technical efficiency and the determinants of inefficiency? The model shows that the inefficiency effects are not a linear function of effluent–influent ratio of various pollutants. It indicates that all the three ratios corresponding to BOD, COD, and SS should be included in the model, as they are all significant at either the 10% or lesser level. The γ parameter defined in Table 8.3 may be interpreted as the amount of unexplained variation in the technical inefficiency effects (Coelli, 1995). This parameter has a value between zero and one. If it is zero, then the variance of effects of inefficiency is zero and the model reduces to the traditional mean response model. On the other hand, a high value for this parameter shows that the model of determinants of inefficiency accounts for the bulk of the variation in technical inefficiency. In our model specification, the absolute value of this parameter is very low, that is, 0.0018, and is statistically significant at the 5% level.

The sign of δ_1 coefficients in Table 8.3 are of particular interest. A negative sign for the estimated coefficient shows that an increase in the value of the variable, that is, ratio of effluent to influent (lower level of regulation) will result in a decrease in the value of the technical inefficiency effect. Thus the more restrictive the regulation, the more inefficient the production process will be. In our estimates, the signs for the BOD and COD ratios are negative and for the SS ratio the sign is positive. This result may be due to the type of regulatory instrument used, for example, command and control versus economic instruments. Since in India today, only command and control measures are used to control water pollution and it is known that the use of such instruments results in the firms using inefficient pollution abatement technologies, the result found above is expected.[6]

However, in a situation using economic instruments (pollution taxes or marketable pollution permits), the result that the stricter regulation leads to the decrease of technical efficiency of polluting firms may not hold good. There are studies arguing that environmental regulation results in improvement of the technical efficiency of firms, a win-win situation explained by the Porter hypothesis (Porter and van der Linde, 1995).

8.4.3 Scale Economies

One more issue of importance in the ongoing debate is about the implications of pollution control requirements for economies of scale and barriers to entry. Although this issue has not been as widely debated, it may have important policy implications. Many industries facing strict pollution control requirements are already characterized by capital intensity and a large minimum efficient size (MES) of plant. A large MES in an industry may act as a barrier to entry, either because of the number of customers that must be pirated away from other suppliers or because of the difficulty in raising the huge sums of money required to build a plant. If entry is difficult, actual and potential competition in the industry may be less vigorous; tacit and explicit collusion may be less difficult; and supercompetitive prices and profits may be easier to achieve. Thus if pollution control requirements increase MES in an industry, they may have harmful allocation effects, and the resulting resource costs should be weighed against the benefits of pollution control in policy decisions.

The measure of scale economies may be estimated for each firm in the sample, and one may then examine whether firms that show a high level of pollution control are those that have economies of scale in production and controlling pollution. If this association is found, one may conclude that pollution control regulations have increased MES in the sample. Table 8.3 (column 3) and Appendix (column 4) provide estimates of scale economies of water-polluting industries and firms in the sample.

Three questions are of interest concerning the results of testing for scale economies of joint production:

1. Are the firms in the sample generally operating under conditions of increasing, neutral, or decreasing economies of scale? In the sample, the average figure for this is 0.823.

---

[6]There are now studies to show that the compliance to the pollution standards by the industries in the developing countries including India are due to both formal regulation (command and controls) and the informal regulation by the local communities (Murty et al., 1999; World Bank, 1999).

2. Does any systematic difference in scale economies exist for different firms/industries in the sample (e.g., are higher levels of turnover/production associated with increasing or decreasing scale economies). In the sample of 60 firms, the correlation coefficient is 0.047.
3. Are higher levels of pollution control associated with increasing or decreasing scale economies? Unfortunately, a correct measure of pollution control is not available for answering this question. A low level of pollution may reflect either a high level of pollution control or merely a general low level of production. Obviously, any measure of pollution control must include both levels of influent and effluent. The measure chosen here is the ratio of effluent to influent; a lower value of the ratio reflects a higher level of control. The correlation coefficients between effluent/influent of BOD and COD and scale economies are −0.197 and −0.098, respectively.

8.5 Conclusion

The distance function in the theory of production helps to characterize the technology of a firm producing a vector of outputs jointly and to define their shadow prices or opportunity costs. In the case of a firm generating air and water pollution, the output distance function can be used to represent the firm's technology as a joint production of good and bad outputs. With the assumption of weak disposability of outputs, the shadow prices of pollutants can be defined in terms of positive output or revenue foregone.

The distance function approach helps to derive firm-specific shadow prices for pollutants. The estimated shadow prices of pollutants have to be equal for all the firms if pollution taxes are levied on all the firms in order to obtain their conformity with the prescribed standards and for all the firms reduced pollution loads to meet the standards. Since there are no pollution taxes in India, command and control instruments are used to compel the firms to meet the set standards, and a majority of firms do not comply with the standards. The shadow prices of estimated pollutants vary across the firms. The estimated shadow prices of pollutants BOD and COD for all the 60 firms in the sample differ across the firms. The estimated sample averages for shadow prices of BOD and COD are Rs. 0.246 and 0.077 per gram of pollutant, respectively. That means, as per the current pollution abatement practices, the Indian water-polluting industry is forgoing revenue amounting to Rs. 246 and 77 for reducing one kilogram of BOD and COD, respectively. Large differences in the firm-specific shadow prices of pollutants reflect the use of inefficient pollution abatement technologies by water-polluting industries in India. The large differences in the estimates of shadow prices of pollutants bring out clearly the case for using economic instruments, like pollution taxes or marketable pollution permits, in India instead of the currently used command and control instruments.

In an economy in which industries are meeting the pollution standards fixed for the sustainable use of environmental resources, the distance function approach in the theory of production can be used to estimate the maintenance cost of environmental resources. This can be a methodology that potentially can be used for estimating the environmentally corrected GDP by making use of the maintenance cost version of the United Nations methodology of "Integrated Environmental and Economic Accounting."

The estimates of production efficiency for water-polluting industries in India reported in this paper explain production efficiency with a joint production of good and bad outputs. For the Indian water-polluting industries as a whole, the estimated efficiency index is approximately 90%. It means that by employing the same set of inputs, the good output can be further increased by 10%. Among the industries for which an efficiency index is estimated, distillery has the lowest, while iron and steel has the highest efficiency in the sample of 60 firms from 17 water-polluting industries in India.

The estimates of economies of scale show that the water-polluting industry as a whole has decreasing returns to scale. Estimates show that three industries, that is, fertilizers, refinery, and drugs, have increasing returns to scale, while others have decreasing returns to scale. There is a positive correlation between the economies of scale and the turnover of a firm. Also, there is a positive association between pollution control and economies of scale (the higher the scale economies, the lower the effluent–influent quality ratio).

The shadow prices of pollutants estimated in this study may be interpreted as the marginal costs of respective pollutants. The result – a negative relationship between pollution load reductions and shadow prices across the firms found in this study – confirms the presence of scale economies in pollution abatement found in the earlier studies on industrial water pollution abatement in India.

Appendix: Estimates of Shadow Prices of BOD and COD and Technical Efficiency and Economies of Scale

Industry	Firm	Efficiency	Scale economies	Estimates of shadow prices	
				BOD	COD
Fertilizer	1	0.997	1.028	−0.086	−0.019
	2	1.000	1.451	−0.061	−0.003
	3	0.388	0.686	−0.083	−0.063
	4	0.828	0.903	−0.024	−0.010
Sugar	5	1.000	1.184	−0.414	−0.047
	6	0.763	1.106	−0.799	−0.264
	7	0.902	1.098	−0.099	−0.055
	8	0.790	1.217	−0.250	−0.152
	9	0.983	0.792	−0.007	−0.007
	10	0.994	0.751	0.000	0.000
	11	0.828	0.803	−0.010	−0.006
	12	0.998	1.035	−0.021	−0.015
	13	0.942	0.99	−0.046	−0.018

(*continued*)

(continued)

Industry	Firm	Efficiency	Scale economies	Estimates of shadow prices	
				BOD	COD
	14	0.821	1.067	−0.066	−0.024
	15	0.983	0.942	−0.035	−0.013
Distillery	16	0.747	0.575	−0.077	−0.035
	17	1.000	0.343	0.000	0.000
	18	0.718	0.338	−0.325	−0.108
	19	0.738	0.281	−0.001	0.003
	20	0.777	0.155	−0.001	0.000
Chemical	21	0.788	0.623	−0.102	−0.017
	22	0.743	0.849	−2.138	−0.406
	23	1.000	1.477	−0.503	−0.217
	24	0.93	0.823	−0.056	−0.016
	25	0.915	0.348	−0.012	−0.035
	26	0.873	0.645	−0.028	−0.003
	27	0.841	0.64	−0.013	−0.007
	28	0.944	0.348	−0.137	−0.015
	29	0.80	0.572	−0.106	−0.013
	30	0.926	0.937	−0.051	−0.004
	31	0.998	0.748	−0.013	−0.003
Refinery	32	0.862	1.469	−0.471	−0.167
	33	0.916	0.877	−0.024	−0.013
Tannery	34	0.887	0.848	−0.293	−0.149
	35	0.793	0.502	−0.016	−0.008
	36	0.962	0.772	0.000	0.000
	37	0.858	0.509	−0.056	−0.071
Iron and steel	38	1.000	0.768	−0.007	−0.001
Paper and paper products	39	0.999	0.575	−0.005	−0.001
	40	0.841	0.54	−0.004	−0.001
	41	0.936	0.481	−0.002	−0.000
	42	0.997	0.402	−0.000	0.000
	43	0.803	0.460	−0.003	−0.002
	44	1.000	0.437	−0.001	0.000
	45	0.802	0.372	−0.002	−0.007
	46	1.000	0.62	−0.006	−0.001
	47	0.888	0.498	−0.012	−0.001
	48	1.000	0.557	−0.002	0.000
	49	0.998	0.386	0.000	0.000
	50	1.000	0.514	−0.003	0.000
	51	1.000	0.62	−0.003	0.000
	52	0.835	0.576	−0.013	−0.001
	53	0.867	0.601	−0.003	−0.001
	54	0.998	0.551	−0.005	−0.001
Drugs	55	0.645	0.418	−0.005	−0.019
	56	1.000	1.115	−1.090	−0.094
	57	0.925	0.787	−0.060	−0.014
	58	1.000	0.657	−0.018	−0.002
Misc.	59	1.000	0.667	−0.088	−0.008
	60	0.987	1.805	−1.091	−0.182

Source: Estimated

Chapter 9
Win–Win Opportunities and Environmental Regulation: Test of the Porter Hypothesis

9.1 Introduction

Environmental regulation makes firms internalize the costs of environmental externality generated by them. It may result in firms complying with the regulation being less competitive in the market than the noncomplying firms. This conventional view about the effects of regulation on the competitiveness of firms has recently been subjected to scrutiny, especially in the context of empirically testing the so-called Porter hypothesis (Porter, 1990, 1991). Porter and van der Linde (1995) argue that properly designed environmental standards can trigger innovation that may partially or more than fully offset the costs of complying with them. Such "innovation offsets," as one can call them, cannot only lower the net costs of meeting environmental regulations, but even lead to absolute advantage (p. 98). The authors further contend that innovation offsets occur mainly because pollution regulation is often coincident with improved efficiency of resource usage; the inference is that stiffer environmental regulation results in greater production efficiency. Many economists (e.g., Palmer et al., 1995) remain skeptical of the widespread existence of this hypothesis or such "win–win" opportunities. Although Palmer et al. clearly do not accept the basic arguments of the Porter hypothesis, they do agree that environmental regulation and production efficiency may be related. According to them, "we acknowledge that regulations have sometimes led to the discovery of cost saving or quality improving innovation; in other words, we do not believe that firms are ever vigilantly perched on their efficiency frontier." However, they indicate that more systematic studies are needed to establish the extent of the effect. Indeed, empirical literature on the relationship between environmental regulation and production efficiency is still rather scarce. The objective of this chapter is to study the effect of environmental regulation relating to water pollution by the manufacturing industry in India on the productive efficiency of firms. The panel (time series–cross section) data of 92 water-polluting firms for the three-year period 1996–1999 are used to test the Porter hypothesis.

S. Kumar and S. Managai, *The Economics of Sustainable Development*, Natural Resource Management and Policy 32,
DOI: 10.1007/978-0-387-98176-5_9,

There are three major approaches used in the literature to measure the effect of environmental regulations on the production efficiency of firms:

1. Adjusting the output of the plant to account for the marginal benefit or cost of the emission reduction or the shadow prices of pollutants (Pittman, 1981, 1983; Färe et al., 1993; Hetemaki, 1996; Coggins and Swinton, 1996; Repetto et al., 1996; Kumar, 1999; Murty and Kumar, 2002)
2. Accounting for the effect of pollution abatement costs on the total factor productivity (Gollop and Roberts, 1983; Barbara and McConnell, 1990; Gray and Shadbegian, 1995) and the plant cost function (Morgenstern et al., 1997)
3. Directly measuring efficiency and computing the changes in inputs and outputs if pollution levels or abatement expenditures were not constrained (Färe et al., 1986, 1989; Boyd and McClelland, 1999)

All of these studies can be further classified in to two types, the first type using conventional approaches, such as production, cost, or profit functions, while a second category employs the theory of distance functions. Here the analysis is carried out using the distance functions approach.

Although the theoretical framework on which the distance functions are based has been known for a long time (Shephard, 1953) it is only recently that their usefulness in empirical applications has come to be appreciated. In particular, the work of Färe and others (Färe et al., 1986, 1989, 1993, 1994; Färe and Primont, 1995) has been influential in popularizing the use of distance functions. Most of the existing applications of distance functions are either nonparametric studies or based on the parametric linear programming approach. It appears that only a few econometric distance functions studies have been carried out (Lovell et al., 1994; Grosskopf and Hayes, 1993; Hetemaki, 1996; Kumar, 1999). Probably the most important reason for the paucity of econometric applications is the fact that the stochastic estimation of distance functions is more involved than the application of linear programming models or the estimation of production, cost, and profit functions without considering the joint production of good output and bad outputs (pollution loads). This results in a potentially misleading comparison of the productive efficiency of firms producing significant amounts of undesirable outputs, such as water and air pollution. When firms divert resources for reducing undesirable outputs, the input/output ratios of the firm are higher and the productivity of the plant appears lower. An output efficiency measure, which is the amount by which desirable output can be increased while maintaining the level of inputs usage, will label the plant as less inefficient than it would be in the absence of this diversion of resources. An input efficiency measure, which is the amount by which the usage of conventional inputs and undesirable outputs can be decreased while maintaining the level of desirable outputs, will similarly label the plant less inefficient than it would be in the absence of this diversion of resources. It is understood that the constraints imposed by environmental regulation on the decisions of the firm will be subsumed within an overall measure of efficiency.

The linear programming approach to compute distance functions is deterministic in that random errors are absent. It is rather a limitation of this approach. Of course, in some cases it may turn out that the random errors are of negligible importance for the final results, but even in these cases, this is usually not known a priori. Consequently, it is important to be able to estimate distance functions stochastically so that random errors are accounted for. As it turns out, the estimation of distance functions is not as straightforward as the estimation of conventional cost, production, or profit functions. Indeed, this may be the reason for the paucity of econometric distance function studies. In the present study the output distance function and the cause and effect relationship between technical inefficiency and environmental regulation are simultaneously estimated employing the framework proposed by Battese and Coelli (1995b) in a production function setting.

9.2 Methodology for Testing Porter Hypothesis

Conventionally, a firm's performance is assessed by a measure of productivity based on the estimate of production function without considering the joint production of good output and bad outputs (pollution loads). This results in a potentially misleading comparison of the productive efficiency of firms producing significant amounts of undesirable outputs, such as water and air pollution. When firms divert resources for reducing undesirable outputs, the input/output ratios of the firm are higher and the productivity of the plant appears lower. An output efficiency measure, which is the amount by which desirable output can be increased while maintaining the level of inputs usage, will label the plant as less inefficient than it would be in the absence of this diversion of resources. An input efficiency measure, which is the amount by which the usage of conventional inputs and undesirable outputs can be decreased while maintaining the level of desirable outputs, will similarly label the plant less inefficient than it would be in the absence of this diversion of resources. It is understood that the constraints imposed by environmental regulation on the decisions of the firm will be subsumed within an overall measure of efficiency.

Consider a firm employing a vector of inputs $x \in \Re_+^N$ to produce a vector of outputs $y \in \Re_+^M$ where $\Re_+^N$ and $\Re_+^M$ are nonnegative N- and M-dimensional Euclidean spaces, respectively. Let $P(x)$ be the feasible output set for the given input vector x and $L(y)$ is the input requirement set for a given output vector y. Now the technology set is defined as

$$T = \{(x,y) \in \Re^{M+N}, y \in P(x), x \in L(y)\}. \tag{9.1}$$

Assumptions about the disposability of outputs become very important in the context of a firm producing both good and bad outputs. The normal assumption of strong or free disposability about the technology implies, if $(y_1, y_2) \in P(x)$ and $0 \leq y_1^* \leq y_1,\ 0 \leq y_2^* \leq y_2 \Rightarrow (y_1^*, y_2^*) \in P(x)$. This means we can reduce some outputs given the other outputs or without reducing them. This assumption may exclude important production processes, such as undesirable outputs. For example, in the case of water pollution, biological oxygen demand (BOD), chemical oxygen demand (COD), and suspended solids (SS) are regulated and the firm cannot freely dispose of them. The assumption of weak disposability is relevant to describe such production processes. The assumption of weak disposability implies that if $y \in P(x)$ and $0 \leq \theta \leq 1 \Rightarrow \theta y \in P(x)$. This means a firm can reduce the bad output only by decreasing simultaneously the good output. Hence one can characterize a world where there are nonpriced outputs in production that the plant manager has an interest in controlling. The assumption of weak disposability about the production technology enables one to consider this behavior of the firm while defining the factor productivity. For the problem considered here it is convenient to decompose the plant's output vector into two subvectors, $y = (g, b)$, which represent desirable output, g, and undesirable outputs, b, of the production process. The difference between these two types of outputs is captured via the disposability assumptions. Here it is assumed that the desirable outputs are freely disposable and the undesirable outputs may only be weakly disposable. That is, the firm may have to expand resources (or reduce "good" output) to reduce the bad outputs.

9.2.1 *Output Distance Function Approach*

The conventional production function defines the maximum output that can be produced from an exogenously given input vector, while the cost function defines the minimum cost to produce the exogenously given output. The output and input distance functions generalize these notions to a multioutput case (Färe et al., 1994a; Färe and Primont, 1995). The output distance function describes "how far" an output vector is from the boundary of the representative output set, given the fixed input vector. The output distance function is defined as

$$D_0(x, y) = \min\{\lambda > 0 : (y/\lambda) \in P(x)\} \quad \forall x \in \Re^N_+. \tag{9.2}$$

Equation (9.2) characterizes the output possibility set by the maximum equiproportional expansion of all outputs consistent with the technology set (9.1). The output distance functions can be used to measure the Debreu–Farrell technical efficiency (DF) (Debreu, 1951; Farrell, 1957). For example, in terms of the above output set, the Debreu–Farrell measure can be defined as $\mathrm{DF}(y, x) = \max\{\lambda : \lambda y \in P(x)\}$; and in terms of the output distance function $\mathrm{DF}(y, x) = 1/D_0(x, y)$. Thus, the DF measure is the reciprocal of the value of the distance function and it gives the factor by which all output could be expanded proportionately if the production units were operating on the frontier. It is clear that $D_0(x, y) \leq 1$. If $D_0(x, y) = 1$, the firm can be regarded as 100% efficient and y is on the boundary of feasible production set. For $D_0(x, y) \leq 1$; y is in the interior of feasible production set and could be characterized as $100 \times D_0$ percent efficient. One should also keep in mind that the output distance function is the dual of the revenue function.

9.2.2 Econometric Estimation of Distance Functions

The procedure for estimating the output distance function econometrically is described as follows. The econometric formulation of the output distance function (9.2) can be expressed as

$$D_0 = f(x, y) \exp \varepsilon, \tag{9.3}$$

where ε is the random disturbance term and is assumed to be independently and identically distributed (IID) as $N(0, \sigma_{\varepsilon}^2)$. In econometric estimation, the basic problem with output distance function is the inability to observe the dependent variable. Further, if the function is assumed to be efficient (i.e., $D_0 = 1$), the left-hand side of the equation is invariant, an intercept cannot be estimated, and the ordinary least squares (OLS) parameter estimates will be biased. To solve this problem, the property used is that the output distance function is homogenous of degree +1 in outputs (Lovell et al., 1994; Grosskopf, 1996; Kumar, 1999):

$$\lambda D_0(x, y) = D_0(x, \lambda y), \tag{9.4}$$

now suppose $\lambda = 1/y_m$; then

$$1/y_m D_0(x, y) = D_0(x, y/y_m). \tag{9.5}$$

From (9.2)

$$1/y_m D_0(x, y) \geq D_0(x, y/y_m). \tag{9.6}$$

Equation (9.6) can be converted into a stochastic frontier model for D_0 and introducing the composed error term:

$$\ln(1/y_{mk}) = \ln D_{0k}(x_k, y_k/y_{mk}) + u_k + v_k, \tag{9.7}$$

where $k = 1, 2, \ldots, K$ denotes kth plant, v refers to random shocks and noise, and u represents the production inefficiency. It is assumed that v_k is IID as $N(0, \sigma_v^2)$; and u is assumed to be distributed independently of v and to satisfy $v_k \leq 0$: After having estimated (9.7), $E\langle u_k | v_k + u_k \rangle$ is calculated for each plant from which plant-specific measures are computed as

$$D_{0k}(x, y) = \exp[-E\langle u_k | v_k + u_k \rangle]. \tag{9.8}$$

The composed error structure was originally formulated in a production function setting by Aigner et al. (1977), and in the context of the output distance function it was first used by Grosskopf and Hayes (1993) and later by Hetemaki (1996). This framework is extended to incorporate a model for u_k which is employed to estimate simultaneously the technical inefficiency and its determinants.

9.2.3 *Relationship Between Technical Inefficiency and Environmental Regulation*

Battese and Coelli (1995) proposed a framework, in a production setting, to estimate simultaneously the magnitude of inefficiency and its determinants. This framework is applied here in the distance function setting. Assuming that u_k in (9.8) be defined as

$$\exp(-u_k) = \exp(-Z_k\delta - w_k), \tag{9.9}$$

where the u_ks are assumed to be independently distributed so that u_k is obtained by truncation (at zero) of the normal distribution with mean $Z_k\delta$ and variance σ^2; Z_k is an $(1 \times h)$ vector of plant-specific variables; δ is an $(1 \times h)$ vector of unknown coefficients of the plant-specific inefficiency variables; and w_k accounts for the residual efficiency and is defined by a truncation of the normal distribution with zero mean and variance σ^2; so that the point of truncation is $-Z_k\delta$; i.e., $-Z_k \leq w_k$.

In this model, the explanatory variables of technical inefficiency may not enter into the distance function directly, but they affect technical inefficiency. The appropriate content and term of the Z vector is not obvious. The Z vector should reflect the reason why inefficiency may arise, that is, why the plants are not operating on the output distance frontier. Here we examine three factors that may contribute to inefficiency. These factors reflect the intensity of the environmental regulation, wastewater per unit of revenue, and time.

The parameters of (9.7) and (9.9) may be estimated simultaneously by the maximum-likelihood method following the approach of Battese and Coelli (1995). The likelihood function is expressed in terms of the variance parameters, $\sigma_s^2 = \sigma_v^2 + \sigma_u^2$ and $\gamma \equiv \sigma_u^2/\sigma_s^2$ (Battese and Coelli, 1993).

9.3 Data and Translog Distance Function

In order to estimate the output efficiency for water-polluting Indian manufacturing industries, the parameters of output and input distance functions must be estimated. The translog functional form is chosen for estimating the distance functions. Many earlier studies for estimating the shadow prices of pollutants have used the translog functional form for estimating the output distance function (Pittman, 1981; Färe and Primont, 1990; Coggins and Swinton, 1996). The distance function in the translog functional form is given as follows:

$$\ln D(x,y) = \alpha_0 + \sum \beta_n \ln x_n + \sum \alpha_m \ln y_m + \frac{1}{2}\sum\sum \beta_{nn'}(\ln x_n)(\ln x_{n'}) + \frac{1}{2}\sum\sum \alpha_{mm'}(\ln y_m)(\ln y_{m'}) + \sum\sum \gamma_{nm}(\ln x_n)(\ln y_m), \tag{9.10}$$

where x and y are, respectively, $N \times 1$ and $M \times 1$ vectors of inputs and outputs. The homogeneity conditions, i.e., $\sum_{m=1}^{M} \alpha_m = 1$; $\sum_{m'=1}^{M} \alpha_{mm'} = \sum_{m=1}^{M} \gamma_{nm} = 0$; $m = 1, 2, \ldots, M$; $n = 1, 2, \ldots, N$; and symmetry conditions, $\alpha_{mm'} = \alpha_{m'm}$; $\beta_{nn'} = \beta_{n'n}$ are imposed.

Detailed questionnaires seeking information about the production and pollution abatement activities were sent to 1,500 water-polluting firms belonging to 18 categories of industries declared as water-polluting industries by the Central Pollution Control Board in India. The panel data during the period 1996–1999 for 92 firms for which full information is available are used in this study. The data consist of sales values, BOD, COD, and SS load as outputs and conventional inputs such as wage bill, capital stock, and materials. For a calculation of the relationship between the technical inefficiency and environmental regulation, the intensity of environmental regulation is measured by the two new variables, regulation index (RI), and water conservation index (CI). The RI variable is constructed by making use of effluent concentrations of BOD, COD, and SS for all the firms. To begin, an index of compliance of firms with respect to a given pollutant is constructed by scaling down each observation of effluent concentration by its maximum value (the value for the firm with least compliance) among 276 observations. Then the regulation index is defined as the geometric mean of the three compliance indices of BOD, COD, and SS. The range of this index is from zero to one. It takes the value one for the firm with the least compliance and approaches zero for the firm with the maximum compliance or zero pollution. The water conservation index is defined as the ratio of wastewater to turnover. The lower is this ratio; the higher is the conservation effort of the firm. Similar types of indices had been used earlier by Gollop and Roberts (1983) and Hetemaki (1996).

Table 9.1 lists the descriptive statistics of variables used in the estimation of translog output distance function. In most instances, the standard deviation is higher than the mean values for almost all the variables. This could be attributed to the fact that the firms in the sample belong to 12 categories of water-polluting industries with widely varying characteristics with respect to pollution and the size of the firm.

9.4 Results

The parameters of estimated output distance function are given in Table 9.2. Most of these parameters are significant either at the 1% level or at the 5% level. The log-likelihood ratio test is also significant at the 1% level with the number of restrictions equal to five. The value of the distance function computed for each observation

Table 9.1 Descriptive statistics of the data used in the study and estimates of technical efficiency

Variable (unit)	Mean	SD	Max	Min
Efficiency	0.511	0.098	0.620	0.087
Turnover (Rs. million)	1,911	3,291	25,190	0
BOD effluent load (kg)	50,634	106,941	813,262	2
COD effluent load (kg)	344,605	830,895	5,635,000	21
SS effluent load (kg)	99,471	252,681	1,481,200	4
Materials (Rs. million)	774	1,382	11,143	0
Wage bill (Rs. million)	169	794	10,080	0
Capital stock (Rs. million)	2,323	7,811	74,538.092	0
Regulation intensity (RI)	0.028	0.097	0.7866671	6.33192×10^5
Wastewater/turnover (ton) (CI)	1,676	4,403	37,172	0

Source: Primary Survey

gives a measure of technical efficiency at a plant level. Table 9.1 reports the descriptive statistics of technical efficiency for the sample. The mean value of technical efficiency for the firms in the sample is 0.51, meaning that the water-polluting industry in India is 49% inefficient. The effect of environmental regulation on the technical efficiency of firms is studied by estimating a relationship between technical inefficiency and the indices of environmental regulation, water conservation, and time – simultaneously with the output distance function.

The model shows that technical inefficiency is not a linear function of regulation and water conservation indices and the time variable. In Table 9.2, the coefficients of RI and CI are significant at the 1% level, but the coefficient of the time variable is not significant even at the 10% level. The γ parameter defined in Table 9.2 may be interpreted as the amount of unexplained variation in the technical inefficiency by its determinants (Coelli, 1995). This parameter varies between zero and one. If it is zero, the determinants do not explain the variation in inefficiency and the model reduces to the traditional mean response model. On the other hand, a high value for this parameter shows that the determinants of inefficiency account for the bulk of the variation in technical inefficiency. In the model specification, the absolute value of this parameter is very low, i.e., 0.05 and it is statistically significant at the 1% level.

The sign of δ_i coefficients in Table 9.2 are of particular interest in the case of testing the Porter hypothesis described in Section 9.1. A positive sign for the estimated coefficients of RI and CI shows that the higher the compliance to regulation (higher RI) and higher the conservation of water (higher CI) by a firm, the higher is its technical efficiency. The signs of estimated coefficients of RI and CI in the model are positive. In other words, the more the industry complies with the regulation, the more efficient it becomes. This result supports the Porter hypothesis. The positive (negative) sign of the coefficient of time implies an increase (decrease) of technical inefficiency over time.

Water conservation results in the saving of costs to the industry and thus contributes to an increase in productive efficiency. There may be potential complementarities between production of conventional output and a reduction of

Table 9.2 Parameter estimates of the output distance function (stochastic estimation)

Parameter	Coefficient	Standard error	t-Ratio
β_0	−21.670	6.077	−20.275
β_1	1.024	0.171	5.973*
β_2	−20.111	0.339	−20.328
β_3	0.012	0.335	0.035
β_4	0.352	0.249	0.014**
β_5	−20.224	0.265	−20.843
β_6	−20.938	0.212	−24.425*
β_7	0.019	0.008	2.307***
β_8	−20.037	0.017	−22.165***
β_9	0.091	0.036	2.541***
β_{10}	−20.169	0.019	−28.942*
β_{11}	0.005	0.031	0.163
β_{12}	−20.040	0.012	−23.253*
β_{13}	−20.008	0.026	−20.305
β_{14}	−20.061	0.027	−22.254***
β_{15}	−20.007	0.043	−20.161
β_{16}	0.092	0.035	2.629*
β_{17}	0.166	0.041	4.054*
β_{18}	−20.367	0.039	−20.946
β_{19}	0.084	0.026	3.188*
β_{20}	−20.015	0.028	−20.546
β_{21}	−20.076	0.019	−23.981*
β_{22}	0.022	0.053	0.410
β_{23}	0.227	0.052	4.343
β_{24}	−20.150	0.040	−23.761*
β_{25}	−20.099	0.053	−21.877**
β_{26}	−20.128	0.053	−22.428***
β_{27}	0.145	0.045	3.183*
β_{28}	0.522	0.269	0.194**
β_{29}	−20.629	0.251	−22.506***
β_{30}	−20.103	0.181	−20.572
β_{31}	0.126	0.183	0.688
β_{32}	−20.692	0.197	−23.520*
β_{33}	0.164	0.198	0.825
β_{34}	0.158	0.311	0.508
β_{35}	0.291	0.196	0.149**
σ-squared	0.183	0.015	11.901*
γ	0.047	0.065	7.156*
δ_0	0.583	6.008	0.097
δ_1	2.203	0.433	5.088*
δ_2	0.0001	0.00001	5.373*
δ_3	−20.012	0.032	−0.368
Log-likelihood function		−2,157.341	
LR test of the one-sided error		42.319*	
With number of restrictions		5	

*Significant at 1% level, **significant at 10% level, ***significant at 5% level
Source: Estimation

pollution loads. With the abatement technologies involving process changes as opposed to the end-of-pipe treatment, the cost of jointly producing conventional output and clean environment may be lower than the cost of producing them separately. Such complementarities might arise, for example, from cost savings associated with recovered or recycled effluents and reuse of wastewater.

The proponents of the Porter hypothesis argue that complementarities between environmental activities and conventional production combined with the induced innovations associated with environmental requirement can partially offset or actually exceed the direct expenditures associated with environmental protection.

9.5 Conclusion

The approach used in this chapter has the advantage of simultaneously measuring efficiency and determining the factors affecting it. Many of the empirical studies about the effect of environmental regulation on the productive efficiency of firms show that the regulation makes the firms less efficient. However, there are a few studies showing the opposite, the study reported in this chapter being one of them.

Environmental regulation could provide incentives to the firms for innovation and resource conservation in environmental management. To study this problem we require firm-specific panel data on the production and environmental management practices of firms. This chapter uses firm-specific data for 3 years for a sample of 92 water-polluting firms in India. The Porter hypothesis about the possibility of win–win opportunities for firms subjected to environmental regulation is tested for the Indian water-polluting industry. This is done by estimating the output distance function jointly with the equation explaining the relationship between technical inefficiency and indices of environmental regulation and water conservation and the time variable. The main empirical result is that the technical efficiency of firms increases with the intensity of environmental regulation and water conservation efforts. This result supports the Porter hypothesis about environmental regulation.

The win–win opportunities from environmental regulation could be found more in some industries and less in others. Similar studies for specific industries could help us to identify the industries with no such opportunities so that monitoring and enforcement could be directed to those industries in which incentives are absent. Given the very high monitoring and enforcement cost of environmental regulation, this could result in the significant cost savings.

Chapter 10
Industrial Water Demand and Shadow Price

10.1 Introduction

Use of water may be broadly classified into three consumption categories: agricultural, industrial, and domestic. While there is substantial literature dealing with the agricultural[1] and domestic[2] uses of water, relatively little has systematically analyzed industrial water use, especially in the context of developing countries.[3] This may partly be due to the lack of reliable information on water consumption at the firm level. There is no consensus on the range of industrial water demand price elasticity and the sensitivity of water demand to other factors such as other input prices and output levels. The question of assessing the economic value (shadow price) of water still remains open.

There are several reasons for analyzing industrial water demand in developing countries. First, although current industrial withdrawal of water in developing countries is quite low in comparison to developed countries, this is expected to increase in comparison to other sectors of the economy, as well in absolute terms, since these countries are expected to have higher growth in industrial production in the near future. Second, in developing countries, toxic and some persistent organic pollution (including toxins from the heavy metals industry) may be present in most effluent emissions, and since a large proportion of the urban population lives in the vicinity of industrial areas, many suffer the ill effects of high-level water pollution. Third, in countries like India, where concentration-based environmental standards are adopted for water pollutants and the financial extraction costs of water are too

---

[1] On survey of agricultural use of water and its pricing, see Varela-Ortega (1998) and Johansson et al. (2002).

[2] On survey of residential water demand, see Arbues et al. (2003).

[3] Frederick et al. (1997) report only seven estimates that deal with industrial water use in 494 estimates of economic value of freshwater in a survey for the U.S. In the context of developing countries we could locate only few studies which are devoted to the analysis of demand for industrial water use; i.e., Onjala (2001), Wang and Lall (2002), Feres and Reynaud (2003), Goldar (2003), etc.

S. Kumar and S. Managi, *The Economics of Sustainable Development*,
Natural Resource Management and Policy 32,
DOI: 10.1007/978-0-387-98176-5_10,

low, firms have incentives to dilute the effluent stream with the excessive use of water (Goldar and Pandey, 2001). Finally, since water is a scarce input, there are conflicts over its allocation for different uses. Thus, the valuation of water in competitive uses (domestic, industrial, and agricultural, as well as within different industries or firms) is a prerequisite for any water resource policy design.

Water enters into the production process of manufacturing firms as an intermediate public good, which reduces the unit cost of production (Wang and Lall, 2002). In estimating demand models (Turnovsky, 1969; Rees, 1969; DeRooy, 1974; etc.), earlier studies on industrial water use have used the ratios of total expenditure to total quantities of water purchased as proxies for prices. In cost function models (Grebenstein and Field, 1979; Babin et al., 1982; Ziegler and Bell, 1984; Williams and Shu, 1986; Renzetti, 1988, 1992, 1993, 2002; Dupont and Renzetti, 2001; Reynaud, 2003; Feres and Reynaud, 2003; etc.), studies were conducted by including water as an input, along with labor, capital, and materials, and the average cost of water consumption was used to determine the price. These studies find that the price elasticities of water are small and industry specific. They also find that water and labor are mostly substitutes, whereas capital and water are complementary inputs. The results of these studies should be considered with caution since they are based on aggregate data and do not take into account the specificity of water as input. Moreover, in these studies, water quantity appeared on both sides of the demand equation, which may introduce a simultaneity bias, and the use of average cost is not consistent with economic theory, since firms respond to marginal prices in their decision-making process.

This chapter contributes to the literature on industrial water use by estimating the industrial water demand for a panel of Indian manufacturing firms observed from 1996/1997 to 1998/1999. We characterize the structure of industrial water demand by estimating a translog input distance function. We model production technology by distinguishing four inputs (material, labor, capital, and water) and one output (sales revenue). We are especially interested in analyzing the following issues:

- What are the complementarity or substitutability relationships between the different inputs?
- What can be said about the price elasticity of industrial water demand in India?
- What can be the per unit shadow price of industrial use of water?

A firm's production technology could be modeled in different ways: the production function, profit function, or the cost function. Then Hotelling's Theorem and Shephard's Lemma allow one to derive compatible input demands and output offers with optimization behavior. Our approach to modeling the production process differs from earlier studies which use cost functions (see, e.g., Reynaud, 2003; Feres and Reynaud, 2003) or production functions (see Wang and Lall, 2002; Goldar, 2003), and instead uses a distance function to measure technology. The input distance function completely describes multiple output technology and is dual to the cost function (Färe and Primont, 1995).

The input distance function has an obvious advantage over production functions in allowing for the possibility of multiple outputs and joint production. One advantage of the input distance function over the cost function is that no information on input prices is required, nor is the maintained hypothesis of cost minimization. In fact, no specific behavior goal is embedded in the input distance function (Grosskopf et al., 1995b). Moreover, the distance functions allow one to calculate the shadow prices of the inputs, as the observed prices of inputs in developing countries are not market-clearing prices, especially for commodities like water. Similar to other analyses of production and technology, we calculate ease of substitution among the various inputs. Using parameter estimates of input distance function, the Morishima and Allen elasticity of substitution are computed. The Morishima elasticity is viewed as a more appropriate measure of substitutability when the production process has more than two inputs (Blackorby and Russell, 1989).

The remainder of the chapter is organized as follows. Section 10.2 presents the economic modeling. Industrial production technologies are represented by the input distance function and are approximated by a translog form. The estimation model is the subject matter of Section 10.3. Then we present an empirical application. The model is applied on a panel data of 92 firms concerning different water polluting industries. The original data come from a survey conducted by the Institute of Economic Growth, Delhi, in 2000 and is presented in Section 10.4. Section 10.4 also presents and discusses the results of the study. The chapter closes in Section 10.5 with some concluding remarks.

10.2 Economic Model

Consider a manufacturing firm employing a vector of inputs $x \in \Re_{+}^{N}$ to produce a vector of outputs $y \in \Re_{+}^{M}$ where $\Re_{+}^{N}$ and $\Re_{+}^{M}$ are nonnegative N- and M-dimensional Euclidean spaces, respectively. Let $P(x)$ be the feasible output set for the given input vector x and $L(y)$ is the input requirement set for a given output vector y. Now the technology set is defined as (Färe et al., 1994a)

$$T = \{(\mathbf{y},\mathbf{x})\hat{I}\ \hat{A}_{+}^{M+N}, \mathbf{y}\hat{I}\ P(\mathbf{x}), \mathbf{x}\hat{I}\ L(\mathbf{y})\}. \tag{10.1}$$

The conventional production function defines the maximum output that can be produced from an exogenously given input vector, while the cost function defines the minimum cost to produce the exogenously given output. The output and input distance functions generalize these notions to a multioutput case. The input distance function describes "how far" an input vector is from the boundary of the representative input set, given the fixed output vector. Formally, the input distance function is defined as

$$D(\mathbf{y},\mathbf{x}) = \min\{\lambda : [\mathbf{x}/\lambda, \mathbf{y}] \in T\}. \tag{10.2}$$

Equation (10.2) characterizes the input possibility set by the maximum *equi*proportional contraction of all inputs consistent with the technology set (10.1). The input distance function can be used to measure Debreu-Farrell technical efficiency. The input distance function is homogeneous of degree one in inputs, concave in inputs, convex in outputs, and nondecreasing in inputs.[4] It is dual to the cost function. That is,

$$\begin{aligned} D(\mathbf{y},\mathbf{x}) &= \min_{w}\{\mathbf{wx} : C(\mathbf{y},\mathbf{w}) \geq 1\}, \\ C(\mathbf{y},\mathbf{w}) &= \min_{x}\{\mathbf{wx} : D(\mathbf{y},\mathbf{x}) \geq 1\}, \end{aligned} \tag{10.3}$$

where **w** is a vector of minimum cost-deflated input prices and C is a unit cost function if the costs are minimized. This implies that the value of input distance function would be equal to one only when the inputs are used in their cost-minimizing proportions, i.e.,

$$C(\mathbf{y},\mathbf{w}) = \mathbf{wx}/D(\mathbf{y},\mathbf{x}). \tag{10.4}$$

Both cost and input distance functions completely describe the production technology, but they have different data requirements. Whereas both require data on output quantities, the distance function requires data on input quantities rather than input prices. Applying the dual Shephard's Lemma, the cost-deflated (i.e., normalized) input shadow prices can be derived from the input distance function. Färe and Primont (1995) show that the cost-deflated shadow price for each input is given by

$$\mathbf{w} = C(\mathbf{y},\mathbf{w})\,\tilde{\mathrm{N}}_x D(\mathbf{y},\mathbf{x}). \tag{10.5}$$

The undeflated (i.e., absolute) shadow prices can be expressed as the product of the cost function and the deflated shadow price. Hence when the cost function is known, the absolute shadow prices can be computed. The difficulty in computing undeflated shadow prices is that cost function depends on these undeflated shadow prices, which are unknown. However, if we assume that the observed price for the input is equal to its undeflated shadow price, then cost function is the ratio of its undeflated and deflated shadow prices. It is assumed that the undeflated shadow price of x_j is equal to its observed market price.[5] The remaining undeflated shadow prices (w_i) are computed as

$$w_i = w_j \frac{\partial D(\mathbf{x},\mathbf{y})/\partial x_i}{\partial D(\mathbf{x},\mathbf{y})/\partial x_j}, \quad i \neq j, \tag{10.6}$$

[4]For the properties of input distance function, see Färe and Primont (1995).

[5]To the extent that markets are imperfectly competitive, or there are subsidies or taxes, the assumption that the shadow price and observed prices are equal is inaccurate.

where w_i and w_j stands for the shadow prices of two different inputs x_i and x_j, respectively. Equation (10.6) states the undeflated shadow price of input (e.g., water) is the product of the actual price of other input (e.g., materials) and the marginal rate of technical substitution (MRTS) between two inputs. According to this equation, the absolute shadow price of the input for an inefficient producer is determined by making a radial projection to the isoquant from the observation.[6] The shadow prices of the inputs associated with that observation are calculated at the point on the isoquant. Hence, the absolute shadow price reflects the actual proportions of inputs used by an inefficient producer.

As the input distance function completely describes the production technology and identifies the boundaries of technology, one may use it to describe the characteristics of the frontier or surface technology, including curvature, i.e., the degree of substitutability along the surface technology (Grosskopf et al., 1995). Therefore, we calculate indirect Morishima elasticity of substitution as defined by Blackorby and Russell (1989). That is,

$$M_{ij}(\mathbf{x},\mathbf{y}) = -\frac{\mathrm{d}\ \ln[D_i(\mathbf{x},\mathbf{y})/D_j(\mathbf{x},\mathbf{y})]}{\mathrm{d}\ \ln[x_i/x_j]} = x_i\left(\frac{D_{ij}(\mathbf{x},\mathbf{y})}{D_j(\mathbf{x},\mathbf{y})}\right) - x_i\left(\frac{D_{ii}(\mathbf{x},\mathbf{y})}{D_i(\mathbf{x},\mathbf{y})}\right), \quad (10.7)$$

where the subscripts on the distance functions refer to partial derivatives with respect to inputs: e.g., $D_{ii}(\mathbf{x},\mathbf{y})$ is the second-order partial derivative of the distance function with respect to x_i. As noted earlier, the first derivatives of the distance function with respect to inputs yield the normalized shadow price of that input; therefore the first line of the definition may be thought of as the ratio of the percentage change in shadow prices brought about by a 1% change in the ratio of inputs. This would represent the change in relative marginal products and input prices required effecting the substitution under cost minimization. High values reflect low substitutability and low values reflect relative ease of substitution between the inputs. We can simplify the Morishima elasticity as follows:

$$M_{ij} = e_{ij}(\mathbf{y},\mathbf{x}) - e_{ii}(\mathbf{y},\mathbf{x}), \quad (10.8)$$

where $e_{ij}(\mathbf{y},\mathbf{x})$ and $e_{ii}(\mathbf{y},\mathbf{x})$ are the constant output cross- and own elasticities of shadow prices with respect to input quantities. The first term provides information on whether pairs of inputs are net substitutes or net complements, and the second term is the own price elasticity of demand for the inputs. Here it should be noted that these elasticities are indirect elasticities. Therefore,e_{ij} greater than zero indicates net complements and less than zero indicates net substitutes. (In contrast, a direct substitution elasticity greater than zero indicates net substitutes and less than zero indicates net complements.)

[6]The radial projection assumed a proportional contraction of all inputs for given output vector until the isoquant is attained.

The Allen elasticity of substitution may be defined in terms of distance function as

$$A_{ij} = [D(\mathbf{y},\mathbf{x})D_{ij}(\mathbf{y},\mathbf{x})/D_i(\mathbf{y},\mathbf{x})D_j(\mathbf{y},\mathbf{x})]. \tag{10.9}$$

Here it should be noted that the Morishima and Allen elasticities yield the same result in the two-input case; when the number of inputs exceeds two, however, they no longer coincide. Moreover, the Morishima elasticities may not be symmetric, i.e., $M_{ij} \neq M_{ji}$. This is as it should be and allows for the asymmetry in the substitutability of different inputs, e.g., substitutability between skilled and unskilled personnel.

The returns to scale RTS measure can be calculated from the input distance function using the formula

$$\mathrm{RTS}(\mathbf{y},\mathbf{x}) = \frac{\partial \ln \varsigma}{\partial \ln \xi} = \frac{-1}{\nabla_y D(\mathbf{y},\mathbf{x})}, \tag{10.10}$$

where ς and ξ are scalars representing equiproportionate changes in the output and in the input vectors, respectively.

10.3 Estimation Model

The distance functions can be computed either nonparametrically using the data envelope analysis (DEA) or parametrically. Here we adopt the parametric approach for the computation of distance functions; the advantage of this approach is that it is differentiable. We employ the translog form of input distance function that is twice differentiable and flexible. The form is given by

$$\begin{aligned}\ln D(\mathbf{x},\mathbf{y}) &= \alpha_0 + \sum_{n=1}^{N} \alpha_n \ln \mathbf{x}_n + \sum_{m=1}^{M} \beta_m \ln \mathbf{y}_m \\ &+ \frac{1}{2}\sum_{n=1}^{N}\sum_{n'=1}^{N} \alpha_{nn'} \ln \mathbf{x}_n \ln \mathbf{x}_{n'} + \frac{1}{2}\sum_{m=1}^{M}\sum_{m'=1}^{M} \beta_{mm'} \ln \mathbf{y}_m \ln \mathbf{y}_{m'} \\ &+ \sum_{n=1}^{N}\sum_{m=1}^{M} \gamma_{nm} \ln \mathbf{y}_m \ln \mathbf{x}_n. \end{aligned} \tag{10.11}$$

To compute the parameters of (10.11), we use the linear programming approach developed by Aigner and Chu (1968), that is,

$$\text{Minimize} \quad \sum_{k=1}^{K} \{\ln D(\mathbf{x},\mathbf{y}) - \ln 1\}, \quad k = 1, 2, \ldots, K. \tag{10.12}$$

Subject to

(i) $\ln D(\mathbf{x}, \mathbf{y}) \geq 0$

(ii) $\dfrac{\partial \ln D(\mathbf{x}, \mathbf{y})}{\partial \ln y_m} \leq 0, \quad m = 1, \ldots, M$

(iii) $\dfrac{\partial \ln D(\mathbf{x}, \mathbf{y})}{\partial \ln x_n} \geq 0, \quad n = 1, \ldots, N$

(iv) $\sum_{n=1}^{N} \alpha_n = 1, \quad \sum_{n'=1}^{N} \alpha_{nn'} = \sum_{n=1}^{N} \gamma_{nm} = 0, \quad n, n' = 1, \ldots, N$

(v) $\alpha_{nn'} = \alpha_{n'n}, \quad \beta_{mm'} = \beta_{m'm}, \quad n, n' = 1, \ldots, N, \quad m, m' = 1, \ldots, M$

where K denotes the number of observations. The restrictions in (i) ensures that the value of input distance function is greater than or equal to one as the logarithm of this function are restricted to be greater than or equal to zero. Restriction in (ii) enforces the monotonocity condition of nonincreasing of input distance function in good outputs, whereas the restriction in (iii) enforces that the input distance function is nondecreasing in inputs. Restriction (iv) and (v) impose the homogeneity and symmetry conditions, respectively, as required by the theory.

From the translog specification, some characteristics of interest may be computed. We focus in particular on the price elasticities on input demands and elasticities of input demands with respect to output levels. The shadow price elasticities with respect to input quantities are obtained as

$$\varepsilon_{ij} = [\alpha_{ij} + S_i S_j]/S_i \qquad \text{if } i^1 \quad j,$$
$$\varepsilon_{ii} = [\alpha_{ii} + S_i(S_i - 1)]/S_i \quad \text{if } i = i.$$

The Allen elasticities of substitution, $A_{ij} = \varepsilon_{ij}/S_i$ and Morishima elasticities of substitution, $M_{ij} = \varepsilon_{ij} - \varepsilon_{jj}$, $i \neq j$ are computed. Where S_i is the first-order derivative of the translog output distance function with respect to input In x_i, i.e., $S_i = \partial \ln D(\mathbf{x}, \mathbf{y}) / \partial \ln x_i$.

10.4 Data and Estimation Results

The data used in this chapter are from a recent survey of water-polluting industries in India.[7] These survey data provide information about characteristics of the main plant for the 3 years 1996/1997 to 1998/1999. The data about the main plant are given for sales value, capital stock, wage bill, other material input costs, and water consumption for a sample of 92 firms. The firms in the sample belong to leather, distillery, chemicals, sugar, paper and paper products, fertilizers, pharmaceuticals,

[7]"A Survey of Water Polluting Industries in India," Research Project on "Environmental and Economic Accounting for Industry," Institute of Economic Growth, Delhi (2000).

Table 10.1 Descriptive statistics of the variables used in the estimation

	Sales revenue (Rs. millions)	Materials (Rs. millions)	Wage bill (Rs. millions)	Capital stock (Rs. millions)	Water (million kL)
Mean	1,911.59	774.23	169.79	2,323.67	1,676.91
Maximum	25,190.00	11,143.58	10,080.00	74,538.09	37,172.41
Minimum	0.38	0.52	0.14	0.33	0.04
Std. Dev.	3,291.95	1,382.05	794.93	7,811.52	4,403.29
Observations	276	276	276	276	276

Source: Primary Survey

drugs, petrochemicals, iron and steel, refining, and other industries. For details on characteristics of the data, see Murty and Kumar (2004). Descriptive statistics of the variables used in the study are given in Table 10.1.

In order to compute absolute (undeflated) producer shadow price and own and cross-price elasticities for water, the input distance function is estimated using (10.11) and (10.12) with data from 1996/1997 to 1998/1999 for 92 manufacturing firms. To capture industry and time effects we have included ten dummy variables. The first two dummy variables are for the time effect, as we have data for 3 years, and the next eight dummy variables are industry specific since the whole data belongs to nine industries.[8] Since a single distance function is estimated, input and output substitution possibilities are constant over time and across industries. The estimation also included tests of regularity conditions. For each observation, monotonicity with respect to inputs and outputs is imposed by the linear programming problem. The distance function satisfies convexity in outputs for most observations, while it also appears to satisfy concavity in inputs for a majority of observations. The parameter estimates are presented in Table 10.2.

Recall that the input distance function is the reciprocal of the input-based measure of technical efficiency. On average the technical efficiency for our sample observations is 0.46. This reflects that on average the firms can produce the same level of output with less than half of the inputs if they were operating at the input frontier. Industry-wide mean and standard deviation of technical efficiency are presented in Table 10.3. Table 10.3 also provides estimates of scale economies of water-consuming industries in the sample. On average, firms are operating under increasing returns to scale.

[8] In our sample of 276 observations and there are 114 observations that belong to sugar industry. Therefore, we have tried to estimate the distance function parameters without sugar industry and only for sugar industry, but the results were not statistically different from the estimates obtained from the whole sample. This may be due to introduction of industry specific dummy variables since linear programming is sensitive to outliers.

Table 10.2 Parameter estimates of translog input distance function

Variable	Coefficient	Variable	Coefficient
Constant	0.415	x_3x_4	−0.002
x_1	0.424	x_4x_1	0.001
x_2	0.373	x_4x_2	3.87×10^{-4}
x_3	0.196	x_4x_3	−0.002
x_4	0.007	x_4x_4	3.67×10^{-4}
y_1	−0.824	y_1y_1	0.018
x_1x_1	0.097	Year 1 dummy	0.135
x_1x_2	−0.062	Year 2 dummy	0.043
x_1x_3	−0.036	Leather dummy	0.205
x_1x_4	0.001	Distillery dummy	1.102
x_2x_1	−0.062	Chemicals dummy	0.448
x_2x_2	0.053	Sugar dummy	0.356
x_2x_3	0.008	Paper and paper products dummy	−0.145
x_2x_4	3.87×10^{-4}	Fertilizer dummy	−0.241
x_3x_1	−0.036	Drug and pharmaceutical dummy	2.136
x_3x_2	0.008	Petrochemicals dummy	−0.427
x_3x_3	0.03		

y_1 sales revenue, x_1 material inputs, x_2 wage bill, x_3 capital stock, x_4 water
Source: *Authors'* calculations

10.4.1 Shadow Price of Water

The undeflated shadow price of water is computed using (10.6). The parameter estimates of input distance function were used to compute the shadow price of water for each observation. Recall that the computation of the shadow price of the industrial use of water requires the assumption that the observed price of one of the inputs is equal to its shadow price. Here we have obtained the shadow price of water relative to the price of materials.[9] Table 10.3 provides estimates of industry-specific shadow prices of water. These shadow prices are positive, reflecting that water is a normal input in the production process of these industries.

The average shadow price of water is Rs. 7.21 per kiloliter. There is a wide variation of shadow prices of water across firms and industries as shown in Table 10.3. This wide variation can be explained by the variation in the degree of water intensity as measured by the ratio of water consumption to sales value. The shadow price of water increases with the degree of water intensity of firms. The correlation coefficient between the shadow price of water and water intensity is 0.32. The correlation coefficient is 0.68 for the firms in which the intensity of water is more

[9]Given that the data set used provides values of materials rather than quantities and prices of various material inputs, materials is considered as a single composite input and firms are price taker for this input, therefore, it will be less restrictive to assume that observed price of materials is equal to its shadow price.

Table 10.3 Technical efficiency, return to scale, and shadow price of water

Name of industry	Number of observations	Technical efficiency	Returns to scale	Shadow price of water
Leather	09	0.637 (0.239)	1.365 (0.037)	1.161 (0.950)
Distillery	18	0.393 (0.229)	1.362 (0.062)	6.752 (6.620)
Chemicals	48	0.343 (0.216)	1.436 (0.033)	3.164 (5.872)
Sugar	114	0.424 (0.235)	1.404 (0.051)	4.862 (8.907)
Paper and paper products	33	0.630 (0.224)	1.435 (0.027)	30.535 (32.632)
Fertilizers	18	0.442 (0.217)	1.465 (0.048)	2.465 (3.192)
Drug and pharmaceuticals	06	0.514 (0.505)	1.337 (0.036)	3.919 (3.609)
Petrochemicals	09	0.516 (0.386)	1.431 (0.023)	1.396 (1.682)
Misc.	21	0.546 (0.285)	1.470 (0.046)	3.026 (4.995)
All	276	0.455 (0.260)	1.418 (0.054)	7.209 (15.611)

Note. Figures in *parentheses* are standard deviations
Source: authors' *calculations*

than 1 kL and it is 0.14 for the firms in which the water intensity is less than 1 kL. It implies that the higher the water intensity, the higher the shadow prices would be. The average shadow price of water is much higher than the (average) price paid for water by the industry, Rs. 1.94 per kiloliter (Goldar, 2003). Since the shadow price of water for industries is quite high as compared with the price charged, it may be concluded that there is ample scope for raising the water price. The high shadow price of water in industries also indicates that water shortage in industries has a significant cost in terms of lost industrial output.[10]

10.4.2 Analysis of Derived Demand for Water

The distance function estimate enables us to derive the cross- and own price elasticities. Here we should recall that we measure indirect elasticities. A higher value implies less responsiveness, and lower values means more responsiveness. Table 10.4 presents the mean of these elasticities. We discuss now these results and more carefully analyze those dealing with water input.

All own price elasticities have the expected negative sign, implying that an inverse relationship exists between the price of an input and the quantity demanded. The derived demand for materials is more elastic in comparison to other inputs. We observe relatively high labor own price elasticity compared to capital. Regarding cross-price elasticities between inputs, labor appears to be a complement to all other inputs, i.e., materials, capital, and water. Just as materials appear to be a complement

[10]The serious adverse affect that water shortage has on industrial production has been analyzed by Bhatia et al. (1994) in the context of India and some other developing countries.

Table 10.4 Mean of cross- and own indirect price elasticity of input demands (ε_{ij})

	Materials	Wage bill	Capital stock	Water
Materials	−0.268 (0.083)	0.144 (0.056)	0.117 (0.040)	0.008 (0.002)
Wage bill	0.292 (0.057)	−0.522 (0.027)	0.220 (0.041)	0.007 (0.003)
Capital stock	0.270 (1.064)	0.328 (0.239)	−0.589 (0.890)	−0.009 (0.060)
Water	0.788 (0.409)	1.239 (1.690)	−0.313 (0.845)	−0.902 (0.158)

Note. Figures in *parentheses* are standard deviations
Source: Authors' *calculations*

to all other inputs, capital appears to be a complement to materials and labor and a substitute for water.

Considering now the water input, water is found to be a substitute for capital and a complement to materials and labor. Substitution between capital and water was also observed by Dupont and Renzetti (2001) and Feres and Reynaud (2003), in contrast with previous results from Grebenstein and Field (1979) and Babin et al. (1982), where water was found to be a substitute for labor and a complement to capital. The substitutability between water and capital implies that as the price of water increases, the industry employs more capital. As the price of water increases, the industry may try to reduce water consumption by investing in water-conserving/recirculation technologies. Water conservation/recirculation is generally accompanied by reduction in energy costs, recapturing valuable raw materials and reduction in effluent stream (Dupont and Renzetti, 2001). Therefore, the complementarity between water and materials found here is in conformity with Dupont and Renzetti (2001).

It should be noticed that the own price elasticity of water is quite high, −0.902 (in conventional sense it is −1.11) at the sample mean with standard deviation (0.16). The result suggests that pricing policies can be a potential instrument for water conservation. This elasticity is close to the one obtained for the Chinese economy by Wang and Lall (2002), who estimate an average price elasticity of approximately −1.0, and for the Brazilian economy by Feres and Reynaud (2003), who estimate an average price elasticity of approximately −1.078. However, since Wang and Lall (2002) adopt a marginal productivity approach and Feres and Reynaud (2003) adopt a cost function approach to derive elasticity estimates, any comparison between elasticity estimates should be made with caution. The estimates of own price elasticity of water for India, China, and Brazil are higher than those obtained by Onjala (2001) for Kenya and Goldar (2003) for India. Onjala estimates water price elasticities ranging from −0.60 to 0.37. Goldar estimates water price elasticities ranging from −0.4 to 0.64. Onjala (2001) adopts a dynamic adjustment model with data on input prices and production levels, whereas Goldar adopts a marginal productivity approach with aggregate data on inputs and outputs and water input data that include only the quantity of water purchased and not water consumed. However, once more, comparison between estimates seems to be difficult to establish, and results should be used with caution.

The water price elasticity estimates for developing countries (India, Brazil, and China) are significantly higher than the ones obtained for developed countries (U.S,

Canada, and France). For example, for the U.S manufacturing sector Grebenstein and Field (1979) find elasticities ranging from −0.80 to −0.33; Babin et al. (1982) find elasticities ranging from −0.66 to +0.14; for Canadian manufacturing, Renzetti (1992) finds elasticities ranging from −0.59 to −0.15; Dupont and Renzetti (2001) find elasticity −0.77; and for French manufacturing, Reynaud (2003) finds elasticity −0.29.

It is quite difficult to attribute these differences between water price elasticities in developed and developing countries to any structurally based explanation; it may be due to the difficulties of getting accurate water-related data in developing countries. Indeed, the water price used in our study corresponds to the marginal cost, whereas the prices paid by Indian firms are far below this level.[11] This may lead to an upward bias in our estimates. The same upward bias could be present in Wang and Lall (2002) and Feres and Reynaud (2003). Moreover, the three samples (Indian, Chinese, and Brazilian) consist of medium and large plants, which tend to have higher water price elasticities than small ones. Since large firms withdraw high volumes of water, they face high incentives to invest in water-recycling activities. Since water recirculation is a substitute for water withdrawal, these firms should have more elastic water withdrawal price elasticity (Reynaud, 2003). In developing countries, it should be noticed that water is not a scarce resource in the sense that firms do not face stringent water resource constraints, since water is often an underpriced or unpriced intermediate input. In such a context, firms are likely to overuse water resources, and the marginal productivity of the water tends to be low, as reported by Wang and Lall (2002). This may result in high responsiveness to water prices, since any increase in water prices would lead to a substantial cut in water withdrawals, although additional research and more accurate data on industrial use of water in developing countries is needed in order to answer these questions.

The indirect Morishima and Allen elasticities of substitution can be computed from own and price elasticities, and they are presented in Tables 10.5 and 10.6, respectively. In terms of Morishima elasticities, all of the inputs appear to be complements to each other, whereas according to Allen elasticities of substitution, water and capital are substitutes to each other, and all other inputs appear to be complements.

Table 10.7 presents the indirect own and cross-elasticities of input-derived demand. These elasticities are computed at the mean sample for each industrial sector. The own price elasticity of water ranges from −0.301 for the drug and pharmaceutical sector to −0.942 for the leather industry, which means that the own price elasticity is not much different across sectors, except for drug and pharmaceutical, and it is price elastic for all the sectors. The own price elasticity for labor is not much different across industries and it ranges between −0.505 and −0.529.

[11] Gupta et al. (1989) has estimated the financial cost of groundwater extraction Rs. 0.25 per kiloliter, therefore, Goldar and Pandey (2001) are of the view that the price of water in India often does not cover the cost of delivery, let alone its opportunity cost or scarcity value. This results in overuse/wasteful use of water.

Table 10.5 Mean of Morishima elasticity of substitution (M_{ij})

	Materials	Wage bill	Capital stock	Water
Materials	0	0.666 (0.068)	0.706 (0.900)	0.910 (0.158)
Wage bill	0.561 (0.101)	0	0.809 (0.900)	0.909 (0.159)
Capital stock	0.539 (1.081)	0.850 (0.233)	0	0.893 (0.167)
Water	1.056 (0.406)	1.762 (1.683)	0.276 (1.224)	0

Note. Figures in parentheses are standard deviations
Source: Authors' calculations

Table 10.6 Mean of Allen elasticity of substitution A_{ij}

	Materials	Wage bill	Capital stock	Water
Materials	0	0.258 (0.465)	0.216 (0.251)	0.016 (0.018)
Wage bill	1.175 (0.384)	0	0.853 (0.215)	0.030 (0.015)
Capital stock	−29.750 (520.918)	9.336 (125.336)	0	−1.848 (29.915)
Water	369.495 (2,403.71)	1,039.308 (8,930.628)	−442.696 (4,417.971)	0

Note. Figures in parentheses are standard deviations
Source: Authors' calculations

Similarly, own price elasticity for materials ranges between −0.203 and −0.320. Concerning cross-price elasticities, water appears to be a substitute for capital and a complement to materials and labor in all the industries.

10.5 Conclusions

This chapter investigates the structure of industrial water demand in India. We have estimated production technology with an input distance function, which is dual to the more generally used cost function. This duality is employed to retrieve the shadow price of water. The advantage of using the distance function approach instead of the cost function approach is that one can calculate elasticities of substitution without the maintained axiom of cost minimization, including Morishima elasticities of substitution. We have estimated derived demand for water using the establishment-level data for 92 firms belonging to different industries over the 3-year period. In our empirical model, water, as well as capital, labor, and materials are treated as input to industrial production (sales revenue). Translog functional form is specified for the input distance function with dummies for year and industry-specific characteristics.

In the literature, cost, production, and demand functions have been used to estimate the derived demand of industrial water use. These three approaches are based on the maintained axioms of optimization and assume that firms are operating at their frontiers, and cost and demand functions require an established market for water and information regarding costs and prices. In the absence of well-established

Table 10.7 Price elasticities by industry ε_{ij}

	Mat	Wage	Cap	Water		Mat	Wage	Cap	Water
		Leather (09)					Fertilizers (18)		
Mat	−0.203 (09)	0.113 (0.04)	0.080 (0.05)	0.010 (0.003)	Mat	−0.279 (0.05)	0.144 (0.03)	0.127 (0.03)	0.007 (0.001)
Wage	0.327 (0.02)	−0.519 (0.03)	0.178 (0.05)	0.010 (0.004)	Wage	0.290 (0.05)	−0.528 (0.02)	0.228 (0.03)	0.007 (0.002)
Cap	−2.123 (6.62)	0.827 (1.46)	1.442 (5.54)	−0.146 (0.37)	Cap	0.344 (0.06)	0.307 (0.05)	−0.645 (0.01)	−0.005 (0.001)
Water	0.778 (0.09)	0.736 (0.16)	−0.133 (0.02)	−0.942 (0.009)	Water	0.745 (0.06)	1.088 (0.32)	−0.228 (0.11)	−0.917 (0.03)
Distillery (18)					Drug and pharmaceutical (06)				
Mat	−0.305 (0.05)	0.164 (0.04)	0.134 (0.03)	0.007 (0.002)	Mat	−0.296 (0.07)	0.135 (0.03)	0.156 (0.04)	0.005 (0.002)
Wage	0.268 (0.06)	−0.516 (0.02)	0.239 (0.04)	0.007 (0.002)	Wage	0.251 (0.08)	−0.523 (0.02)	0.266 (0.06)	0.004 (0.003)
Cap	0.308 (0.06)	0.336 (0.06)	−0.639 (0.01)	−0.005 (0.001)	Cap	0.330 (0.09)	0.306 (0.06)	−0.630 (0.03)	−0.006 (0)
Water	0.711 (0.11)	1.159 (0.39)	−0.235 (0.17)	−0.913 (0.04)	Water	2.389 (2.45)	7.607 (10)	−3.558 (5.08)	−0.301 (0.94)
Chemicals (48)					Petrochemicals (09)				
Mat	−0.283 (0.05)	0.155 (0.04)	0.120 (0.02)	0.008 (0.002)	Mat	−0.221 (0.08)	0.105 (0.06)	0.109 (0.04)	0.008 (0.003)
Wage	0.283 (0.06)	−0.517 (0.03)	0.225 (0.03)	0.007 (0.002)	Wage	0.281 (0.06)	−0.505 (0.04)	0.212 (0.04)	0.008 (0.003)
Cap	0.325 (0.08)	0.326 (0.07)	−0.646 (0.01)	−0.005 (0.001)	Cap	0.388 (0.08)	0.259 (0.07)	−0.640 (0.02)	−0.007 (0.002)
Water	0.732 (0.16)	1.117 (0.6)	−0.236 (0.28)	−0.915 (0.05)	Water	0.827 (0.16)	1.021 (0.45)	−0.251 (0.22)	−0.917 (0.04)
Sugar (114)					Misc. (21)				
Mat	−0.247 (0.1)	0.132 (0.07)	0.107 (0.05)	0.008 (0.002)	Mat	−0.263 (0.05)	0.125 (0.03)	0.131 (0.03)	0.006 (0.002)
Wage	0.307 (0.06)	−0.529 (0.03)	0.211 (0.04)	0.008 (0.002)	Wage	0.289 (0.04)	−0.530 (0.01)	0.232 (0.04)	0.006 (0.002)
Cap	0.347 (0.07)	0.303 (0.05)	−0.644 (0.02)	−0.005 (0.001)	Cap	0.370 (0.05)	0.280 (0.04)	−0.644 (0.007)	−0.006 (0.001)
Water	0.766 (0.2)	1.059 (0.89)	−0.236 (0.41)	−0.918 (0.08)	Water	0.912 (0.33)	1.596 (1.3)	−0.505 (0.65)	−0.867 (0.12)
Paper and paper products (33)									
Mat	−0.320 (0.04)	0.187 (0.03)	0.124 (0.04)	0.009 (0.003)					
Wage	0.272 (0.05)	−0.510 (0.02)	0.228 (0.05)	0.008 (0.003)					
Cap	0.276 (0.06)	0.365 (0.05)	−0.636 (0.02)	−0.004 (0.002)					
Water	0.681 (0.16)	1.146 (0.67)	−0.223 (0.32)	−0.915 (0.06)					

Mat materials cost, *Wage* wage bill, *Cap* capital stock, *Water* water consumption. Figures in parentheses are standard deviations

Source: Authors' calculations

water market and information about prices and cost, the distance function approach can be used to assess the shadow prices of water for industrial use if information about quantities of inputs and outputs is available when firms are not operating at their frontiers. Thus, the distance function also provides estimates of a firm's efficiency and returns to scale.

The main results of our analysis are the following: We first have shown that there is high variability in the production efficiency of Indian manufacturing industries; they can produce the same level of output with less than half of the quantities of inputs than they are using on average. There are increasing returns to scale in our sample of firms, with an average of 1.42. Returns to scale is positively associated with turnover and water intensity. The estimated average shadow price of water is Rs. 7.21 per kiloliter. We observe a wide variation across industries and firms in these shadow prices. The shadow price varies from Rs. 1.40 per kiloliter for petrochemicals to Rs. 30.54 per kiloliter for paper and paper products industry.

We have also estimated own and cross-price elasticities of water for other inputs. We find that water is a complement to labor and materials and a substitute for capital. We find a price elasticity of water demand about -0.902 (in the conventional sense -1.11) at the sample mean. This high value is similar to what has been found by other researchers working on developing countries (e.g., China and Brazil). Thus, given the high responsiveness of water demand to price, water charges may act as an effective instrument for water conservation.

Part III
Environmental Productivity, Oil Prices and Induced Innovations

Chapter 11
Environmental Productivity and Kuznets Curve

11.1 Introduction

It has been a tough trade-off decision between economic growth and environmental protection, especially in developing countries. Tireless efforts to accelerate economic growth had kept environmental considerations as secondary objectives in policy making in these countries. This indifference towards environmental protection has led to serious environmental problems in developing countries and has threatened their sustainable future. For example, damage caused by pollution in India is estimated to cost $14 billion annually, amounting to close to 4.5%–6% of GDP (Economic Survey of India, 1998–1999). In response, many developing countries have begun to enact and implement environmental policies that regulate air and water pollution and solid waste disposal in order to limit the severity of environmental degradation. The stringency of these regulations has been increasing over the years.

It has been increasingly recognized that technological progress can play a key role in maintaining a high standard of living in the face of these increasingly stringent environmental regulations. However, the extent of the contribution of technological progress depends on how well environmental policies are designed and implemented. Successful environmental polices can contribute to technological innovation and diffusion (Jaffe et al., 2003), while poor policy designs can inhibit innovation.

On the other hand, successful implementation of environmental regulations may crucially be linked with the pattern of economic growth. This argument is the basis of the environmental Kuznets curve (EKC) hypothesis, which has gained tremendous popularity among the researchers over the past decade. EKC draws its roots from the pioneering study by Grossman and Krueger (1993), which established the empirical relationship between measures of environmental quality and national income. An inverted U-shaped relationship of the EKC implies that environmental degradation increases with income at low levels of income and then decreases once a threshold level of per capita income is reached.

S. Kumar and S. Managi, *The Economics of Sustainable Development*, Natural Resource Management and Policy 32,
DOI: 10.1007/978-0-387-98176-5_11,

After the study by Grossman and Krueger (1993), many studies, such as those by Seldon and Song (1994) and Holtz-Eakin and Selden (1995), investigated this relationship for alternative measures of environmental degradation with levels of pollutants or pollutant intensities (see Dinda [2004], Stern [2004], and Managi [2006] for recent literature). Their studies supported the EKC relationship between pollution and per capita national income. Their argument for such a finding was that after a certain level of income, concern for environmental degradation becomes more relevant and a mechanism to reduce environmental degradation is put in place through necessary institutional, legal, and technological adjustments.

However, a major criticism of these studies is that they have adopted a reduced form approach to examine the relationship between per capita income and pollution emissions (Stern, 1998). These two variables are merely the outcomes of a production process, and they do not explain the underlying production process, which converts inputs into outputs and pollutants. In fact, the transformation of this production process may lead to environmental improvement at a higher level of income (Zaim and Taskin, 2000). Therefore, studies that examine the transformation of the production process by quantifying the opportunity cost of adopting alternative environmentally superior technologies are more relevant to our study.

The more efficient utilization of pollution abatement technologies at least in part influences the cost of alternative production and pollution abatement technologies (e.g., Jaffe et al., 2003). An extensive body of theoretical literature examines the role of environmental policy in encouraging (or discouraging) productivity growth. On the one hand, abatement pressures may stimulate innovative responses that reduce the actual cost of compliance below those originally estimated. On the other hand, firms may be reluctant to innovate if they believe regulators will respond by "ratcheting-up" standards even further. Therefore, in addition to the changes in environmental regulations and technology, management levels also affect environmental performance level or environmental productivity, which explains how efficiently pollutions are treated, as defined by Managi et al. (2005). Thus, whether environmental productivity increases over time is an empirical question.[1]

Against this backdrop, the objective of this chapter is twofold: First, attempts are made to measure data for the technological/productivity change for environmental (nonmarket) outputs of sulfur dioxide (SO_2), nitrogen dioxide (NO_2), and suspended particular matter (SPM) in India using state-level industry data over the period 1991–2003. Second, the change in environmental productivity in different states is linked with their respective per capita income in order to find an EKC-type relationship. We intend to measure environmental productivity following the

[1]Most current empirical studies focus on developed countries (Managi et al., 2005). To the authors' knowledge, there are few studies that have estimated the efficiency changes of environmental technology or management in the context of developing countries. See Murty et al. (2006) for recent application to the Indian Sugar industry.

traditional productivity literature.[2] Regulations requiring more stringent pollution abatement do not necessarily change environmental productivity since the linear expansion of pollution abatement costs and pollution reduction does not necessarily change the pollution reduction per abatement cost.

The chapter is structured as follows. Section 11.2 briefly reviews environmental policies in India. The empirical model and data are explained in Section 11.3, while the results are presented in Section 11.4. Concluding remarks and further discussions are provided in the final section.

11.2 Environmental Policies in India

To combat the problem of environmental degradation, several environmental polices were initiated by the Government of India from the late 1970s. India was the first country to insert an amendment into its Constitution allowing the state to protect and improve the environment for safeguarding public health, forests, and wildlife. The 42nd amendment was adopted in 1976 and went into effect January 3, 1977. The *Directive Principles of State Policy* (*Article 47*) requires not only a protectionist stance by the state but also compels the state to seek the improvement of polluted environments.

The Air (Prevention and Control of Pollution) Act was passed in 1981, and the Parliament passed the Environmental Protection Act in 1986. The responsibility for administering new legislation fell on the central and state pollution control boards. The Department of Environment (DOE) was created in 1980, and was supposed to appraise the environmental aspects of development projects, to monitor air and water quality, to establish an environmental information system, to promote environmental research, and to coordinate activities between federal, state, and local governments. The DOE was criticized, however, by environmental groups for its small political and financial base. Environmentalists recognized quickly that the DOE would essentially serve as an advisory body with few enforcement powers.

This deficiency was soon recognized, and a Ministry of Environment and Forests (MoEF) was created in 1985. It continued the same functions that the DOE originally had, such as monitoring and enforcement, conducting environmental assessments and surveys, but also provided promotional work about the environment. The MoEF's implementation of a monitoring system was noteworthy (see MoEF, 2001). In 1984, there were 28 monitoring stations for air pollution in India. The number increased to 290 stations by 1994 and included 51 stations from the Global Environmental Monitoring System (GEMS).

In December 1993, the MoEF completed its Environmental Action Plan to integrate environmental considerations into developmental strategies, which,

[2] There are several studies that measures market productivity, For example, Pallikara (2004) finds 2.8% annual increase of market TFP using Solow residual type total factor productivity over 1992 and 2001.

among other priorities, included industrial pollution reduction. However, control of environmental pollution had not been found to be satisfactory, mostly because of growth-oriented economic policies. Since the adoption of reform policies in India in 1991, the economy has achieved a higher-trajectory growth rate. Between 1993–1994 and 1997–1998, the Indian economy has averaged more than a 7% growth rate per annum (Economic Survey of India, 1998–1999). The growth of industrial production and manufacturing has averaged at 8.4% and 8.9%, respectively, during these years. This expansion of economic activities placed a heavy toll on the country's environmental quality. Furthermore, lack of properly functioning markets for environmental goods and services and market distortions created by price controls and subsidies have aggravated environmental problems.

The weakness of the existing system lies in the enforcement capabilities of environmental institutions, both at center and state levels. There is no effective coordination among various ministries/institutions regarding integration of environmental concerns at the inception/planning stage of the project (Economic Survey of India, 1998–1999). Further, current policies are fragmented across several government agencies with differing policy mandates. Lack of trained personnel and a comprehensive database delay many projects. Most state government institutions are relatively small and suffer from inadequate technical staff and resources.

Although it was claimed by the Central Pollution Control Board (CPCB, 2001) that the overall quality of the Environmental Impact Assessment (EIA) process has improved over the years, little is known about how environmental productivity has changed over time in India. By considering the divergence of policy intention and actual implementation in each province/state, this study measures the efficiency of environmental management in India using two techniques explained in the following section.

11.3 Models

11.3.1 Measurement of Productivity

We measure productivity change in a joint production model, with a vector of market and nonmarket outputs using production frontier analysis (see Kumar [2006] for the literature). This approach uses the Luenberger productivity index, which is dual to the profit function and does not require the choice of an input-output orientation (Chambers et al., 1996a)[3]. In contrast, the more commonly used Malmquist productivity

[3]Though Luenberger Productivity is theoretically well developed, there is very little empirical work in the literature (Boussemart et al., 2003). A commonly used technique in productivity measurement is growth accounting, which forms a residual after taking the impact of changes in capital and labor inputs out of changes in real output. Compared with the approach used, however, this approach has a number of disadvantages, including an assumption of constant returns-to-scale and zero inefficiency.

index requires the choice between of an output or input orientation corresponding to whether one assumes revenue maximization or cost minimization as the appropriate behavioral goal (Färe et al., 1985). Since the Luenberger productivity index can be applied with either an output- or input-oriented perspective, it is a generalization of, and superior to, the Malmquist productivity index (Luenberger, 1992a, b; Chambers et al., 1998; Boussemart et al., 2003). In this study, we estimate the Luenberger productivity index.

Following Managi et al. (2005), this study uses two datasets, one of which includes only market input/output and TFP_{Market}; the other includes environmental input/output in addition to the market input/output, TFP_{Joint}, consideration of the maximum expansion of good outputs, and contraction of bad outputs. The total factor productivity (TFP) associated with environmental outputs, TFP_{Env}, or environmental productivity, is then calculated as:

$$TFP_{Env} = TFP_{Joint} - TFP_{Market}, \tag{11.1}$$

where TFP is Luenberger indices, which take the difference of the two models. This is because Luenberger indices employ the difference method (see Chambers et al., 1998). The TFP includes not only the change in technology, but also the effect of management-level changes in institutions, including environmental regulations. Thus, even though the technology level remains constant, there are cases where there are changes in TFP.

Production frontier analysis yields the Luenberger index (e.g., Luenberger, 1992a), which can then be used to quantify productivity change. The index-based approach measures the TFP change between two data points by calculating the ratio or difference of two associated distance functions or shortage functions (e.g., Caves et al., 1982; Luenberger, 1992a). This approach has several advantages. One advantage is immediate compatibility with multiple inputs and outputs. This is important for environmental applications, since pollutants, as the by-product of market outputs, can be multiple. This technique estimates the weight given to each observation, such as the weight or shadow price for each item, e.g., environmental pollution data, and implicitly combines these into one index. In addition, this approach can incorporate the inefficient behavior of the decision maker and avoid the need for the explicit specification of the production function (see Managi et al. [2005] for further details).

Using the distance function specification, our problem can be formulated as follows. Let **x**, **b**, **y** be vectors of inputs, environmental output (or undesirable output), and market outputs, respectively, and then define the production possibilities set by;

$$\mathbf{P}^t \equiv \{(\mathbf{x}^t, \mathbf{b}^t, \mathbf{y}^t) : \mathbf{x}^t \text{can produce}(\mathbf{y}^t, \mathbf{b}^t)\}, \tag{11.2}$$

which is the set of all feasible production vectors. We assume that $\mathbf{P}^{\mathrm{t}}$ satisfies standard axioms, which suffice to define meaningful directional distance functions. The directional distance function is defined at t as;

$$D^t(\mathbf{y}^t, \mathbf{x}^t, \mathbf{b}^t; \mathbf{g}^t) = \sup\{\delta : (\mathbf{y}^t, \mathbf{x}^t, \mathbf{b}^t) + \delta \mathbf{g}^t \in P^t\}, \tag{11.3}$$

where $\mathbf{g}$ is the vector of directions by which outputs are scaled. For this directional distance function, we define $\mathbf{g} = (\mathbf{y}, \mathbf{0}, -\mathbf{b})$, i.e., desirable outputs are proportionately increased, inputs are held fixed, and environmental outputs (pollution) are proportionately decreased. In contrast to traditional market productivity measurements, which simply seek to maximize good production, this directional distance function is able to credit the reduction of pollution at the same time.

The Data Envelopment Analysis (DEA) formulation calculates the Luenberger productivity index under variable returns-to-scale (VRS) by solving the following optimization problem (Chambers et al., 1996b):

$$\begin{aligned} D^t(\mathbf{y}^t, \mathbf{x}^t, \mathbf{b}^t) &= \max_{\delta,\lambda} \delta \\ \text{s.t.} \qquad \mathbf{Y}^t\lambda &\geq (1+\delta)\, y_i^t \\ B^t\lambda &\geq (1-\delta)\, b_i^t \\ \mathbf{X}^t\lambda &\leq x_i^t \\ N1'\lambda &= 0 \\ \lambda &\geq 0, \end{aligned} \tag{11.4}$$

where $N1$ is an identity matrix, λ is a $N \times 1$ vector of weights, $\mathbf{Y}^t$, $\mathbf{X}^t$, $\mathbf{B}^t$ are the vectors of market outputs, $\mathbf{y}^t$, inputs, $\mathbf{x}^t$, and environmental outputs, $\mathbf{b}^t$.

As in the Malmquist indices, several different proportional distance functions are necessary to estimate the change in productivity over time. For the mixed period distance function, we have two years, t and $t+1$. For example, $D^t(\mathbf{y}^{t+1}, \mathbf{x}^{t+1}, \mathbf{b}^{t+1})$ is the value of the distance function for the input-output vector of period $t+1$ and technology at t. The Luenberger productivity index defined by Chambers et al. (1996a) and Chambers (2002) is as follows:

$$\begin{aligned} \text{TFP} = \frac{1}{2}[(\mathbf{D}^t(\mathbf{y}^t, \mathbf{x}^t, \mathbf{b}^t) - \mathbf{D}^t(\mathbf{y}^{t+1}, \mathbf{x}^{t+1}, \mathbf{b}^{t+1})) + (\mathbf{D}^{t+1}(\mathbf{y}^t, \mathbf{x}^t, b^t) - \mathbf{D}^{t+1} \\ \times (\mathbf{y}^{t+1}, \mathbf{x}^{t+1}, \mathbf{b}^{t+1}))]. \end{aligned} \tag{11.5}$$

This is an arithmetic mean of period t (the first difference) and period $t+1$ (the second difference) Luenberger indices, as an effort once again to avoid any arbitrary selection of base years (e.g., Balk, 1998). This study measures the TFP index of market outputs ($\text{TFP}_{\text{Market}}$) and TFP of both market and environmental output ($\text{TFP}_{\text{Joint}}$) in a joint production analysis. These two TFP indices are then used to estimate the TFP of environmental output (TFP_{Env}).

TFP includes all categories of productivity change, which can be decomposed into two components including technological change and efficiency change. Technological Change (TC) and Efficiency Change (EC) have additive relations to compose TFP. TC measures shifts in the production frontier, while EC measures changes in the position of a production unit relative to the frontier so-called “catching up” (Färe et al., 1994a).

11.3.2 Kuznets Curve Relationship: Environmental Productivity and Income Level

According to the EKC literature, successful implementation of environmental regulations depends upon the pattern and stages of their growth. It is expected that higher-income regions would be more sensitive about implementing environmental regulations and thereby curbing pollution. Recent work by Zaim and Taskin (2000) undertakes an efficiency approach in the EKC literature. They measure the environmental efficiency of Organization for Economic Co-operation and Development (OECD) countries over 1980–1990 using DEA with a proxy for environmental quality as the EKC-dependent variable. Finding the determinants of the factors underlying the changes in the environmental efficiency is their main concern. They find a Kuznets curve in the efficiency.

We attempt to find a relationship between state-wise per capita income and their respective environmental productivity indices. To analyze the determinants of the productivity change, several variables are used as independent variables, e.g., per capita gross state product (GSP), population density, education level, and urbanization. The following equation is estimated is this study:

$$E_{\text{kit}} = \beta_1 + \beta_2 \text{GSP}_{it-1} + \beta_3 \text{GSP}^2_{it-1} + \beta_4 \text{PO}_{it} + \beta_5 \text{UR}_{it} + \beta_6 \text{ED}_{it} + \varepsilon_{it} \quad (11.6)$$

where, E_{kit} is the environmental productivity index (environmental TFP) or joint TFP of the pollution parameter k in state i in year t; GSPL is the log of gross state product (GSP) per capita[4]; PO is the population density; UR is the urbanization index; ED is an education index; β_s are the coefficients and ε_{it} is the random error term.

It is expected that per capita income has a negative relation with environmental productivity. This is because an increase in income in the initial phase of growth

[4]The TFP indices of environmental variables are transformed into their logarithmic form. Since most of the observations are in negative values and a simple log-transformation was not possible, TFP data are converted into (1 + TFP) form to make the observations positive. Grossman and Krueger had done such transformation in their 1993 paper to avoid the negative values of data series. Furthermore, since the dependent variable for one year is the difference of the productivity between current year and the base year. Therefore, the TFP for year t is affected by the per capita income of year $t-1$. Therefore, log TFP are regressed on 1-year lagged values of log GSP.

raises pollution, which would eventually reduce productivity. Therefore, β_2 should bear a negative sign from our regressions. However, this negative effect might be reversed, and therefore we expect a positive relation between per capita income square and environmental productivity. After a sufficiently high per capita income is reached, a further increment in income is expected to increase environmental productivity, i.e., β_3 is positive.

The population density variable may bear a negative sign since there might be more pressure on the environment in more densely populated areas. A positive association is expected between the education index variable and environmental productivity. The education level of a society affects the level of environmental awareness among people. Some studies have considered a time variable to capture this unobservable factor in their models (Grossman and Krueger, 1995; Seldon and Song, 1994; and Antweiler et al., 2001). They have argued that an increase in environmental awareness and knowledge over time would lead to reduction in environmental degradation. However, it is meaningful to consider an observable variable, which can capture the relevant character of this factor. Therefore, an awareness level index is used to represent environmental awareness in our study. The level of education is one indicator that shows the awareness level among people regarding environmental degradation and the need for its protection. Therefore, an education proxy index is constructed by taking all persons who have passed at least matriculation in a particular year in a state. Finally, urbanization, which is measured as urban population as a percentage of total population, is expected to bear a negative sign due to its spiraling effect on environmental quality.

We employ panel regression techniques to estimate (11.6). The panel data approach encompasses data across cross-sections and over time series, and thus provides a comprehensive analysis to examine variables of interest. However, this type of two-step approach, where productivity measures are estimated by DEA in the first step and regressed on explanatory variables in the second step, should be treated with caution. Following Simar and Wilson (2007), productivity measures estimated by DEA are serially correlated. They argue that a bootstrapping method should be used. However, the use of panel data and dynamic specifications make this problem more complex. Alternatively, to eliminate the serial correlation problem, Zhengfei and Oude Lansink (2006) suggest the use of a dynamic generalized method of moments (GMM) model to analyze TFP measures estimated by DEA. Therefore, in addition to the sensitivity analysis of the OLS method and fixed effects model, we employ dynamic GMM to analyze productivity change as descried in Zhengfei and Oude Lansink (2006).

The previous year's productivity change affects the current year's productivity change because further improvement of productivity after high growth in the previous year might be more difficult. To address the dynamics, two lags of the dependent variable are included in (11.6). Furthermore, the error term of ε consists of an individual effect η and random disturbance v i.e., $\varepsilon_{it}=\eta_i + v_{it}$. In the first-differenced model we estimate, all observations of the dependent variable before $(t - 2)$ are valid instruments. Arellano and Bond (1991) proposed a difference GMM estimator, in which all the valid historical instruments are used in the equation. Arellano and Bover (1995) and Blundell and Bond (1998) propose a system GMM in which the moment conditions in the differenced model and levels model are combined. In their study, it is shown that the system GMM can dramatically improve the problem of weak instruments. Therefore, the system GMM is used in this article.

The dataset consists of annual data for the period 1991–2003 for 16 states in India. For conventional market output, state-level manufacturing data are from *Annual Survey of Industries* (ASI) constructed by the Central Statistical Organization (CSO, 1995, 2004). This study uses real gross manufacturing output as market output in the model. Capital stock and labor as number of workers from ASI are employed as inputs. Data of gross state product are collected from various issues of the *Economic Survey of India* reports, and data of the control variables, such as urbanization, education level and pollution densities, are collected from various editions of the *Statistical Abstract of India.* On the other hand, environmental output is treated as a by-product from the industries in the production process in this study. To account for environmental outputs, data of SO_2, NO_2, and SPM are extracted from the various yearly reports of *National Air Quality Monitoring Programme* (NAMP) (see CPCB, 1995, 1998, 2003). The emissions from all pollution data are estimated from stations located in 46 cities/towns of 16 states of India. The name of the states is provided in the Appendix.

11.4 Results

11.4.1 Productivity Analysis

Separate frontiers are estimated for each year, and shifts in the frontiers over time are used to measure technological change. The arithmetic mean of the Luenberger productivity indices for each state in each year[5] are estimated under the assumptions of VRS production technologies. Note that we also estimate the productivities under the assumptions of constant returns to scale and find similar results.

Arithmetic mean values of TFP, TC, and EC across the states for each period are presented in Tables 11.1 and 11.2[6]. In these tables, the study period (1991–2002) is divided into three subperiods of 1991–1994 (1st periods), 1995–1998 (2nd periods), and 1999–2002 (3rd periods). The purpose of this division is to compare productivity indices between the subperiods to assess how changes in productivity have taken place vis-à-vis policy changes.

[5] See Balk (1998) for theoretical reasoning underlying the use of arithmetic means to average data.

[6] Note that the Luenberger TFP technique is difference based technique and therefore minus value implies that productivity decreases compared to base period. On the other hand, a plus value reflects a positive increase.

Table 11.1 Market and joint productivity changes (average changes in each periods)

Periods	Market productivity			Joint productivity		
	TFP	EC	TC	TFP	EC	TC
1991–1994	−0.022	0.007	−0.029	−0.008	−0.003	−0.005
1995–1998	0.013	0.000	0.013	−0.012	−0.001	−0.011
1999–2002	0.004	0.004	0.000	−0.010	−0.003	−0.007
Mean	−0.001	0.004	−0.005	−0.010	−0.002	−0.008

Source: Authors' calculations

Table 11.2 SO_2, NO_2, and SPM productivity changes (average changes in each periods)

Periods	SO_2			NO_2			SPM		
	TFP	EC	TC	TFP	EC	TC	TFP	EC	TC
1991–1994	−0.028	−0.022	−0.005	−0.011	−0.006	−0.005	−0.008	−0.002	−0.005
1995–1998	−0.007	0.003	−0.011	−0.017	−0.005	−0.011	−0.012	−0.001	−0.012
1999–2002	0.005	0.012	−0.006	−0.031	−0.022	−0.009	−0.010	−0.003	−0.007
Mean	−0.010	−0.002	−0.007	−0.020	−0.011	−0.009	−0.010	−0.002	−0.008

Source: Authors' calculations

11.4.2 Market Productivity

The results of market productivity are presented in Table 11.1. The results of TFP_{Market} have two different phases: 1991–1996 and 1997–2002. In the initial phase the productivity index has negative values, showing a decline in productivity from the base period. However, although the absolute value of the index has decreased during this period, the rate of decline has narrowed down by 60%, i.e., −0.025 in 1991 to 0.010 in 1996. In the latter phase, changes in TFP_{Market} change value are positive, indicating a net productivity gain.

Overall, the movement of the index suggests that the productivity of the market declined in the initial years of economic reforms in India. In fact, the country went through a transition phase in the early 1990s following a massive policy change in 1991 that has resulted in a turbulent period in the industrial sector. The growth rates in both GDP and manufacturing output are low during 1991–1992 and 1992–1993. However, during the mid-1990s, the industrial sector recovered from the early shocks of the reform process and registered reasonable growth rates. This is reflected in the positive changes in TFP later in the decade. The value of EC decreased from 0.007 in first period to 0 in the second period and finally increased to 0.004 in the third period. On the other hand, the TC increased from –0.029 in first period to 0.013 in the second period, and it decreased to 0.00 of third period.

11.4.3 Joint Output Productivity

Joint output productivity indices are constructed using a joint output production technology in which both a desired output (conventional good) as well as undesired outputs (environmental pollutions) of SO_2, NO_2, and SPM are jointly produced, the latter being the by-product. The Luenberger productivity index uses directional distance functions that attempt to maximize market output while minimizing the undesired by-products and minimizing inputs.

The results in Table 11.1 show that TFP_{Joint} has negative values in almost all the years, showing consistent decline in productivity. The TFP_{Joint} declines from −0.008 to −0.012, a 50% deceleration while moving from the first period to the second period and then it remains steady with a mean value of −0.010 in the third period. This shows that the productivity of joint output does not show any improvement in the postreform periods in India. Moreover, combining the market output productivity and joint output productivity indices, it can be suggested that while the former starts increasing from the mid-1990s, the latter consistently declines throughout our study periods. This finding indicates that the productivity of environment declines continuously. Technological progress increases market productivity and simultaneously creates possible threats to society, which are unknown in the early phase of the implementation of technology. Currently, India may be facing this problem. However, it is difficult to say which of the three pollutants – SO_2, NO_2, or SPM – is the main cause of the overall environmental productivity decrease from these results. Therefore, each specific environmental productivity of these pollutants is estimated and the indices are provided in Table 11.2.

11.4.4 Environmental Productivity

The environmental productivity indices in our study are calculated by taking the difference of market productivity indices and joint productivity indices. We have estimated separate productivity indexes for the three pollution variables of SO_2, NO_2, and SPM, respectively. For example, the environmental productivity of SO_2 is represented as TFP_{SO_2}, i.e., SO_2 pollution productivity. The TFP_{SO_2} given in Table 11.2 shows that productivity declines from 1991 to 1999. Although the first two periods show a negative sign, the deteriorating rate decreases. In the third period, the index shows a positive sign. These results indicate that the implementation of environmental regulations to control and prevent emissions of sulfur dioxide improves over the years in India – more particularly in recent years because of the increase in EC. This indicates that the externalities of SO_2 are identified as social institutions formulate laws and regulations to consider SO_2 pollution. The generation of new technologies to reduce SO_2 is more efficiently implemented in catching up to the frontier states, which is reflected by the increase in EC.

In contrast, the changes in TFP_{NO_2} (also both of TC and EC) are monotonously negative over the whole study period and show a continuous decline in productivity. Moreover, the alarming feature of the trend is that the rate of this decline is actually increasing over the years. The mean value of the index declines from −0.011 in the first periods to −0.017 in the second periods, with a 55% decline in productivity. It has further gone down to −0.031 in the third periods, with an 82% decline in productivity. This is quite significant and seriously questions the implementation efficiency of government pollution control boards in controlling the emission and concentration of nitrogen oxides in India. The CPCB annual report (2003–2004) also raises concerns about the unabated spiraling of nitrogen oxide in industrial cities in the country.

Finally, the estimated productivity indexes of the third pollutant in our study, i.e., SPM, show that the performances of TFP, EC, and TC are not any better than the NO_2 case. The index has been negative in all the years, indicating a net decrease in productivity. The mean values show that the index has decreased from -0.008 in first periods to -0.012 in second periods, thus registering a 50% decline in productivity. The rate of decrease in the third periods is smaller than that of NO_2; nevertheless, it raises serious concerns for policy makers in the country.

The discussion above reveals that although market productivity recovered after the mid-1990s from a slump in the early stages of economic reforms, environmental productivity, on the other hand, has deteriorated constantly. Except for the productivity of SO_2, which has shown some improvement after 1999, the abatement of other forms of air pollution has been worse.

11.4.5 Environmental Kuznets Curve Test

Furthermore, the analysis of environmental productivity in the individual states suggests that there is variation among the states in terms of productivity. For example, productivity of SO_2 improves in states like Andhra Pradesh, Gujarat, Haryana, Madhya Pradesh, Maharashtra, Punjab, Rajasthan, Uttar Pradesh, and West Bengal after 1999; whereas in other states productivity declines monotonously (see Appendix). A similar pattern is found for the productivity of NO_2 and SPM. In general, environmental productivities decrease more in high-income states than in low-income states.

To examine how income levels are associated with environmental productivity at the state level, we provide the panel analysis estimates of the TFP of SO_2 in Table 11.3. A perusal of the estimates shows that both fitness and the coefficient values improve while moving from the OLS to fixed effects models. The coefficients are estimated using White heteroscedasticity-consistent standard errors and covariances. To correct the existing autocorrelation problem in the model, AR terms with appropriate lags are incorporated in the estimation process. The fixed effects estimates show that GSP has a statistically significant negative relationship with productivity.

However, the potential of serial correlation, proposed by Simar and Wilson (2007), makes us more careful in evaluating the effects. To obtain more robust results, GMM estimation is also applied. Following Zhengfei and Oude Lansink (2006), GMM estimation is a valid solution to the serial correlation problem.

A 1% increase in linear term of GSP reduces the TFP of SO_2 by 0.034%. The TFP of the environmental parameter reflects, on one hand, the technology used in the production process that emits this kind of pollutants; on the other hand, it shows the management efficiency of pollution control boards to control and prevent emission of pollutants. Therefore, an increase in TFP would mean both employment of greener technologies by industry and more efficient implementation of environmental regulations. The coefficient of GSP is in fact, the scale effect; an increase in income would raise the pollution level and thus would decrease environmental productivity. The term of GSP^2, on the other hand, shows the technique effect; that is, an increase in per capita income induces technological as well as managerial changes leading to reduction in pollution level and increase in productivity. There is 0.029% increase in productivity due to 1% increase in technique effect.

The regression estimates of TFP of NO_2, SPM, and joint output are given in Tables 11.3 and 11.4, respectively. The signs of the estimated coefficients of GSP and GSP^2 with these variables are similar to those of SO_2, with the former having negative sign and the latter having positive sign. Therefore, the scale effect is negative, and the technique effect is positive across all the environmental variables. In the case of NO_2, a 1% increase in per capita income reduces the TFP of NO_2 by 0.087% and, at the margin, a 1% increase in per capita income square, increases productivity by 0.078%. With SPM, the coefficients are −0.02 and 0.015 with GSP and GSP^2, respectively; on the other hand, the coefficients of these variables with joint output are −0.03 and 0.009, respectively.

These elasticities can be added together to arrive at a net effect of income on productivity. For example, in the case of SO_2, scale elasticity is −0.034 and technique elasticity is 0.029; adding them together we find the net elasticity of −0.005. Similarly, the net elasticity for NO_2 and SPM are −0.009 and −0.005, respectively. It can be noticed that the net effect of income on environmental productivity is negative. Although at the margin, increase in per capita income (technique effect) has the potential to improve productivity, this effect is too insignificant to offset the dominant scale effect. Therefore, negative scale effect outperforms the positive technique effect to make productivity decline. The lower TFP index values in high-income states articulate these results. The scale effect of income in the states has been stronger than the technique effect, and thus increases in per capita income are fuelled by higher-growth trajectory leads. In the case of joint output results, net elasticity is 0.014, implying that increase in income level induces better performance, including both market and environmental outputs. Negative results in environmental productivities are caused by higher market productivity.

Among other control variables, population density has negative coefficients, with productivity indices implying environmental performances are adversely affected in densely populated areas. The urbanization variable has also been found to be negatively associated with environmental productivity. The education index, which measures the level of environmental awareness, has positive coefficients, though the magnitude of these coefficients is small. These findings suggest that regions with a higher level of education seem to have experienced less environmental degradation.

Table 11.3 Productivity determinants of SO_2 and NO_2

Dependent variable	SO_2			NO_2		
Estimation technique	OLS	Fixed effects	Dynamic GMM	OLS	Fixed effects	Dynamic GMM
Intercept	0.018^{***} (13.04)	–	–	0.056^{***} (4.719)	–	–
Gross state product (GSP)	-0.016^{***} (−11.78)	-0.029^{***} (−17.17)	-0.034^{***} (−17.74)	-0.058^{***} (−4.41)	-0.078^{***} (−6.76)	-0.087^{***} (−7.23)
GSP^2	0.004^{***} (11.82)	0.005^{***} (11.93)	0.006^{***} (11.99)	0.011^{***} (3.45)	0.019^{***} (6.60)	0.016^{***} (6.01)
Population density	$-1.11e-06^{***}$ (− 4.11)	$-9.08e-06^{***}$ (−3.13)	$-9.83e-06^{***}$ (−3.52)	$1.49e-07$ (0.119)	$-1.45e-05^{***}$ (− 3.16)	$-4.72e-03^{****}$ (−4.43)
Urbanization	$7.79e-06$ (0.32)	0.003^{***} (10.67)	0.009^{***} (11.41)	-0.0002^{**} (−2.20)	-0.0007^{*} (−1.75)	-0.0011^{***} (−2.53)
Education index	$-4.60e-05^{***}$ (−4.45)	$7.00e-07$ (0.03)	$2.72e-02^{***}$ (3.09)	0.0002^{***} (6.00)	0.0005^{***} (5.40)	0.0013^{***} (6.41)
R^2	0.01	0.18	0.21	0.01	0.13	0.16
F-statistic	0.35	4.26^{***}	4.35^{***}	1.25	4.89^{***}	4.95^{***}

*** Significant at 1%, ** Significant at 5%, * Significant at 10%. *t* ssstatistics are in parentheses
Source: Authors' calculations

Table 11.4 Productivity determinants of SPM and joint outputs

Dependent variable	SPM			Joint outputs		
Estimation technique	OLS	Fixed effects	Dynamic GMM	OLS	Fixed effects	Dynamic GMM
Intercept	0.014^{***} (49.16)	–	–	0.026^{***} (8.85)	–	–
Gross state product (GSP)	-0.01^{***} (−39.20)	-0.01^{***} (−37.03)	-0.02^{***} (−41.05)	-0.02^{***} (−7.32)	-0.02^{***} (− 6.25)	-0.03^{***} (−6.97)
GSP^2	0.002^{***} (25.96)	0.002^{***} (41.22)	0.003^{***} (36.52)	0.004^{***} (4.65)	0.008^{***} (10.60)	0.009^{***} (11.46)
Population density	1.59e − 08 (0.285)	$-4.80e-06^{***}$ (− 23.05)	$-3.47e-03^{***}$ (− 20.52)	$-9.41e-07^{*}$ (−2.57)	$-9.30e-06^{***}$ (−4.68)	$-4.88e-04^{***}$ (−5.71)
Urbanization	$-7.04e-05^{***}$ (−20.62)	-0.001^{***} (−35.61)	-0.003^{***} (−25.53)	6.35e − 05 (1.37)	-0.002^{***} (−8.11)	-0.011^{***} (−8.74)
Education index	$1.32e-05^{***}$ (5.53)	$2.40e-05^{***}$ (7.41)	$1.42e-03^{***}$ (6.94)	$2.63e-05^{*}$ (2.36)	$7.81e-05^{**}$ (2.99)	$2.74e-03^{***}$ (3.14)
R^2	0.04	0.14	0.19	0.01	0.34	0.36
F-statistic	1.37	6.25^{***}	6.29^{***}	0.35	5.43^{***}	5.73^{***}

*** Significant at 1%, ** Significant at 5%, * Significant at 10%. *t* statistics are in parentheses

Source: Authors calculations

11.5 Concluding Remarks

As a result of India's extremely rapid economic growth, the scale and seriousness of environmental problems are no longer in doubt. The use of more efficient pollution abatement technologies is crucial in the analysis of environmental management because such technologies influence the cost of alternative production and pollution abatement technologies, at least in part (e.g., Jaffe et al., 2003). Using recently developed productivity measurement techniques, we show that overall environmental productivity decreases over time in India. At present, existing environmental management is not sufficient to bring sustainable development to the country. However, once we disaggregate the pollutants to specific pollution by SO_2, NO_2, and SPM, we find environmental productivity recently increases in SO_2. The results for NO_2 and SPM are the main causes of the productivity reduction over the study periods.

Furthermore, we analyze the determinants of environmental productivity and find an EKC-type relationship exists between environmental productivity and income. However, environmental productivities in general decline more in high-income states than in low-income states. Panel analysis results show that the scale effect is negative and dominant over the positive technique effect. Therefore, a combined effect of income on environmental productivity is negative, which answers the question of why productivity has declined faster in developed states than in their underdeveloped counterparts.

We conclude that if the ongoing pace of industrialization is not met with effective environmental management, there will be untoward consequences in India. Society is required to create environmental practices based on incentives for industries to perform well in their environmental management and to simultaneously formulate economic and environmental policies in order to achieve a sustainable growth process.

Appendix. TFP of SO_2 of Indian States

States	1992	1993	1994	1995	1996	1997	1998	1999	2000	2001	2002	2003
Andhra Pradesh	−0.002	0.018	−0.004	−0.049	−0.03	0.008	0.031	0.011	−0.007	0.003	0.041	0.002
Delhi	−0.007	−0.0003	−0.011	−0.007	0.014	−0.001	−0.002	1.5e − 05	0.002	0.013	−0.01	−0.034
Goa	−0.164	0.069	−0.283	0.039	−0.213	−0.259	−0.126	−0.071	−0.153	−0.066	−0.012	−0.087
Gujarat	−0.016	−0.188	−0.044	−0.58	−0.003	0.54	−0.316	−0.219	0.24	−0.006	0.049	0.068
Haryana	0.008	−0.011	−0.02	−0.02	−0.019	−0.012	−0.009	0.069	−0.079	0.041	0.011	0.064
Himachal Pradesh	−0.0005	0.0003	−0.001	0.0002	5.8e − 05	−0.002	0.0006	0.002	−0.001	−0.002	0.0002	0.0004
Karnataka	−0.012	−0.02	−0.001	−0.039	−0.029	−0.021	−0.037	−0.005	0.058	0.021	−0.003	−0.002
Kerala	−0.002	−0.004	−0.003	0.003	−0.002	−0.003	−0.003	−0.013	−0.026	0.018	0.003	−0.006
Madhya Pradesh	−0.004	−0.034	0.021	−0.066	−0.043	−0.0004	0.022	0.083	−0.098	0.051	0.036	0.021
Maharashtra	0.08	−0.12	−0.12	−0.14	0.11	−0.20	0.15	0.08	0.0009	−0.04	0.065	0.067
Orissa	−0.004	−0.011	0.003	0.003	0.003	−0.004	−0.007	0.022	0.003	−0.001	−0.002	0.007
Punjab	−0.0004	−0.013	0.003	0.007	0.018	−0.023	−0.002	0.002	0.013	0.01	0.01	0.002
Rajasthan	−0.0003	−0.012	0.003	−0.008	−0.004	0.005	0.003	0.01	−0.025	0.012	0.0005	0.016
Tamil Nadu	−0.007	−0.007	−0.037	−0.073	0.127	−0.082	0.013	0.066	−0.061	−0.082	−0.104	0.063
Uttar Pradesh	0.006	−0.022	0.044	0.098	−0.008	−0.022	−0.058	0.069	−0.033	0.003	0.071	0.035
West Bengal	0.043	−0.013	−0.052	0.013	0.018	−0.048	−0.11	0.069	0.09	0.046	0.033	−0.007
Mean	−0.005	−0.023	−0.031	−0.051	−0.004	−0.008	−0.028	0.011	−0.005	0.001	0.012	0.013

Source: Authors' calculations

Chapter 12
A Global Analysis of Environmentally Sensitive Productivity Growth

12.1 Introduction

Concerns about the impact of climate policy on "productivity" or "economic growth" have made countries hesitant to reduce CO_2 emissions. Climate policy has different dimensions: economic, technological, and ecological. The economic dimension offers solutions in terms of price signals, and the technological dimension sees solutions in terms of appropriate technological development and adoption. The ecological dimension adopts a more holistic view of the human-nature relationship and calls for "green accounting" or "sustainable development." This chapter tries to present an extension of the economic approach that includes aspects of technological development and adoption as well as green accounting.

Productivity has acted as a significant engine of growth, allowing living standards in the world to advance rapidly throughout the 20th century. However, its traditional measures do not account for production of harmful by-products such as CO_2, which may lead to environmental damage. CO_2 is conventionally measured using index numbers, which requires data on prices of all outputs and inputs; information for bad outputs does not exist. The distance function approach can help overcome such problems, as it requires data only on quantities of inputs, outputs, and pollutants. Unlike this study, others estimating productivity using the distance function approach have focused on desirable outputs only (e.g., Färe et al., 1994a; Lall et al., 2002). Some studies have used microeconomic data rather than the macroeconomic data used by the present study while estimating total factor productivity (TFP) in the presence of bad outputs (e.g., Yaisawarng and Klein, 1994; Ball et al., 1994; Chung et al., 1997; Hailu and Veeman, 2000).

The Malmquist indexes are an increasingly popular method of measuring TFP using the distance function. However, incorporation of bad outputs into the Malmquist indexes can be problematic. As the Malmquist indexes are based on Shepherd distance functions, which are radial in nature, firms cannot be credited with the reduction of bad outputs. This does not allow for changes in technology that reduce the amount of pollution generated while increasing production of good outputs.

S. Kumar and S. Managi, *The Economics of Sustainable Development*, Natural Resource Management and Policy 32,
DOI: 10.1007/978-0-387-98176-5_12,

It does not capture any "decoupling" of the production of good outputs with bad outputs. If there has been a decoupling of pollution and production, then there may be computational problems using the Shepherd distance function (Chapple and Harris, 2003).

There are several studies on the measurement of productivity changes in industries which produce good and bad outputs simultaneously during the production process. Some of these studies have treated the bad outputs as inputs (Cropper and Oates, 1992; Pittman, 1981; Haynes et al., 1993, 1994; Boggs, 1997; Kopp, 1998; Reinhard et al., 1999; Murty and Kumar, 2004) etc., while the others have treated these as synthetic output such as pollution abatement (e.g., Gollop and Robert, 1983). Murty and Russell (2002) have pointed out that the treatment of bad outputs as inputs is not consistent with the materials balance approach. The approach adopted by Gollop and Robert to treat reduction in bad output as good output creates a different nonlinear transformation of the original variable in the absence of base-constrained emission rates (Atkinson and Dorfman, 2002). To overcome this problem, Pittman (1983) proposed that good and bad outputs should be treated nonsymmetrically. He suggested the maximal radial expansion of good outputs and contraction of bad outputs. Chung et al. (1997) have used the directional distance function to calculate production relationships involving good and bad outputs that treats good and bad outputs asymmetrically. This study follows Chung et al. (1997) and uses the directional distance function to measure Malmquist-Luenberger (ML) productivity index and its components.

The components of the productivity index – technical and efficiency changes – are analogous to the notions of technological innovation and adoption, respectively. The ML index credits producers for simultaneously increasing good outputs and reducing the production of bad outputs such as CO_2. It also offers an alternative way of assigning weights on the relative importance of bad outputs, which can be interpreted as if consumers have preferences for reducing bad outputs regardless of the actual damage resulting from these products (Färe et al., 2001). Although the ML index does not directly relate to changes in welfare level, it does provide a complete picture of productivity growth under environmental regulations of emissions that are of concern to society.

The measures of productivity are often obtained under alternative assumptions about the disposability of CO_2. That is, it could be either strongly or weakly disposable. While strong disposability implies that a country can reduce CO_2 emissions without incurring any abatement costs, weak disposability assumes diversion of resources from the production of good outputs. Thus the ML index encompasses green accounting while accounting for undesirable outputs.[1]

This chapter uses the nonparametric linear programming method to estimate directional distance function. Thus for each year the same "meta" best practice frontier is constructed based on the data for 41 countries for the period 1971–1992.

---

[1] Hailu and Veeman (2000) termed the measurement of productivity under weak disposability of pollutants as environmentally sensitive productivity.

Each country is then compared to this best practice frontier to provide the performance scores.

Productivity analysis helps to explain the level of economic prosperity, standard of living, and the degree of competitiveness of a country, although it is not the only determinant of economic growth and welfare. Therefore, it is important to find which factors determine productivity growth in countries in the presence of reduced carbon emissions. Though there are various theories that explain productivity growth in countries, two are of particular interest.[2] The convergence hypothesis states that in low-income countries productivity tends to converge towards those of high-income countries, (Baumol, 1986; Baumol et al., 1989). The rationale behind the convergence hypothesis is the concept of diminishing returns to capital. In developed countries the capital-labor ratio is found to be high in comparison to developing countries, and therefore the marginal productivity of capital is low.

Two, the endogenous growth theory advocates that the difference in productivity between developed and developing countries remains constant or even diverges over time (Arrow, 1962). The foundation of endogenous growth theories lies in the concept of increasing returns to scale, which are generated from externalities associated with the acquisition of technical knowledge. However, there are institutions and policies that determine the development process of a country (Olson, 1996). This chapter tries to extend this literature by empirically examining the causes of productivity changes while accounting for carbon emissions.

The remainder of the chapter is structured as follows: In Section 12.2, we discuss the theoretical approach of the chapter. Section 12.3 discusses the data used in the study and its results. The data set is richer than the past examinations of efficiency and productivity analyses in that it includes energy as input. The addition allows for a more thorough assessment of the production processes that generate carbon emissions from the use of energy. The chapter closes in Section 12.4 with some concluding remarks.

12.2 Measuring Environmentally Sensitive Productivity

Suppose that a country employs a vector of inputs $x \in \Re^N_+$ to produce a vector of good outputs $y \in \Re^M_+$, and bad outputs $b \in \Re^I_+$. Let $P(x)$ be the feasible output set for the given input vector x and $L(y, b)$ is the input requirement set for a given output vector (y, b). Now the technology set is defined as:

$$P(x) = \{(y, b) : x \text{ can produce}(y, b)\}, x \in \Re^N_+. \tag{12.1}$$

We assume that the good and bad outputs are null-joint; a country cannot produce good output in the absence of bad outputs, i.e., if $(y, b) \in P(x)$ and $b=0$ then $y=0$. The output is strongly or freely disposable if

[2]for a brief summary of growth theories, see Lall et al. (2002).

$$(y,b) \in P(x) \text{ and } \hat{y} \leq y \text{ imply}(\hat{y},b) \in P(x). \tag{12.2}$$

This implies that if an observed output vector is feasible, then any output vector smaller than that is also feasible. It excludes production processes that generate undesirable outputs that are costlier to dispose. In contrast, concerns about CO_2 and other greenhouse gases imply that these should not be considered to be freely disposable. In such cases bad outputs are considered as being weakly disposable and

$$(y,b) \in P(x) \text{ and } 0 \leq \theta \leq 1 \text{imply}(\theta y, \theta b) \in P(x) \tag{12.3}$$

This implies that pollution is costly to dispose of, and abatement activities would typically divert resources away from the production of desirable outputs and thus lead to lower good output with given inputs.

Following Färe et al. (1994) a DEA model can be constructed which satisfies the above conditions. For each time period $t = 1,\ldots,T$ there are $k = 1,\ldots,K$ observations for of inputs and outputs ($\mathbf{x}^{k,t}$,$\mathbf{y}^{k,t}$,$\mathbf{b}^{k,t}$). Using these data in a DEA framework, an output set can be constructed that satisfies the above conditions:

$$\begin{aligned} P^t(x^t) = \{(y^t,b^t) : & \sum_{k=1}^{K} z_k^t y_{km}^t \geq y_m^t & m = 1,\ldots,M \\ & \sum_{k=1}^{K} z_k^t b_{ki}^t = b_i^t & i = 1,\ldots,I \\ & \sum_{k=1}^{K} z_k^t x_{kn}^t \leq x_n^t & n = 1,\ldots,N \\ & z_k^t \geq 0 & k = 1,\ldots,K\} \end{aligned} \tag{12.4}$$

where z_k^t are the intensity variables or weights assigned to each observation in constructing the production possibility frontier. Non-negativity of the intensity variable has the effect of imposing constant returns to scale.

Furthermore, to incorporate the null jointness of outputs, the following conditions are imposed on the DEA model:

$$\sum_{k=1}^{K} b_{ki}^t > 0 \qquad k = 1,\ldots,K \tag{12.5}$$

$$\sum_{i=1}^{I} b_{ki}^t > 0 \qquad i = 1,\ldots,I \tag{12.6}$$

These state that every bad output is produced by some country, k, and that every country, k, produces at least one bad output (in multiple bad output situations).

12.2.1 Directional Distance Functions

As stated earlier, the directional distance function seeks to increase the good outputs while simultaneously reducing the bad outputs. Formally it is defined as:

$$\vec{D}_o(x, y, b; g) = \sup\{\beta : (y, b) + \beta g \in P(x)\} \tag{12.7}$$

where g is the vector of directions in which outputs can be scaled. Following Chung et al. (1997), the direction taken is $g = (y, -b)$, so that as good outputs are increased, bad outputs are decreased. The difference between the output distance function and the directional distance function is illustrated in Fig. 12.1. In contrast to the output distance function, which places output vector A on the boundary at point C, expanding both good and bad outputs simultaneously, the directional distance function starts at A and scales in the direction of increase in good outputs and decrease in bad outputs to point B on the boundary. At point B, the output vector is $(y^t + \beta^* g_y, b^t - \beta^* g_b)$ where, $\beta^* = \vec{D}_o^t(x^t, y^t; g_y, -g_b)$ with $\beta^* g_y$ has been added to the good output and $\beta^* \mathbf{g}_b$ has been subtracted from the bad output.

12.2.2 Malmquist–Luenberger Productivity Index

Using directional distance functions, we define the ML productivity index. The ML index is very much based on the traditional Malmquist indexes – the main difference being that they are constructed from directional distance functions rather than Shepherd distance functions. The ML index requires a definition of the directional distance function with respect to two different periods, i.e.,

$$\vec{D}_o^{t+1}(x^t, y^t, b^t; g) = \sup\{\beta : (y^t, b^t) + \beta g \in P^{t+1}(x^t)\} \tag{12.8}$$

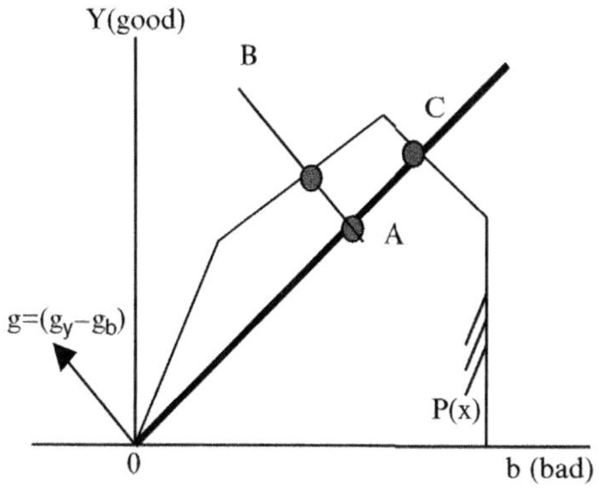

Fig. 12.1 Shephard output and directional distance functions

This version of the directional distance function measures observations at time t based on the technology at time $t+1$. Chung et al. (1997) define the ML index of productivity between period t and $t+1$ as:

$$\mathrm{ML}_t^{t+1} = \left[\frac{(1+\vec{D}_o^{t+1}(x^t,y^t,b^t;y^t,-b^t))}{(1+\vec{D}_o^{t+1}(x^{t+1},y^{t+1},b^{t+1};y^{t+1},-b^{t+1}))}\frac{(1+\vec{D}_o^{t}(x^t,y^t,b^t;y^t,-b^t))}{(1+\vec{D}_o^{t}(x^{t+1},y^{t+1},b^{t+1};y^{t+1},-b^{t+1}))}\right]^{\frac{1}{2}}. \tag{12.9}$$

The index can be decomposed into two component measures of productivity change:

$$\mathrm{ML}_t^{t+1} = \underbrace{\left[\frac{1+\vec{D}_o^{t}(x^t,y^t,b^t;y^t,-b^t)}{1+\vec{D}_o^{t+1}(x^{t+1},y^{t+1},b^{t+1};y^{t+1},-b^{t+1})}\right]}_{\mathrm{MLEFFCH}_t^{t+1}}$$
$$\times \underbrace{\left[\frac{(1+\vec{D}_o^{t+1}(x^t,y^t,b^t;y^t,-b^t))}{(1+\vec{D}_o^{t}(x^t,y^t,b^t;y^t,-b^t))}\frac{(1+\vec{D}_o^{t+1}(x^{t+1},y^{t+1},b^{t+1};y^{t+1},-b^{t+1}))}{(1+\vec{D}_o^{t}(x^{t+1},y^{t+1},b^{t+1};y^{t+1},-b^{t+1}))}\right]^{\frac{1}{2}}}_{\mathrm{MLTECH}_t^{t+1}}. \tag{12.10}$$

The first term, MLEFFCH, represents the efficiency change component, a movement towards the best practice frontier, while the second, MLTECH, the technical change, i.e., a shift. If there have been no changes in inputs and outputs over two time periods, then $\mathrm{ML}_t^{t+1} = 1$. If there has been an increase in productivity, then $\mathrm{ML}_t^{t+1} > 1$, and finally, a decrease when $\mathrm{ML}_t^{t+1} < 1$. Changes in efficiency are captured by $\mathrm{MLEFFCH}_t^{t+1}$, which gives a ratio of the distances the countries are to their respective frontiers, in time periods t, and t+1. If $\mathrm{MLEFFCH}_t^{t+1} > 1$, then there has been a movement towards the frontier in period $t+1$. If $\mathrm{MLEFFCH}_t^{t+1} < 1$, then it indicates that the country is further away from the frontier in $t+1$, and hence has become less efficient. If technical change enables more production of good and less production of bad outputs, then $\mathrm{MLTECH}_t^{t+1} > 1$. Whereas if $\mathrm{MLTECH}_t^{t+1} < 1$, there has been a shift of the frontier in the direction of fewer good outputs and more bad outputs (Färe et al., 2001).

12.2.3 Computation of Directional Distance Function

The technique of linear programming is used to compute directional distance functions. Four programs need to be solved for each observation. Two programs use observations and technology for the time period t, or $t+1$, and the other two programs use mixed periods, using, for example, technology calculated from period t with the observation $t+1$. The directional distance function for observation k' in period t, using period t technology, can be calculated by solving the following LP problem.

$$\vec{D}_o^t(x^t, y^t, b^t; y^t, -b^t) = \max \beta$$
$$s.t.$$
$$\sum_{k=1}^{K} z_k^t y_{km}^t \geq (1+\beta) y_{k'm}^t, m = 1, \ldots, M$$
$$\sum_{k=1}^{K} z_k^t b_{ki}^t = (1-\beta) b_{k'i}^t, i = 1, \ldots, I \tag{12.11}$$
$$\sum_{k=1}^{K} z_k^t x_{kn}^t \leq x_{k'n}^t, n = 1, \ldots, N$$
$$z_k^t \geq 0, k = 1, \ldots, K.$$

The mixed period problems can cause difficulties in calculation, whereby the observed data in period $t+1$ are not feasible in period t. For example, the observation ($y^{t+1,k'}$, $b^{t+1,k'}$) may not belong to the output set $P^t(x^{t+1})$. To minimize this problem, we follow Färe et al. (2001), whereby multiple year "windows" of data are the reference technology. All frontiers are constructed from 3 years of data – hence the frontier for 1973, for example, would be constructed from data in 1973, 1972, and 1971, which reduces the likelihood of "nonsolutions."

12.3 Data and Results

We obtain the data on five variables namely, GDP, CO_2, labor force, capital stock, and commercial energy consumption for 41 countries,[3] a mix of developed and developing countries for the period 1971–1992.[4] Out of these five variables the first two, GDP and CO_2 are considered as proxies of good and bad outputs respectively,

[3]We have grouped all the countries in two categories- Annex I and Non-Annex I Countries. We have 21 Annex I countries (Australia, Austria, Belgium, Canada, Denmark, Finland, France, Greece, Iceland, Ireland, Italy, Japan, Luxembourg, Netherlands, New Zealand, Norway, Spain, Sweden, Switzerland, United Kingdom, United States) and 20 Non-Annex I countries (Bolivia, Chile, Colombia, Ecuador, Guatemala, Honduras, Hong Kong, India, Israel, Kenya, Mexico, Morocco, Nigeria, Peru, Philippines, Syrian Arab Republic, Thailand, Venezuela, Zambia, and Zimbabwe). The Annex-1 parties to the United Nations Framework Convention on Climate Change are those developed countries, or regional organizations (the EU), that are listed in the Annex-I of the Climate Convention.

[4]The choice of period and countries is based on the availability of data particularly on capital stock. The Penn World Tables provide capital stock data up to 1992 especially for developing countries.

Table 12.1 Descriptive statistics of the variables used in the study (1971–1992)

Variable	Mean	Std. dev.	Minimum	Maximum
Annex-1 countries				
GDP growth rate	2.662	0.750	1.398	3.900
CO_2 growth rate	0.414	1.700	−2.934	4.148
Labor growth rate	1.184	0.602	0.464	2.409
Capital growth rate	4.142	1.164	2.476	6.941
Energy consumption growth rate	1.293	2.164	−6.405	3.855
Capital per labor (US$)	29,227.530	11,088.051	9,721.000	76,733.000
GDP per capita (US$)	20,755.585	7,628.875	8,039.286	45,951.950
Energy/GDP (metric tons/million US$)	0.285	0.141	0.006	0.761
Openness (Export + Import as % GDP)	65.601	40.745	11.239	233.537
Non-Annex-I countries				
GDP growth rate	3.708	1.870	0.820	7.880
CO_2 growth rate	3.911	2.920	−2.755	9.143
Labor growth rate	3.030	0.483	2.052	4.033
Capital growth rate	4.312	2.252	−1.218	8.245
Energy consumption growth rate	3.884	1.811	0.682	8.698
Capital per labor (US$)	7,352.907	5,988.130	349.000	22,307.000
GDP per capita (US$)	2,775.676	3,813.108	205.645	20,963.160
Energy/GDP (metric tones/million US$)	0.972	0.882	0.127	4.57
Openness (Export + Import as % GDP)	56.278	38.467	8.365	274.955

Source: Authors' calculations

and the remaining three are taken as inputs. Data on GDP, labor force, and energy consumption are collected from the World Development Indicators (WDI, World Bank), whereas data on CO_2 are taken from the website of World Resources. Capital stock[5] data are obtained from the Penn World Tables (Mark 5.6). GDP and capital stock are measured in 1985 US dollars, whereas CO_2 and energy consumption are measured in 1,000 metric tons. The labor force data are in millions of workers.

The descriptive statistics of all the variables used in the study for both of the groups, i.e., Annex-I and Non-Annex-I countries, are presented in Table 12.1, which bring into focus the contrast between these two groups.

The highest growth rate with respect to CO_2 was observed in the Syrian Arab Republic (9.14%). A newly industrialized country, Hong Kong, registered the highest growth rate of GDP (7.9%) during the period studied and also had a high growth rate for CO_2 emissions (6.5%). Thailand had the second highest growth rate of GDP (7.3%) but a more rapid rate of growth in the production of CO_2 (8.3%). An examination of the overall growth rates reveals that developing countries witnessed higher growth in GDP, CO_2, and commercial energy consumption than did developed countries. Sweden, Luxembourg, France, Belgium, United Kingdomm and

[5]Capital stock does not include residential construction but does include gross domestic investment in producers' durables, as well as nonresidential construction. These are the cumulated and depreciated sums of past investment.

Denmark registered negative CO_2 growth rates of 2.93%, 1.96%, 1.71%, 1.57%, 0.66%, and 0.32%, respectively. Thus we observe that in Non-Annex-I countries not only were growth rates of income and emissions higher, but there was also a higher degree of variability within the group. The emission intensity of output measured as a ratio of CO_2 emission to GDP was quite high in developing countries in comparison to developed countries. However, it should be noted that the per capita GDP, capital, CO_2 emissions, and commercial energy consumption were substantially higher in developed countries than in developing economies.

The approach outlined in Section 12.2 constructs a best-practice frontier from the data.[6] Table 12.2 sums up the main results, which describe the average[7] annual performance of each country and each group.[8] Recall that index values greater (less) than one denote improvements (deterioration) in the relevant performance. Here we have calculated the ML index and its components for both cases: weak and strong disposability of CO_2 emissions.

To examine the relationship between productivity and its determinants, the study considers variables such as GDP per capita, technical inefficiency in the previous year, capital per labor, and energy intensity of output measured by the use of commercial energy per unit of GDP. We also included the openness index as a determinant of productivity. The openness index could be a proxy for a country's institutional and policy framework. The source of data on the openness index is the WDI.

If one could establish a positive relationship between (a) GDP per capita and level of productivity and (b) productivity and capital-labor ratio, the findings would favor endogenous growth theories. Higher productivity growth in lower capital-labor ratios countries would favor convergence theory because the marginal product of capital would be low in high-income countries that exhibit a high capital-labor ratio.[9] Moreover, the convergence theory could be restated in the relationship between productivity and lagged technical inefficiency. This relationship would

[6]In the computation of ML index although we followed multiple year "windows" of data as the reference technology to minimize the problem of infeasible solutions, even then there exists the problem for some countries, i.e. Hong Kong (2), Luxembourg (8), and Netherlands (13) observed infeasible linear programming for at least one of the mixed period when the carbon emissions are included as bad outputs (the values in the parentheses indicate the number of years for each country).

[7]Since the total factor productivity index is multiplicative, these averages are also multiplicative (i.e., geometric means) and the average is simply the geometric mean from those years for which the index could be computed.

[8]Disaggregated results for each country are available from the author on request.

[9]Here it is important to differentiate between σ- and β-convergences. When the dispersion of income across a group of economies falls over time, there is σ-convergence and the negative partial correlation between growth in income over time and its initial level favors β-convergence. Young et al. (2003) show that β-convergence is necessary for σ-convergence. They show that "σ^2_t can be rising even if β-convergence is the rule. Intuitively, economies can be β-converging towards one another while, at the same time, random shocks are pushing them apart" (Page 5). The present analysis relates to later category.

Table 12.2 Decomposition of average annual changes, 1973–1992

Country	Environmentally sensitive measure			Conventional measure		
	ML	MLEFFCH	MLTECH	M	MLEFFCH	MTECH
Bolivia	0.9977	0.9955	1.0022	0.9866	0.9965	0.9900
Chile	1.0031	1.0045	0.9986	1.0072	1.0129	0.9943
Colombia	1.0008	0.9986	1.0022	1.0234	1.0237	0.9997
Ecuador	0.9939	0.9922	1.0018	0.9992	1.0112	0.9881
Guatemala	0.9941	0.9975	0.9966	1.0049	1.0038	1.0011
Honduras	0.9996	0.9973	1.0023	1.0044	0.9995	1.0049
Hong Kong	1.0138	1.0077	1.0062	1.0238	1.0375	0.9868
India	0.9987	0.9978	1.0009	0.9892	0.9943	0.9949
Israel	0.9996	1.0001	0.9995	1.0505	1.0000	1.0505
Kenya	1.0044	1.0028	1.0016	1.0158	1.0161	0.9998
Mexico	0.9981	0.9964	1.0017	0.9836	0.9894	0.9941
Morocco	0.9981	0.9982	0.9999	0.9984	1.0037	0.9947
Nigeria	0.9963	0.9970	0.9994	1.0004	0.9981	1.0023
Peru	1.0001	0.9980	1.0021	0.9974	1.0038	0.9937
Philippines	1.0012	0.9987	1.0024	0.9947	1.0041	0.9907
Syrian A. R.	0.9977	0.9968	1.0009	1.0109	1.0285	0.9829
Thailand	0.9982	0.9959	1.0023	1.0303	1.0371	1.0062
Venezuela, RB	0.9993	0.9981	1.0012	0.9923	0.9991	0.9932
Zambia	1.0058	1.0048	1.0010	0.9965	0.9940	1.0025
Zimbabwe	0.9997	0.9988	1.0009	0.9767	0.9841	0.9925
Australia	0.9983	1.0000	0.9982	1.0008	0.9968	1.0040
Austria	1.0035	1.0028	1.0008	1.0023	0.9979	1.0044
Belgium	1.0034	1.0001	1.0033	1.0059	1.0017	1.0042
Canada	0.9983	0.9995	0.9988	0.9937	1.0037	0.9900
Denmark	0.9968	0.9954	1.0015	0.9812	0.9901	0.9910
Finland	0.9996	1.0026	0.9971	1.0048	1.0207	0.9844
France	1.0091	1.0081	1.0010	1.0092	1.0123	0.9969
Greece	0.9930	0.9950	0.9980	1.0029	0.9976	1.0053
Iceland	0.9915	0.9938	0.9977	1.0132	1.0071	1.0060
Ireland	1.0088	1.0079	1.0009	1.0142	1.0178	0.9965
Italy	1.0010	1.0024	0.9986	0.9830	0.9802	1.0029
Japan	1.0072	0.9995	1.0077	0.9681	0.9811	0.9868
Luxembourg	0.9787	1.0000	0.9820	1.0052	1.0014	1.0038
Netherlands	1.0132	1.0000	1.0143	0.9788	0.9892	0.9894
New Zealand	0.9906	0.9939	0.9967	0.9956	0.9899	1.0057
Norway	1.0088	1.0095	0.9993	0.9944	0.9991	0.9926
Spain	0.9932	0.9937	0.9994	1.0022	0.9987	1.0035
Sweden	1.0055	1.0070	0.9985	0.9825	0.9879	0.9946
Switzerland	1.0054	0.9994	1.0059	0.9833	0.9851	0.9981
U.K.	1.0047	1.0022	1.0025	0.9913	0.9857	1.0057
U.S.	0.9985	0.9984	1.0000	0.9962	1.0015	0.9947
Non-Annex-I	1.0000	0.9988	1.0012	1.0042	1.0068	0.9981
Annex-I	1.0004	1.0005	1.0001	0.9956	0.9973	0.9981
All	1.0002	0.9997	1.0006	0.9998	1.0019	0.9981

ML Malmquist–Luenberger index, *MLEFFCH* Malmquist–Luenberger efficiency change, *MLTECH* Malmquist–Luenberger technical change, *M* Malmquist index, *MEFFCH* Malmquist efficiency change, *MTECH* Malmquist technical change

Source: Authors' calculations

state that those countries that were near the production frontier would see a lower level of productivity growth than those were farther away. Therefore, the positive relationship between productivity level and lagged technical inefficiency level would indicate the presence of the convergence hypothesis (Lall et al., 2002).

A number of factors affect the pattern of CO_2 emissions, including technical change, economic growth, and changes in the composition of GDP. Technical progress can yield reductions in CO_2 emissions by increasing the ratio of good output to bad output. A change in the composition of a country's GDP country can also affect the level of CO_2 emissions. For example, presumably a shift away from the energy-intensive sector would yield a decline in CO_2 emissions. Therefore, in the determinants of productivity we have included the energy intensity of production.

The relationship with the openness variable will determine the impact of international trade on productivity growth. The openness variable can show both positive and negative effects of an increased volume of trade on the environmentally sensitive measure of productivity growth (ML index). On the negative side, it captures the environmentally deteriorating effects that stem from the increased volume of trade. On the positive side, it captures the environmentally beneficial effects that arise due to harmonization of environmental policies. The sign and significance of the openness variable help one to select among the competing hypotheses on environment and international trade (Etkins et al., 1994; Taskin and Zaim, 2001).

12.3.1 Conventional Measurement of Productivity

The average Malmquist index value of 0.9998 indicates that the annual productivity decline for the sample countries was 0.002%. On average, this decline was due to technical change; the world witnessed an average technical regression of 0.02% over the study period. This progress in TFP is 0.42% per annum for Non-Annex-I countries, whereas in Annex-I countries it declined by 0.44% per year. From these overall average figures of stagnation in TFP changes in countries it may be argued that effectively all GDP growth in the post-1970 period was due to high rates of input accumulation.

Zimbabwe experienced the highest decline (2.33% per year) in TFP among the sample developing countries (Table 12.2). Israel experienced the highest growth in TFP followed by Thailand and Hong Kong in Non-Annex-I countries. In the Annex-I countries, Japan experienced the highest decline in TFP growth, followed by the Netherlands. Ireland experienced the highest growth rate in TFP in the amount of 1.42% per annum.

Letting the Malmquist index, M_{it} represent the conventional measure of productivity of country i in year t, the equation below specifies a possible form of relation between the conventional measure of productivity and its determinants.

$$M_{it} = \beta_{1i} + \beta_2 \text{GDPPC}_{it} + \beta_3 \vec{D}_{it-1} + \beta_4 \text{CAPLAB}_{it} \\ + \beta_5 \text{ENGDP}_{it} + \beta_6 \text{OPEN} + \beta_7 \text{ANNEX}_{it} + \varepsilon_{it},$$

where i is country index; t is time index; σ is the disturbance term such that $\varepsilon \sim N(0, \sigma_\varepsilon)$; GDPPC is the GDP per capita; $\vec{D}_{it-1}$ is the value of technical inefficiency in the lagged period (when the emissions of carbon are strongly disposable); CAPLAB is capital per labor; ENGDP is use of commercial energy per unit of GDP; OPEN, openness index defined as the ratio of total exports and imports to GDP; and ANNEX is the dummy variable for the group of countries (its value is equal to one if the country belongs to Annex-I countries and zero otherwise).

Table 12.3 (last columns) provides the estimated parameters of the regressions for the M_{it} index under alternative specifications. An LM test performed on the alternative specifications of the fixed effect model rejects the null hypothesis of a common intercept in favor of one with country-specific intercept terms. The choice between the fixed effect and random effect models can be made using the Hausman test. We find in the present study the fixed effect model to be the appropriate specification.

In the fixed effect model, the positive relationship between the lagged technical inefficiency and productivity index favors the existence of the convergence hypothesis. We also find that the sign of capital per labor is negative, while statistically significant at the 15% level; this again favors the convergence hypothesis. These findings concur with Lall et al. (2002), who find the existence of the convergence hypothesis for a sample of 30 countries in the western hemisphere for the period 1978–1994 using the Malmquist index. Moreover, we find that the openness of a

Table 12.3 Determinants of productivity change

Variable	Environmentally sensitive measure Random effect model	Conventional measure Fixed effect model
GDPPC	6.34E−06 (2.883)*	1.33E−06 (0.408)
$\vec{D}_o^{t-1}$	1.90E−01 (2.473)**	2.46E−02 (2.263)**
LN(CAPLAB)	−4.17E−02 (−2.154)**	−5-35E−02 (−1.493)$^+$
ENGDP	−4.35E−02 (−2.06)**	−1.22E−02 (−0.313)
OPEN	7.00E−04 (2.714)*	2.10E−03 (4.904)*
ANNEX	6.79E−04 (0.023)	1.35E−01 (1.96)**
Constant	1.0219 (68.26)*	
Adj. R^2	0.07	0.14
LM Test (p value)		0.00
Hausman Test (p value)	0.118	
N	751	776

Notes. Values in parentheses represent "t-statistics." *, **, ***, and + show the level of significance at 1%, 5%, 10%, and 15% respectively. *GDPPC* GDP per capita, $\vec{D}_o^{t-1}$ lagged period value of directional distance function, *LN(CAPLAB)* natural log of capital-labor ratio, *ENGDP* energy consumption to GDP ratio; *OPEN* openness index; and *ANNEX* Dummy variable 1 if the country belong to Annex-I countries and 0 otherwise

Source: Authors' calculations

country contributes positively to productivity growth. The coefficient of the dummy variable that represents the group of countries is positive and statistically significant. It implies that the growth rate of productivity was higher in Annex-I countries in comparison to Non-Annex-I countries. The other variables included in the study are not statistically significant.

12.3.2 Environmentally Sensitive Measurement of Productivity

The average change in the ML productivity index, when CO_2 was weakly disposable, was 0.02%. This average TFP measure was the product of a positive change in innovation of 0.06% and a negative efficiency change of 0.03%. In Non-Annex-I counties Hong Kong experienced the highest growth in TFP when CO_2 was considered an undesirable output and Ecuador experienced the highest decline in the index. But in Annex-I countries, the Netherlands had the highest growth and Luxembourg experienced the highest decline in the ML index. However, it was technological change that governed the change in overall productivity index in most of the countries.

The ML index had a higher value in comparison to the standard Malmquist index for India, Mexico, Peru, Philippines, Venezuela, Zambia, and Zimbabwe in Non-Annex-I countries. In Annex-I countries Austria, Canada, Denmark, Italy, Japan, Netherlands, Norway, Sweden, Switzerland, the U.K, and the U.S experienced the higher average value of TFP index when we account for CO_2 emissions in comparison to the situation when these emissions are freely disposable. On average in Non-Annex-I countries, the value of standard Malmquist is higher in comparison to the ML index, but the reverse is the situation in Annex-I countries. This shows that Non-Annex-I countries had lower productivity growth when carbon emissions were weakly disposable. This finding confirms that of Kopp (1998). Kopp finds that developed countries experienced technical progress in a way that economizes on CO_2 emissions, but that developing countries did not during 1970–1990.

We run a basic "*t* test" to test the null hypothesis as to whether the two productivity measures and their components were the same. It was found that the TFP index value does not change in either scenario, and the null hypothesis is not rejected. But for technical and efficiency changes, the null hypothesis cannot be accepted for either of the groups of countries (Table 12.4). The relative growth rates of the conventional productivity measure and the productivity measure adjusted for the inclusion of carbon emissions depend on the relative growth rates of the desirable and undesirable outputs.[10]

It should be noted that the technical change index for any one particular country between two adjacent years is merely an index of the shift in the production frontier. A value of this factor greater than unity does not necessarily imply that the country under consideration did actually push the overall frontier outward. Therefore, in

[10]For a rigorous explanation of this relationship see appendix E in Fare et al. (2001).

Table 12.4 Hypothesis testing using basic "*t*" test (paired *t* test on average)

Null hypothesis	*p* value	Result
All countries		
ML = M	0.8563	Accepted
MLEFFCH = MEFFCH	0.0916	Rejected
MLTECH = MTECH	0.0252	Rejected
Non-Annex-I counties		
ML = M	0.9826	Accepted
MLEFFCH = MEFFCH	0.0068	Rejected
MLTECH = MTECH	0.0000	Rejected
Annex-I counties		
ML = M	0.7960	Accepted
MLEFFCH = MEFFCH	0.3054	Accepted
MLTECH = MTECH	0.1100	Accepted

ML Malmquist–Luenberger index, *MLEFFCH* Malmquist–Luenberger efficiency change, *MLTECH* Malmquist–Luenberger technical change, *M* Malmquist index, *MEFFCH* Malmquist efficiency change, *MTECH* Malmquist technical change

order to determine which countries were shifting the frontier or were "innovators," the following three conditions are required for a given country (see Färe et al., 2001, p. 400):

(a) $\text{MLTECH}_t^{t+1} > 1$,
(b) $\vec{D}_0^t(x^{t+1}, y^{t+1}, b^{t+1}; y^{t+1}, -b^{t+1}) < 0$,
(c) $\vec{D}_0^{t+1}(x^{t+1}, y^{t+1}, b^{t+1}; y^{t+1}, -b^{t+1}) = 0$.

The condition (a) indicates that the production possibility frontier shifts in the more good and fewer bad outputs direction. With a given input vector, in period $t+1$ it is possible to increase GDP and reduce CO_2 emissions relative to period t. This measures the shift in the relevant portions of the frontier between periods t and $t+1$ for a given country when the good and bad outputs are treated asymmetrically. The condition (b) indicates that production in period $t+1$ occurs outside the production possibilities frontier of period t (i.e., technical change has occurred). It implies that the technology of period t is incapable of producing the output vector of period $t+1$ with the input vector of period $t+1$. Hence the value of the directional distance function relative to the reference technology of period t is less than zero. The condition (c) specifies that the country must be on the production frontier in period $t+1$.

Table 12.5 lists the innovator countries. Out of 19 two-year periods, the Netherlands shifted the frontier 15 times when carbon emissions were freely disposable and three times when these emissions are accounted for in the measurement of productivity. Japan seemed to be innovative 15 times when carbon emissions were taken into consideration and 14 times when these emissions were strongly disposable. Japan shifted the frontier after 1976–1977. A newly industrialized country, Hong Kong was an innovator during 1983–1984 to 1991–1992 under both scenarios.

Table 12.5 Innovative countries

Years	Environmentally sensitive measure	Conventional measure
1973–1974	Netherlands, Switzerland	Netherlands, Switzerland
1974–1975	–	Netherlands
1975–1976	–	–
1976–1977	Iceland, Japan	Iceland, Japan, Netherlands
1977–1978	Iceland, Japan, Luxembourg	Iceland, Japan, Netherlands
1978–1979	Iceland, Japan, Switzerland	Iceland, Japan, Netherlands, Switzerland
1979–1980	Iceland, Japan, Switzerland	Iceland, Japan, Switzerland
1980–1981	Iceland, Japan, Netherlands, Switzerland	Iceland, Japan, Netherlands, Switzerland
1981–1982	Iceland, Japan, Netherlands, Switzerland	Japan
1982–1983	Japan	–
1983–1984	Hong Kong, Japan	Hong Kong, Japan, Netherlands
1984–1985	Japan, Switzerland	Japan, Netherlands, Switzerland
1985–1986	Japan, Switzerland	Hong Kong, Iceland, Japan, Netherlands
1986–1987	Iceland, Japan, Luxembourg, Switzerland	Hong Kong, Iceland, Japan, Netherlands
1987–1988	Hong Kong, Japan, Switzerland	Hong Kong, Japan, Luxembourg, Netherlands, Switzerland
1988–1989	Japan, Switzerland	Japan, Luxembourg, Netherlands, Switzerland
1989–1990	Hong Kong, Japan, Switzerland	Hong Kong, Japan, Luxembourg, Netherlands, Switzerland
1990–1991	Hong Kong, Japan	Hong Kong, Japan, Luxembourg, Netherlands
1991–1992	Hong Kong	Hong Kong, Luxembourg, Netherlands

Source: Authors' calculations

Out of 41 countries 6 were innovators and all were high-income countries. None of the developing countriees in any 2-year period was shifting the frontier under either scenario.

Letting ML_{it} represent the environmentally sensitive measure of productivity of country i in year t, the equation below specifies a possible form relation between the environmentally sensitive measure of productivity and its determinants.

$$\begin{aligned} ML_{it} = {} & \beta_{1i} + \beta_2 \text{GDPPC}_{it} + \beta_3 \vec{D}_{it-1} + \beta_4 \text{CAPLAB}_{it} \\ & + \beta_5 \text{ENGDP}_{it} + \beta_6 \text{OPEN} + \beta_7 \text{ANNEX}_{it} + \varepsilon_{it}, \end{aligned}$$

where i is country index; t is time index; ε is the disturbance term so that $\varepsilon \sim N(0, \sigma_\varepsilon)$; GDPPC is the GDP per capita; $\vec{D}_{it-1}$ is the value of technical inefficiency in the lagged period; CAPLAB is capital per labor; ENGDP is use of commercial energy per unit of GDP; OPEN, openness index defined as the ratio of total exports and imports to GDP; and ANNEX is the dummy variable for the group of countries – its value is equal to one if the country belongs to Annex-I countries and zero otherwise.

Table 12.3 (second column) provides the estimated parameters of the regressions for the ML index under alternative specifications. An LM test performed on the alternative specifications of the fixed effect model rejects the null hypothesis of a common intercept in favor of the one with country-specific intercept terms. Furthermore, the choice between fixed effect and random effect models can be made using the Hausman test. We reject the null hypothesis and find the random effect model as the appropriate specification.

We find that all coefficients, except that of ANNEX, are statistically significant. It is found that the environmentally sensitive measure of productivity is higher in those countries, which have higher GDP per capita. The positive relationship between the technical inefficiency and productivity index when the disposal of carbon emissions is costly favors the existence of the convergence hypothesis. We also find a negative relationship between the productivity index and capital labor ratio, which again accepts the convergence hypothesis in these countries. It is found that the energy intensity of production contributes negatively to the environmentally sensitive measure of productivity.

Similar to the standard measures of productivity, which are positively related to the openness of a country, the openness of a country contributes positively to the environmentally sensitive measure of productivity also. This finding is similar in spirit to the findings of Hettige et al. (1992), who point out that "...outward oriented, high-growth LDCs have slow growing or even declining toxic intensity of manufacturing...." An OECD report explains this phenomenon by the ability of dynamic and fast-growing developing countries with higher turnover rates of manufacturing capital stock to invest more in new processes based on cleaner techniques.

12.4 Conclusions

Bad outputs are ignored by traditional measures of productivity and hence have limited use with regards to policy evaluation. However, there are environmental regulations and resources that are diverted from traditional productive activities to pollution abatement. As a result, these traditional measures of productivity found that environmental regulations have an adverse effect on productivity. These findings ignore the key feature of environmental regulations, which is that diverting the resources in abatement activities leads to a reduction in environmentally bad outputs. The traditional measures of productivity ignore the reduction in bad outputs due to abatement activities since typically no prices are available for undesirable outputs such as CO_2 emissions, except for situations in which tradable permits are used to restrict the emissions.

This study presents an extended view of TFP growth measured through the ML index using the directional distance function. The index provides insight into the sources of productivity growth and estimates an adjusted rate of TFP growth while accounting for CO_2 emissions minimization activities. Through an asymmetrical treatment of good and bad outputs, the TFP index is decomposed into efficiency and technical changes. This index provides a common dialog on different perspectives about the climate change debate by expanding the basic economic concept of productivity to identify the combined role of technological innovation and adoption and green accounting.

The ML index is calculated using the nonparametric directional distance function for a group of 41 countries consisting of 21 Annex-I countries and 20 Non-Annex-I countries during the 1971–1992 period. On average for either group of countries, the value of the standard Malmquist is not different from the ML index. But for the components of TFP, technical, and technical efficiency changes, the null hypothesis of whether the different indexes are same when emissions are ignored and when they are accounted for cannot be accepted for either group of countries. Out of 41 countries only six – Iceland, Hong Kong, Japan, Luxembourg, the Netherlands, and Switzerland – were innovators. None of the developing countries was shifting the frontier under either scenario.

Subsequent regression analyses find that the environmentally sensitive measure of productivity is higher in those countries which have higher GDP per capita. The value of the ML index is negatively associated with technical efficiency and the capital-labor ratio, implying the presence of the convergence hypothesis. Moreover, it also finds that the energy intensity of production is negatively related to the environmentally sensitive measure of productivity. However, the conventional measure of productivity remains unaffected by the composition of output growth. The openness of a country increases its TFP, whether it is measured by the standard Malmquist index or the ML index.

Beyond measuring environmentally sensitive productivity growth, the present analysis demonstrates the richness of the technique that permits investigation of important research questions on the underlying processes that influence productivity growth. Notwithstanding the striking feature of the techniques used here, data limitations involved in estimation remain an important factor. It is therefore necessary to be cautious while applying these results to policy formulation.

Chapter 13
Macroeconomic Effects of Oil Price Shocks

13.1 Introduction

After reaching a 25-year low in February 1999, oil prices have sharply been rising over the next more than a half decade. Recently, the international price of oil has breached the US$150 mark. Given the macroeconomic developments that followed the oil shocks of the 1970s, the substantial rise in oil prices since 1999 has generated concerns about the prospects for growth and inflation and raised related questions about the appropriate way for monetary and energy policies to respond.

Much of the empirical literature is concerned with the developed countries, particularly the United States and Western Europe.[1] In an international context, an oil price shock may have differential impact on each of the countries, due to variables such as their sectoral composition, their relative position as oil importers or exporters, or their differential tax structure. We analyze the effects of oil price shocks in an oil-importing developing economy – India. To our knowledge this study is the first to assess the impact of the oil price shock on real economic activity in a developing country.

India is the seventh largest consumer of oil in the world. In 2003–2004, India spent about $US 20 billion to meet 70% of its needs. During the decade 1991–2001, oil consumption increased by 68% to reach 2.07 million barrels per day (mbpd) in India – behind only South Korea (78%) and China (109%). Oil imports accounted for 3.7% of gross domestic product (GDP) during 2003–2004. It is estimated that India's fuel consumption will rise to 3.2 million barrels per day by 2010. In the process, India will emerge as the fourth-largest consumer after the United States, China, and Japan.

The Reserve Bank of India (RBI) has estimated that every $1US dollar rise in the international price per barrel of crude oil adds $600 million to the country's oil

[1] To our knowledge there is only one published study, Cunado and Gracia (2005) which studies the impact on Asian countries namely Japan, Singapore, South Korea, Malaysia, Thailand, and Philippines.

S. Kumar and S. Managi, *The Economics of Sustainable Development*, Natural Resource Management and Policy 32,
DOI: 10.1007/978-0-387-98176-5_13,

import bill. Since the mid-1990s, the country's dependence on imported crude oil and petroleum products jumped from barely 30% of total demand in 1991 to over 70% at present, since domestic output has failed to keep pace with demand. These trends are not encouraging for the Indian economy, as higher oil prices are supposed to spur inflation besides straining the balance of payments.

The present study is intended to analyze the oil price-macroeconomy relationship by applying the vector autoregressive (VAR) approach for the Indian economy using quarterly data for the period 1975Q1–2004Q3. In order to account for asymmetry and nonlinearities between oil prices and macroeconomic variables, we use different transformations of oil price data, each one suggesting a different channel through which oil prices may affect real economic activities.

The study is organized as follows. Section 13.2 provides an idiosyncratic synthesis of what can be viewed as the key issues in the oil price-macroeconomy relationship debate and insights gained over the last three decades. In Section 13.3 we briefly present the main features of the oil price market in order to justify the proxy variables of oil price shocks we use in the study. Section 13.4 describes the methodology. Section 13.5 discusses the empirical results. Concluding remarks are offered in Section 13.6.

13.2 Oil Prices-Macroeconomy Relationship

A rich intellectual history since the mid-1970s investigates the economic response to oil price shocks.[2] The theoretical investigation has focused on the channels through which the impact of oil price shocks can be transmitted to economic activities. The empirical research has gone beyond establishing a relationship between oil price movements and aggregate economic activities – such as why rising oil prices appear to retard GDP growth by a greater amount than falling oil prices stimulate it. Some empirical papers have even tried to investigate channels such as monetary policies, through which oil price shocks are transmitted to the economy, and some have tried to examine the possibility of a weakening relationship between oil prices and macroeconomic variables. Section 13.2.1 provides the summary view of theoretical channels, while Section 13.2.2 summarizes the empirical findings from the literature.

13.2.1 Channels Through Which Oil Price Shocks May Affect the Macroeconomy

Oil prices may impact economic activity through various channels. *First*, there is the classic supply-side effect, according to which rising oil prices are indicative of

[2]For literature surveys, see Jones and Leiby (1996), Brown and Yucel (2002), Jones et al. (2004) etc.

the reduced availability of a basic input to production, leading to a reduction of potential output (see, among others, Barro, 1984; Brown and Yucel, 1999). Consequently, there is a rise in costs of production, and the growth of output and productivity are slowed. The decline in productivity growth negatively affects real wages and employment. *Second*, there is a wealth transfer from oil-importing countries to oil-exporting ones, leading to a fall of the purchasing power of firms and households in oil-importing countries that is greater than the increase in oil-exporting countries. Thus, an increase in oil prices deteriorates terms of trade for oil-importing countries (see Dohner, 1981). *Third*, according to the real balance effect (Pierce and Enzler, 1974; Mork, 1994), an increase in oil prices would lead to increase in money demand. Due to the failure of monetary authorities to meet growing money demand with increased supply, there is a rise of interest rates and a retardation in economic growth (for a detailed discussion on the impact of monetary policy, see Brown and Yücel, 2002). *Fourth*, a rise in oil prices generates inflation. The latter can be accompanied by indirect effects, called second-round effects, giving rise to price-wage loops. *Fifth*, an oil price increase may have a negative effect on consumption, investment, and stock prices. Consumption is affected through its positive relation with disposable income, and investment by increasing firms' costs. *Sixth*, if the oil price increase is long-lasting, it can give rise to a change in the production structure and have an impact on unemployment. Indeed, a rise in oil prices diminishes the rentability of sectors that are oil-intensive and can incite firms to adopt and construct new production methods that are less intensive in oil inputs. Given capital and labor inflexibility, oil price changes alter the relative cost of goods and services, which shifts demand and raises unemployment in those sectors most affected (Loungani, 1986). Moreover, it is becoming increasingly clear that in addition to oil price levels, oil price volatility creates uncertainty that reduces wealth and stifles investment.

13.2.2 Oil Prices-Macroeconomy Relationship: Empirical Evidences

Although oil price increases alone are not a necessary or sufficient condition for recessions, considerable research, from an empirical point of view, finds that oil price shocks have affected output and inflation. See, for example, Hamilton (1983, 1988, 1996, 2003), Hooker (1996, 1999, 2002), Huntington (1998), Kahn and Hampton (1990), Mork (1989, 1994), Mork et al. (1994), Jimenez-Rodriguez and Sanchez (2005), etc. Most of the empirical estimates are derived from the impulse response functions of oil price shocks in the GDP equation of a multivariate VAR model. The root of the empirical literature on the oil price-macroeconomy relationship is often traced to Hamilton's (1983) influential study, which establishes a negative association between oil prices and macroeconomic variables. Hamilton's results essentially confirmed those of Darby (1982) and Bruno and Sachs (1982).

The macroeconomic effects of oil price changes are not limited to the United States, but have been measured for other countries also, for example, by Papapetrou (2001) in Greece, Mork et al. (1994) in Canada, Japan, West Germany, France, and the United Kingdom, Jimenez-Rodriguez and Sanchez (2005) in the United States, Japan, Canada, France, Italy, Germany, Norway, and United Kingdom, and Cunado and Gracia (2005) in Asian countries. While the effects vary from country to country, they are overall less dissimilar than expected, even for the oil-exporting United Kingdom, where the GDP impact is similar to that of oil-importing countries.

Earlier studies, using data up to the 1980s, found symmetry in the oil-GDP relationship. However, by the mid-1980s, the estimated linear relationship between oil prices and macroeconomic activities began to lose significance. Although rising oil prices reduced economic growth during the 1980s, the sharp 1986 price declines were found to have a smaller positive effect on economic activity than predicted by linear models. Some authors, such as Mork (1989), Lee et al. (1995) and Hamilton (1996, 2003) proposed nonlinear transformation of oil prices to re-establish the inverse relationship between real economic activities and increase in oil prices.

Mork (1989) separated oil price increases from decreases and found that increases influence economic growth while decreases have only very small effects, if any. Mork, et al. (1994), studying the United States, Japan, Germany, France, United Kingdom, and Norway, found that oil price increases and decreases both produce negative economic impacts in the U.S. and to a lesser extent in Germany and Canada, with ambiguous results for the other countries. Other studies (e.g. Lee et al., 1995) confirmed asymmetry for the U.S. Ferderer (1996) suggests an asymmetric relationship, arguing that positive oil price changes have twice as much influence on industrial production than do negative ones. Cunado and Gracia (2005) confirm this asymmetric relationship for Asian countries also.

Theoretical supply side arguments cannot explain asymmetry of oil price shock impacts. Accordingly, the empirical literature emphasizes other channels, such as monetary policy, adjustment costs and asymmetry in petroleum product prices, as possible explanations for the asymmetry.

Monetary policy could be a possible explanation for the asymmetric response of GDP to oil price shocks. If nominal wages are sticky downward and monetary authority fails to keep nominal GDP constant, then an increase in oil prices will aggravate GDP losses through unexpected inflation. But a fall in oil prices is not stimulating through unexpected disinflation since nominal wages can adjust upward freely. Tatom (1993) and Bernanke et al. (1997) find that monetary policy followed after changes in oil prices caused asymmetry. But Ferderer (1996) and Balke et al. (2002) show that monetary policy cannot account for the asymmetry in impacts.

Lilien (1982) and Hamilton (1988) argue that a change in oil price alters the equilibrium allocation across various sectors. According to this argument, an increase (decrease) in oil price would require a contraction (expansion) of oil-intensive sectors and an expansion (contraction) of oil-efficient sectors. These realignments in production require adjustments which cannot be achieved in the short-run; this is known as *dispersion hypothesis*, and there are negative effects of change in oil prices (either positive or negative) on real economic activities.

Balke et al. (1998) find an asymmetric relationship between crude oil prices and petroleum product prices. Huntington (1998) takes this asymmetric relationship between crude and product prices to the macroeconomic level. He finds that both other energy (other than oil) prices and GDP are symmetrically related to changes in petroleum product prices, while product prices respond asymmetrically to changes in crude price. The result is an asymmetric relationship between crude prices and real economic activities. But according to Jones et al. (2004), "In the crude-product relationship, the asymmetry is in the speed of response, while in the oil price–GDP relationship, it is in the magnitude of response" (p. 10).

13.3 Oil Price Data

The effective oil prices that a country faces have been influenced by many characteristics, such as price controls, taxes on petroleum products, exchange rate fluctuations, and variations in the domestic price index. These characteristics raise great difficulty in measuring the appropriate oil price variable. Most of the empirical literature use the US$ world real price of oil as a common indicator of the world market disturbance (see, for example, Jimenez-Rodriguez and Sanchez, 2005) to analyze the effects of oil price shocks on macroeconomic activities. Some studies use the world oil price converted into the currency of the country, for which analysis is made by means of exchange rate (see, e.g., Mork et al., 1994, for OECD countries; Cunado and Gracia, 2005, for Asian countries). The differential in these two prices reflects whether the oil price shock is due to evolution of world oil prices or due to other factors, such as exchange rate fluctuations or national price index variations. In the present study we use world oil prices converted into Indian Rupees (INR) by the market rate of exchange deflated by the domestic wholesale price index (WPI) to analyze to effect of oil shocks on Indian Economy.[3]

Figure 13.1 shows the evolution of both the real oil price expressed in US$ and in INR over the period 1970Q1–2004Q4. In both the series we observe the effects of the five main negative oil shocks (1973–1974, 1978–1979, 1990, 1999–2000, 2003–2004) and the fall in oil price in 1986 and 1998–1999. However, there is a different evolution of oil prices when they are expressed in US$ and INR.

Until 1986, oil prices were unidirectional in change, but since then they have been characterized by large declines and high volatility (Fig. 13.2). This differential behavior of oil price movements and apparent asymmetric response of the macroeonomy to oil price shocks in the U.S. and Western European economies have led researchers to explore different oil price–GDP specifications in order to re-establish the relationship between these variables (see, for example, Mork, 1989; Hamilton,

[3]In measuring the value of oil price excluding taxes we follow the existing literature since there is no database for tax-including end-use prices of oil products over the sample period.

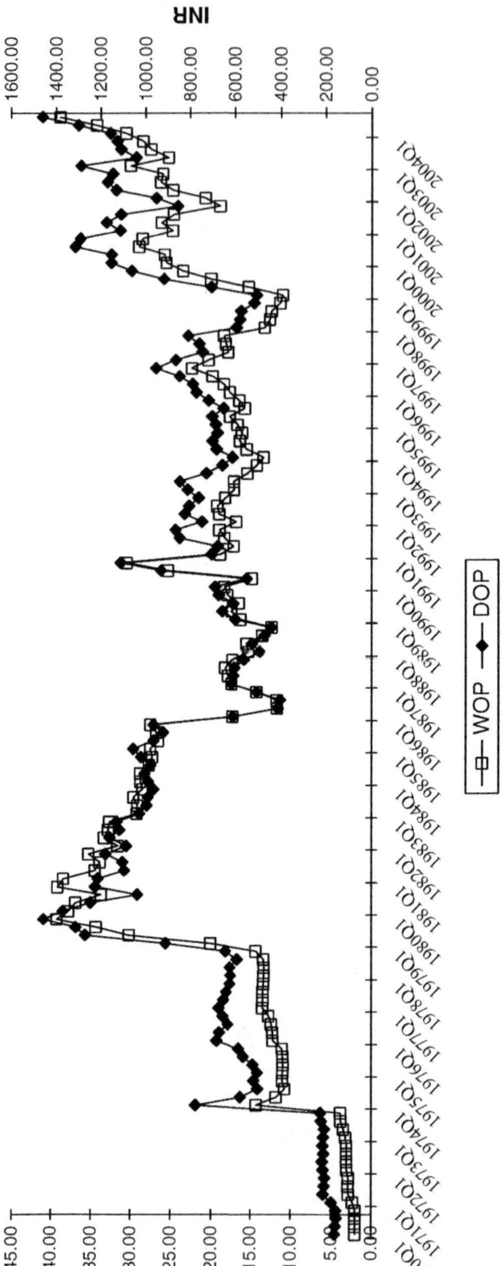

Fig. 13.1 Real oil prices (*Source*: Authors' calculations based on International Financial Statistics data)

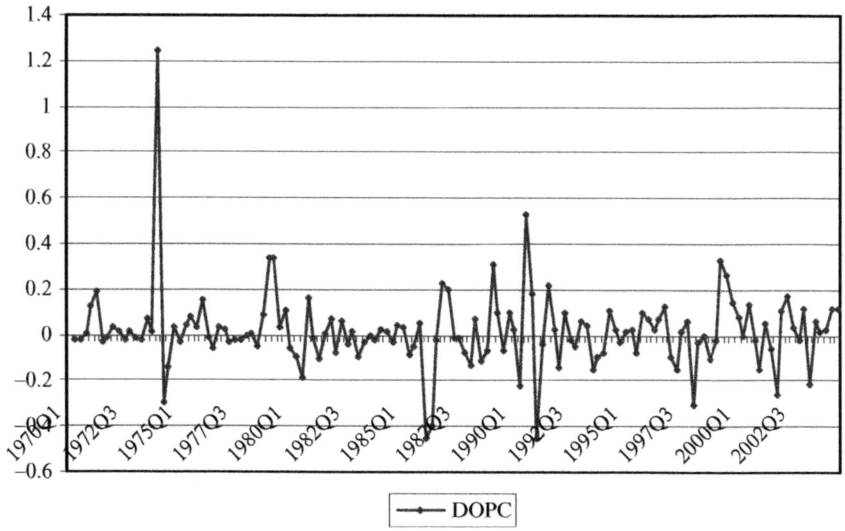

Fig. 13.2 Oil Price Changes in INR
Source: Authors' calculations

1996, 2003; Lee et al., 1995). Following this literature, we define the next four variables for oil price changes expressed both in \$US and INR:

$\triangle \text{oil}_t$: quarterly changes of real oil prices, that is, the conventional first difference transformation of oil price variables (in logs):

$$\Delta \text{oil}_t = \ln \text{oil}_t - \ln \text{oil}_{t-1},$$

where oil_t is the real oil price in period t in \$US or in INR, as defined above.

A significant relationship between this variable and economic activity would lead to a linear oil–output relationship. An asymmetric specification distinguishes between the positive rate of change in oil price oil_t^+ and its negative rate of changeoil_t^-, which are defined as follows:

$$\Delta \text{oil}_t^+ : \text{ real oil price increases, } \Delta \text{oil}_t^+ = \max(0, \Delta \text{oil}_t), \text{ and}$$

$$\Delta \text{oil}_t^- : \text{ real oil price decrease, } \Delta \text{oil}_t^- = \min(0, \Delta \text{oil}_t).$$

In this case, we treat in a different way oil price increases and decreases, that is, we separate oil price changes into negative and positive changes in a belief that oil price increases may have a significant effect on macroeconomic variables even though this might not occur for oil price decreases. The asymmetric model can be rationalized in terms of the dispersion hypothesis described in Section 13.2.

Hamilton (1996) proposed a different nonlinear specification by using the explanatory variable that he calls *net oil price increase* (NOPI). NOPI (expressed in real terms) is defined as the quarterly percentage change in real oil price levels from the past four (and 12) quarters' high – if that is positive – and zero otherwise (NOPI4 and NOPI12). Hamilton (1996) argues that if one wants a measure of how unsettling an increase in the price of oil is likely to be for the spending decisions of consumers and firms, it seems more appropriate to compare the current price of oil with where it has been over the previous years rather than during the previous quarter alone. Hamilton thus proposes to use the amount by which the log oil price in quarter t exceeds its maximum value over the previous periods; if oil prices are lower than they have been at some point during the most recent years, no oil shock is said to have occurred. That is,

$$\mathrm{NOPI4}_t = \max\left(0,\ (\ln(\mathrm{oil}_t) - \ln\left(\max(\mathrm{oil}_{t-1},\ \mathrm{oil}_{t-2},\ \mathrm{oil}_{t-3},\ \mathrm{oil}_{t-4})\right))\right),$$

$$\mathrm{NOPI12}_t = \max\left(0,\ (\ln(\mathrm{oil}_t) - \ln\left(\max(\mathrm{oil}_{t-1}, \ldots,\ \mathrm{oil}_{t-12})\right))\right).$$

Lee et al. (1995) proposed *scaled oil price increases* (SOPI) (where oil price is expressed in real terms). They focus on volatility, arguing that an oil shock is likely to have greater impact in an environment where oil prices have been stable than in an environment where oil price movements have been frequent and erratic because price changes in a volatile environment are likely to be soon reversed. In order to put this idea into practice, Lee et al. (1995) proposed the following AR(4)-GARCH (1,1), representation of oil prices:

$$\Delta\mathrm{oil}_t = \alpha + \sum_{j=1}^{k} \beta_j \Delta\mathrm{oil}_{t-j} + \varepsilon_t, \varepsilon_t | I_t \rightarrow N(0, h_t)$$

$$h_t = \gamma_0 + \gamma_1 \varepsilon_{t-1}^2 + \gamma_2 h_{t-1}$$

$$\mathrm{SOPI}_t = \max\left(0, \frac{\widehat{\varepsilon_t}}{\sqrt{\widehat{h_t}}}\right)$$

$$\mathrm{SOPD}_t = \min\left(0, \frac{\widehat{\varepsilon_t}}{\sqrt{\widehat{h_t}}}\right),$$

where SOPI stands for scaled oil price increases, while SOPD stands for scaled oil price decreases. A significant relationship between this variable and economic activity implies that a "certain" oil price increase will cause a decrease in economic activity, while a price increase in a period of high volatility is less likely to cause it.

The oil price shock proxies (e.g., oil price increases, positive oil price increases, NOPI4 and SOPI) defined in INR are plotted in Figures 13.3, 13.4, and 13.5.[4] As we can see in the figures, the oil price shock proxies detect quite well all the main oil shocks in the period 1970Q1–2004Q4. However, we can also detect some differences

[4] Although all these variables are also constructed in US$, we do not plot them but are available by request from the author.

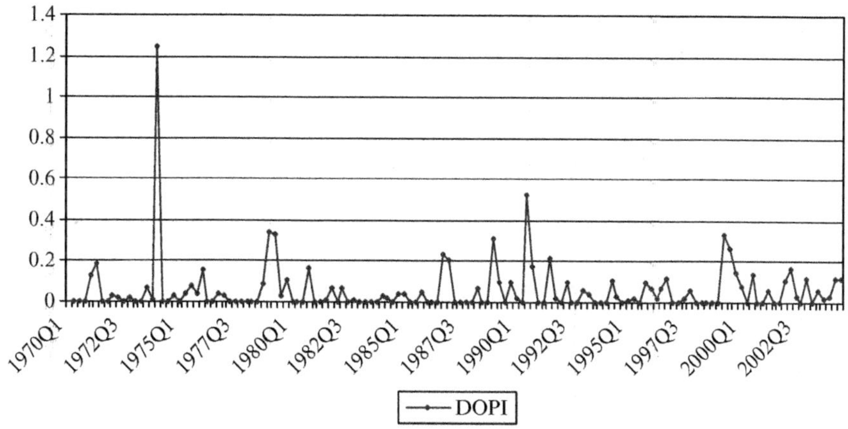

Fig. 13.3 Oil Price Increases in INR (Real Oil Price Increase)
Source: Authors' calculations

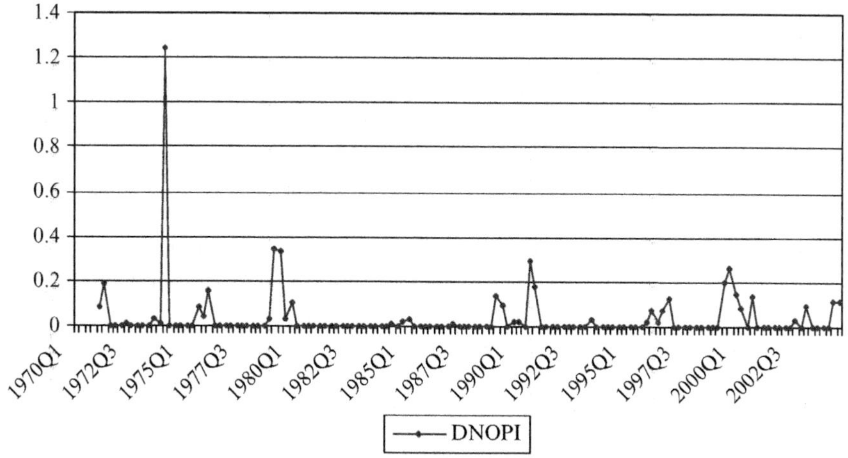

Fig. 13.4 Net Oil Price Increases in INR (NOPI4)
Source: Authors' calculations

between each of the variables. For example, we can observe that the variable Δoil_t^+ takes a much higher value after the increase in oil prices in 1990Q3 than the NOPI variable, a difference which is due to the decrease in oil prices occurred in 1990Q2

13.4 Measurement of Impact of Oil Prices on Macroeconomy

We consider the following vector autoregression model of order p (or simply, VAR(p)):

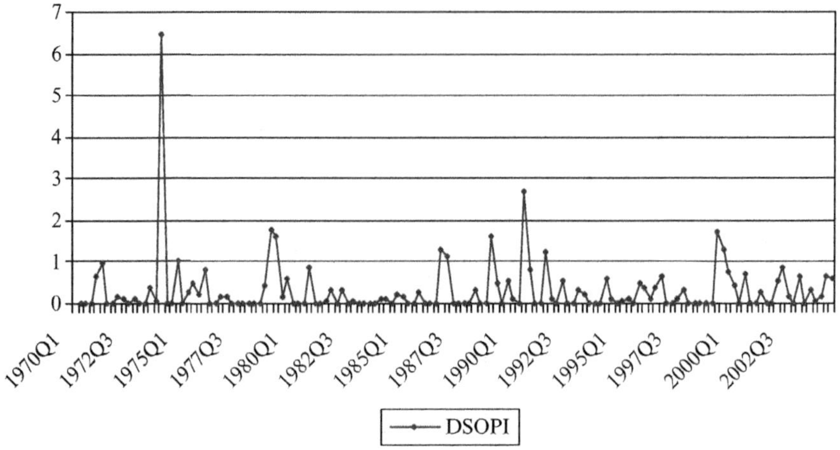

Fig. 13.5 Scaled Oil Price Increase in INR (SOPI)
Source: Authors' calculations

$$y_t = c + \sum_{t=1}^{p} \phi_i y_{t-1} + \varepsilon_t, \tag{13.1}$$

where y_t is a $(n \times 1)$ vector of endogenous variables, $c - (c_1, \ldots c_n)'$ is the $(n \times 1)$ intercept vector of the VAR, $\phi_i$is the i th $(n \times n)$ matrix of autoregressive coefficients for $i = 1, 2, \ldots, p$, and $\varepsilon_t = (\varepsilon_{1t}, \ldots, \varepsilon_{nt})'$ is the $(n \times 1)$ generalization of a white noise process.

In this chapter we use a quarterly five-variable VAR for India. The variables considered for the model are the following: index of industrial production (IIP),[5] real effective exchange rate (REER),[6] real oil price, inflation,[7] and short-term interest rate.[8] Some variables (IIP, REER, and real oil price) are expressed in logs, while the remaining ones are simply defined in levels. We include real oil prices and industrial growth[9] since our main objective is to analyze the effects of the former variable on the latter. We use only one measure of economic activity, namely, industrial growth, while the remaining variables are included to capture some of the most important transmission channels through which oil prices may affect economic activity indirectly, in part by inducing changes in economic policies. Those channels include effects of oil prices on inflation and exchange

[5]The aggregate economic activity is proxied by IIP since the quarterly GDP series in India is available since 1996–1997 only.

[6]REER is defined such that a decrease means a real depreciation of the INR. A depreciation of the REER is expected to increase India's external competitiveness.

[7]Inflation is defined as the change in consumer price index (CPI), i.e. $\triangle CPI = CPI_t - CPI_{t-1}$.

[8]Money market interest rate is considered as the short-term interest rate.

[9]Industrial growth is defined as the change in logarithmic value of IIP, i.e., Industrial Growth = $\ln(IIP_t) - \ln(IIP_{t-1})$.

rates, which then induce changes in real economic activity. Our VAR model also incorporates a monetary sector (by means of short-term interest rate rather than money supply indicators), which can react to inflationary pressures. As is customary in studies focusing on the impact of oil prices, we do not use import prices as a whole but only oil prices, while also allowing for the exchange rate to capture part of the pass-through from import prices (in foreign currency) into domestic prices.

Before studying the effects of oil shocks on economic activity, we proceed to investigate the stochastic properties of the series considered in the model by analyzing their order of integration on the basis of a series of unit root tests. Specifically, we perform the Augmented Dickey–Fuller (ADF) and Phillips–Perron (PP) tests. Results of these formal tests are summarized in Table 13.1, indicating that the first differences of all five variables are stationary. We therefore follow the related literature in defining the vector y_t in (13.1) to be given by the first log-differences of the first three aforementioned variables (IIP, REER, and real oil price), along with the first differences of the remaining ones (inflation, and short-term interest rate).

In order to assess the impact of shocks on endogenous variables, we examine the orthogonalized impulse-response functions, using Cholesky decomposition, as well as the accumulated responses. To do so, we should choose an ordering for the variables in the system, since this method of orthogonalization involves the assignment of contemporaneous correlation only to specific series. Thus, the first variable in the ordering is not contemporaneously affected by shocks to the remaining variables, but shocks to the first variable do affect the other variables in the system; the second variable affects contemporaneously the other variables (with the exception of the first one), but it is not contemporaneously affected by them; and so on. In our case, we have assumed the following ordering: industrial growth, real oil price, inflation, short-term interest rate, and REER. This ordering assumes, as in much of the related literature, that industrial growth does not react contemporaneously on impact to the rest of the variables. The oil price variable is also ranked as a largely exogenous variable, which has an immediate impact on the rate of inflation. The latter is then allowed to feed into changes in the short-term interest rate, while the exchange rate, close the system.[10]

The VAR model in (13.1) is estimated for both a linear specification[11] and the three main nonlinear specifications as defined above. The latter are the following (1) *asymmetric specification*, in which increases and decreases in oil prices are considered as separate variables; (2) *net specifications*, where the relevant oil price variable is defined to be the net amount by which these prices in quarter t exceed the

[10] As a robustness check, other possible ordering are also considered, including the case of an alternative ordering that only differs from the baseline model in that one allows for the contemporaneous influence of real oil price innovation on industrial growth. It was verified that the impulse responses do not change considerably with the baseline specification.

[11] Quarterly changes in real oil prices are used in the linear approach to VAR estimation, and are transformed, as discussed in Sect. 13.3, for their use in non-linear models.

Table 13.1 Unit root test

ADF test						
	Level			First difference		
	(i)	(ii)	(iii)	(i)	(ii)	(iii)
Log (IIP)	4.22	−0.05	−2.69	−1.67***	−4.05*	−4.03**
Log (oil price) INR	0.68	−2.64***	−2.69	−5.86*	−5.92*	−5.90*
Log (oil price) US$	−0.69	−0.71	−2.39	−5.64*	−5.67*	−5.68*
Log (REER)	−2.00**	−0.17	−2.01	−3.43*	−4.03*	−4.04*
CPI	2.94	1.98	−2.42	−1.82***	−3.25**	−4.74*
MMR	−1.35	−3.41**	−3.46**	−5.18*	−5.17*	−5.15*
Phillips-Perron (PP) test						
Log (IIP)	4.63	−0.27	−7.68*	−14.92*	−20.17*	−20.05*
Log (oil price) INR	0.56	−2.57	−2.66	−9.28*	−9.27*	−9.23*
Log (oil price) US$	−0.66	−0.73	−2.28	−10.81*	−10.79*	−10.78*
Log (REER)	−2.84*	−0.13	−1.58	−8.99*	−9.53*	−9.50*
CPI	8.60	3.50	−2.43	−6.22*	−8.53*	−9.61*
MMR	−1.59	−5.40*	−5.47*	−18.76*	−18.71*	−18.64*

Note: (i) with no regressors, (ii) with an intercept, (iii) with an interecept and a linear time trend. *, ** and *** indicate that the test statistics is statistically significant at 1%, 5%, and 10% level respectively

Source: Authors' calculations

maximum value reached in the previous four and 12 quarters; and (3) *scaled specification*, which takes the volatility of oil prices into account.

The sample period runs from 1975Q1 to 2004Q3, for a total of $T = 119$ available quarterly observations (see Appendix for details on data). To select the suitable lag length, different tests are considered, the modified Likelihood Ratio test (Sims, 1980), as well as the Akaike, Schwarz, and Hannan-Quinn tests. Wherever there is conflict among different tests, the optimal lag length is chosen using the Likelihood Ratio test.

13.5 Empirical Results

This section analyzes the empirical results for all the models described in the Section 13.3. In Section 13.5.1 we test the significance of different oil price variables and analyze the Granger-causality in a multivariate context. In the next subsection we estimate the model. In Section 13.5.2, we compare the performance of different specifications under consideration. Then the effects of oil price shocks on macroeconomic variables are examined.[12] The results on impulse-response functions and

[12] Although the analysis of impulse response functions and variance decomposition is also conducted by using the oil price variable in US$, we do not present them as the results are not qualitatively different from using oil price variable in Indian rupees but are available by request from the author.

accumulated responses are first presented; the results of variance decomposition are next discussed. The cases of both impulse response and variance decomposition analysis, for all linear and nonlinear specifications, are examined while focusing on the preferred specification.

13.5.1 Testing for Significance and Granger-Causality

We carry out different tests to investigate the relationship between oil prices and other variables of the model, focusing on the significance of the impact of oil prices on real activities approximated by industrial growth.

First, the Wald test statistics is performed to test the null hypothesis that all oil price coefficients are jointly zero in the industrial growth equation of the VAR model. Table 13.2 displays the χ^2 and p values of the Wald test statistics. The results indicate that we cannot reject the null hypothesis when the oil price variable is decreasing, but the null hypothesis is rejected when the oil prices are increasing in most of the variables. This implies that oil price increases appear to have a significant direct impact on real activities, but the decreases in oil prices do not appear to influence the real activities directly. These results support the asymmetric impact hypothesis of oil price changes on real economic activities.[13]

Second, we test the significance of the oil price variable for the VAR system as a whole. We hypothesize that all of the oil price coefficients are jointly zero in all equations of the system but its own equation (see Table 13.3). This Likelihood Ratio (LR) test provides the information that the oil price variable not only affects real activities directly (as assessed through the Wald test), but also through third variables in the system. It is found that the oil price variable in the linear model, the positive changes in the asymmetric model, the NOPI measured over the previous four quarters (when the oil prices are measured in US dollars), the NOPI measured over the previous 12 quarters, scaled oil price, and SOPI are significant for the system. The negative changes in the oil price variable are not statistically significant in any of the models. The price decrease variable is subsequently eliminated from those models in which it is not significant.

Finally, we perform the so-called test of block exogeneity. A block exogeneity test is useful for detecting whether to incorporate a variable into a VAR (Table 13.4). We test whether an oil price variable Granger-causes the remaining variables of the system. We find that oil price change or increase variable generally Granger-cause the remaining variable of the system at the 1% significance level.

---

[13]The null hypothesis that the sum of positive and negative real oil price variable coefficients is equal in VAR framework has been tested, obtaining the rejection of null hypothesis in all cases.

Table 13.2 Wald test

Model	Oil price in Indian rupees	Oil price in US dollars
$\triangle oil_t$	4.2076[0.040]**	4.8879[0.027]**
Δoil_t^+	5.3402[0.021]**	5.4992[0.019]**
Δoil_t^-	0.14663[0.702]	0.62921[0.428]
NOPI4	5.1911[0.023]**	8.3977[0.004]*
NOPD4	1.9987[0.157]	1.1881[0.276]
NOPI12	12.4496[0.000]*	10.7450[0.001]*
NOPD12	2.1667[0.141]	1.3991[0.237]
SOPC	4.2694[0.039]**	4.9789[0.026]**
SOPI	5.2349[0.022]**	5.4760[0.019]**
SOPD	0.20145[0.654]	0.75492[0.385]

$\triangle oil_t$ real oil price change, Δoil_t^+ increase in real oil prices, Δoil_t^- decrease in real oil prices, *NOPI4* increase in real oil prices over previous four quarters, *NOPD4* decrease in real oil prices over previous four quarters, *NOPI12* increase in real oil prices over previous 12 quarters, *NOPD12* decrease in real oil prices over previous quarters, *SOPC* scaled real oil price change, *SOPI* scaled real oil price increase, and *SOPD* scaled oil price decrease. Values in parentheses are *p* values of the asymptotic distribution Chi-squared for the different models considered. H_o: the oil price coefficients are jointly equal to zero in the IIP growth equation of the VAR model. *, **, *** asterisks mean a *p* value less than 1%, 5%, and 10% respectively
Source: Authors' calculations

Table 13.3 Likelihood ratio test

Model	Oil price in Indian rupees	Oil price in US dollars
$\triangle oil_t$	9.7469[0.045]**	12.2309[0.016]**
Δoil_t^+	10.0428[0.040]**	13.0313[0.011]**
Δoil_t^-	4.9952[0.288]	7.4244[0.115]
NOPI4	7.3431[0.119]	11.4006[0.022]**
NOPD4	7.0485[0.133]	7.4828[0.112]
NOPI12	15.6486[0.004]*	13.7186[0.008]*
NOPD12	5.9627[0.202]	5.9212[0.205]
SOPC	9.6345[0.047]**	12.0660[0.017]**
SOPI	9.8780[0.043]**	12.7647[0.012]**
SOPD	5.6185[0.230]	8.1028[0.088]***

$\triangle oil_t$ real oil price change, Δoil_t^+ increase in real oil prices, Δoil_t^- decrease in real oil prices, *NOPI4* increase in real oil prices over previous four quarters, *NOPD4* decrease in real oil prices over previous four quarters, *NOPI12* increase in real oil prices over previous 12 quarters, *NOPD12* decrease in real oil prices over previous quarters, *SOPC* scaled real oil price change, *SOPI* scaled real oil price increase; and *SOPD* scaled oil price decrease. H_o All oil price coefficients are jointly zero in all equations of the system but its own equation. *, **, *** asterisks mean a *p* value less than 1%, 5%, and 10% respectively
Source: Authors' calculations

13.5.2 Macroeconomic Impacts of Oil Price Shocks

This section assesses the impact of oil shocks on real macroeconomic activities using the different linear and nonlinear models described in Section 133. To facilitate the description of the results, we first evaluate the relative performance

Table 13.4 LR test of block Granger noncausality in the VAR

Model	Oil price in Indian rupees	Oil price in US dollars
Δoil_t	47.1234[0.000]*	47.3433[0.000]*
Δoil_t^+	39.0555[0.001]*	39.1276[0.001]*
Δoil_t^-	45.7280[0.000]*	47.0713[0.000]*
NOPI4	20.8415[0.185]	21.9508[0.145]
NOPD4	33.5816[0.006]*	35.6992[0.003]*
NOPI12	33.7852[0.006]*	23.7708[0.095]***
NOPD12	22.4802[0.128]	36.9883[0.002]*
SOPC	47.6201[0.000]*	47.3668[0.000]*
SOPI	38.9706[0.001]*	38.2700[0.001]*
SOPD	46.5717[0.000]*	48.6510[0.000]*

Δoil_t: real oil price change, Δoil_t^+ increase in real oil prices, Δoil_t^- decrease in real oil prices, *NOPI4* increase in real oil prices over previous four quarters, *NOPD4* decrease in real oil prices over previous four quarters, *NOPI12* increase in real oil prices over previous 12 quarters, *NOPD12* decrease in real oil prices over previous quarters, *SOPC* scaled real oil price change, *SOPI* scaled real oil price increase, and *SOPD* scaled oil price decrease. H_o oil price variable Granger-causes the remaining variables of the system. *, **, *** asterisks mean a p value less than 1%, 5%, and 10% respectively
Source: Authors' calculations

Table 13.5 Relative performance of the models

	Oil price in Indian rupees		Oil price in US dollars	
Model	AIC	SBC	AIC	SBC
oil_t	120.5876	−23.5214	120.9693	−23.1397
Δoil_t^+	163.0239	18.9149	173.9208	29.8119
Δoil_t^-	173.9277	29.8187	168.2319	24.1229
NOPI4	195.9130	51.8041	220.5335	76.4246
NOPD4	199.7075	55.5986	175.3543	31.2454
NOPI12	221.0581	76.9491	248.7439	104.6350
NOPD12	231.8209	87.7120	193.2329	49.1239
SOPC	−68.1654	−212.2744	−69.2026	−213.3115
SOPI	−26.4226	−170.5315	−16.5021	−160.6110
SOPD	−14.0370	−158.1459	−20.7109	−164.8199

Δoil_t: real oil price change, Δoil_t^+ increase in real oil prices, Δoil_t^- decrease in real oil prices, *NOPI4* increase in real oil prices over previous four quarters, NOPD4 decrease in real oil prices over previous four quarters, *NOPI12* increase in real oil prices over previous 12 quarters, *NOPD12* decrease in real oil prices over previous quarters, *SOPC* scaled real oil price change, *SOPI* scaled real oil price increase, and *SOPD* scaled oil price decrease, *AIC* Akaike's information Criterion, *SBC* Schwarz Bayesian Information Criterion
Source: Authors' calculations

of the different linear and nonlinear specifications for the whole VAR system of equations. The goodness of fit of the different model specifications is assessed. We look at the *Akaike Information Criterion* (AIC) and *Schwarz Bayesian Information Criterion* (SBC), since the models are non-nested. Table 13.5 reports the AIC and SBC obtained from each econometric specification. On the basis of these two criteria, we find that the *scaled specification*, i.e., SOPI, performs somewhat better than the other approaches used in the present study.

13.5.2.1 Impulse Response Functions and Accumulated Responses

We examine the impact of oil price shocks on macroeconomic activities in terms of both orthogonalized impulse response functions and accumulated responses for the linear and nonlinear specifications of the model. Impulse response function is a dynamic function comprising the partial derivatives of industrial growth at a given time with respect to the oil price shock at each of a number of periods in the past, possibly beginning with the contemporaneous period. The sum of the impulse response coefficients for a shock at a specific time yields the equivalent of cumulative oil price-industrial growth elasticity for a single period shock.

Figures 13.6–13.11 present the orthogonalized impulse response functions of industrial growth to one standard deviation oil price shock for the specifications used in the study. Table 13.6 reports the accumulated responses of macroeconomic variables to an oil price shock normalized to correspond to a 1% increase in all linear and nonlinear specifications. In order to understand the mechanism behind the impulse and accumulated responses of industrial growth, impulse and accumulated

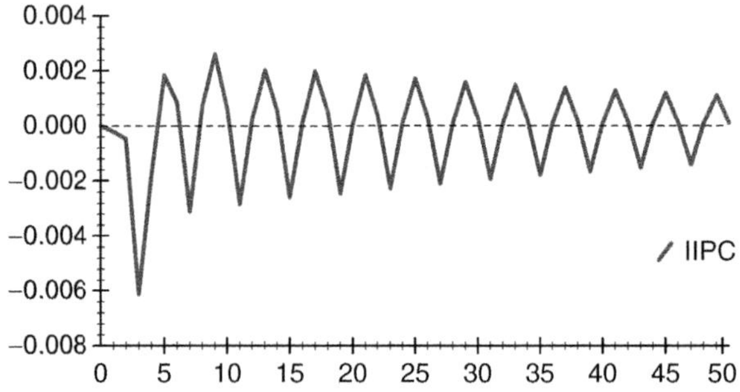

Fig. 13.6 Orthogonalized impulse-response function of industrial growth to a one-standard-deviation oil price innovation (real oil price change)
Source: Authors' calculations

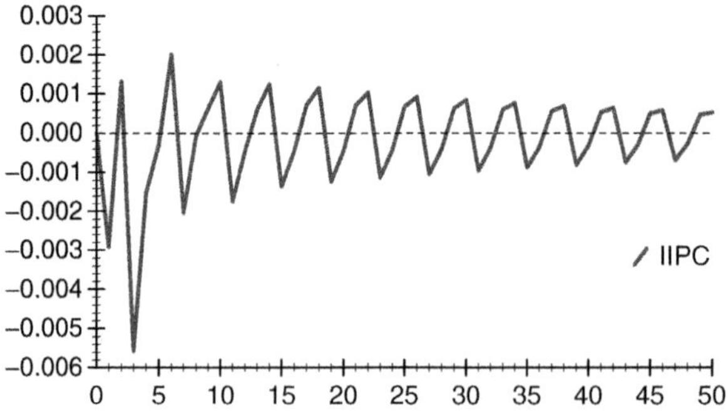

Fig. 13.7 Orthogonalized impulse-response function of industrial growth to a positive one-standard-deviation oil price innovation (real oil price increase)
Source: Authors' calculations

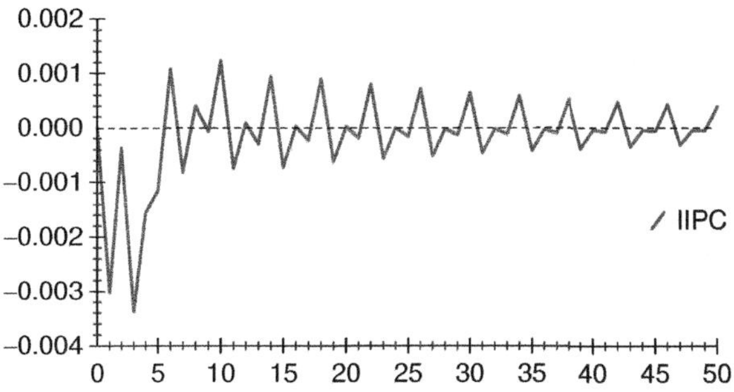

Fig. 13.8 Orthogonalized impulse-response function of industrial growth to a positive one-standard-deviation oil price innovation (net oil price increase, NOPI4)
Source: Authors' calculations

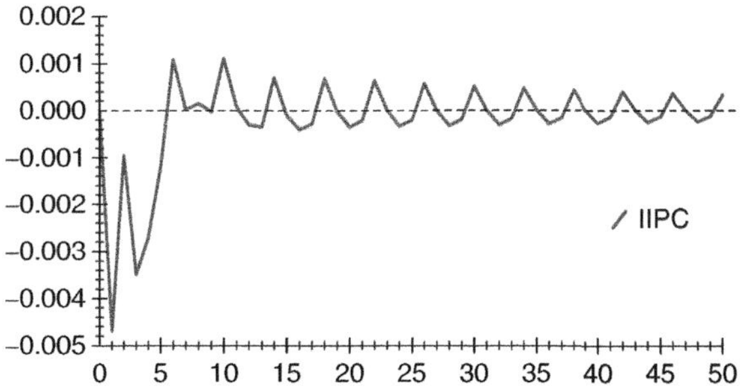

Fig. 13.9 Orthogonalized impulse-response function of industrial growth to a positive one-standard-deviation oil price innovation (net oil price increase, NOPI12)
Source: Authors' calculations

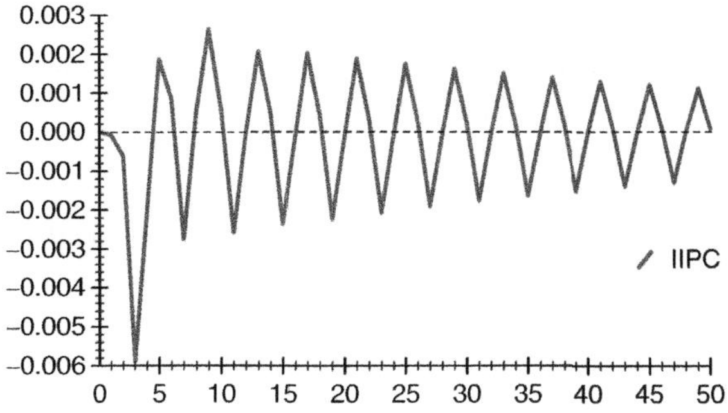

Fig. 13.10 Orthogonalized impulse-response function of industrial growth to a one-standard-deviation oil price innovation (scaled oil price change, SOPC)
Source: Authors' calculations

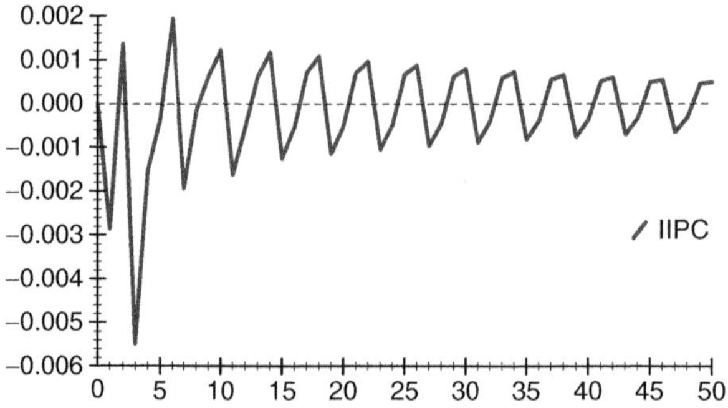

Fig. 13.11 Orthogonalized impulse-response function of industrial growth to a positive one-standard-deviation oil price innovation (scaled oil price increase, SOPI)
Source: Authors' calculations

responses of other variables have been analyzed. It is found that one of the key channels playing a role in the effect of oil prices on real activity is related to the REER.

It is found that the results of the linear specification and that of real oil price increase, NOPI and SOPI are qualitatively similar; however, the results of all the specifications are described at the same time, stressing the results obtained for the preferred model. While the linear model supposes that the impacts of an oil price increase and those of a decline are totally symmetric, nonlinear specifications allow for differential effects of oil shocks of the same magnitude and opposite sign. It was reported in Section 13.5.1 that the negative movements of oil prices in nonlinear specifications are not statistically significant; therefore, we describe the effects of positive oil price shocks for all specifications (Figs. 13.6–13.11).

In the case of positive movements in oil prices, it is observed that the real impact of oil prices is negative in the short-term. The largest negative short-term influence takes place within the year of the shock, being reached in the third quarter after the shock in most of the specifications. Then the impact of the shock becomes smaller, dying out almost completely after 3 years.

Table 13.6 indicates that the accumulated responses of industrial growth to a positive oil price shock in the linear and nonlinear specifications are qualitatively similar. An oil price shock has a negative accumulated effect on industrial growth. It is seen that the accumulated loss to industrial growth for a 100% oil price shock is about 1%. One important mechanism that helps explain this small amount of impact is the depreciation of the REER, which partially offsets the negative impact of oil price increases.[14]

[14] According to Huntington (1998) the crude oil price shocks are essentially energy price shocks that are transmitted to the economy through changes in refined petroleum products. In India, the prices of petroleum products are administered (although theoretical dismantled in 2002 but not in practice) and do not change according to changes in the prices of crude oil.

Table 13.6 Accumulated Impulse response functions

Industrial growth										
	Δoil_t	Δoil_t^+	Δoil_t^-	NOPI4	NOPD4	NOPI12	NOPD12	SOPC	SOPI	SOPD
4Q	−0.0087	−0.0087	−0.0036	−0.0083	−0.0075	−0.0118	−0.0078	−0.0087	−0.0086	−0.0038
6Q	−0.0062	−0.0072	−0.0012	−0.0086	−0.0042	−0.0122	−0.0055	−0.0060	−0.0072	−0.0014
8Q	−0.0084	−0.0091	−0.0039	−0.0088	−0.0085	−0.0117	−0.0103	−0.0081	−0.0090	−0.0040
10Q	−0.0054	−0.0074	−0.0003	−0.0079	−0.0039	−0.0109	−0.0055	−0.0051	−0.0074	−0.0002
12Q	−0.0079	−0.0094	−0.0028	−0.0083	−0.0080	−0.0109	−0.0097	−0.0076	−0.0094	−0.0028
Consumer price index (CPI)										
	Δoil_t	Δoil_t^+	Δoil_t^-	NOPI4	NOPD4	NOPI12	NOPD12	SOPC	SOPI	SOPD
4Q	−0.3088	−0.0425	−0.4307	−0.2026	−0.1953	−0.0097	−0.2052	−0.3034	−0.0446	−0.4199
6Q	−0.2126	0.1192	−0.4451	−0.1131	−0.1054	0.1519	−0.1336	−0.2101	0.1170	−0.4399
8Q	−0.2221	0.0774	−0.4114	−0.1115	−0.0646	0.1124	−0.1083	−0.2252	0.0725	−0.4130
10Q	−0.2793	0.1019	−0.5059	−0.1000	−0.1556	0.1450	−0.1682	−0.2819	0.0976	−0.5042
12Q	−0.2616	0.1070	−0.4885	−0.1069	−0.1455	0.1279	−0.1788	−0.2647	0.1023	−0.4893
Money market interest rate (MMR)										
	Δoil_t	Δoil_t^+	Δoil_t^-	NOPI4	NOPD4	NOPI12	NOPD12	SOPC	SOPI	SOPD
4Q	0.7004	1.6860	−0.5929	1.0384	0.3673	1.0085	0.4671	0.5663	1.6169	−0.7130
6Q	0.1432	1.6542	−1.3156	1.0751	0.1027	1.0209	0.3175	−0.0633	1.5568	−1.5099
8Q	−0.1328	1.8901	−1.9282	1.1247	−0.1972	1.1277	−0.0052	−0.3914	1.7681	−2.1843
10Q	−0.2511	1.9427	−2.1990	1.0644	−0.3318	1.0672	−0.2297	−0.5084	1.8135	−2.4812
12Q	−0.3647	1.9309	−2.3755	1.0114	−0.3979	0.9918	−0.3251	−0.6215	1.7994	−2.6647
Real effective exchange rate (REER)										
	Δoil_t	Δoil_t^+	Δoil_t^-	NOPI4	NOPD4	NOPI12	NOPD12	SOPC	SOPI	SOPD
4Q	0.0296	−0.0168	0.0576	0.0066	0.0397	0.0029	0.0467	0.0287	−0.0173	0.0574
6Q	0.0434	−0.0371	0.0976	0.0028	0.0606	−0.0107	0.0716	0.0430	−0.0367	0.0979
8Q	0.0560	−0.0592	0.1385	−0.0040	0.0797	−0.0275	0.0954	0.0565	−0.0579	0.1397
10Q	0.0713	−0.0796	0.1811	−0.0097	0.0995	−0.0422	0.1200	0.0727	−0.0775	0.1833
12Q	0.0876	−0.0995	0.2248	−0.0143	0.1209	−0.0554	0.1466	0.0900	−0.0965	0.2281

Δoil_t: real oil price change, Δoil_t^+ increase in real oil prices, Δoil_t^- decrease in real oil prices, *NOPI4* increase in real oil prices over previous four quarters, *NOPD4* decrease in real oil prices over previous four quarters, *NOPI12* increase in real oil prices over previous 12 quarters, *NOPD12* decrease in real oil prices over previous quarters, *SOPC* scaled real oil price change, *SOPI* scaled real oil price increase, and *SOPD* scaled oil price decrease

Source: Authors' calculations

Table 13.7 Estimated orthogonal variance decomposition

Real oil price change	Industrial growth	Oil price	CPI	MMR	REER
Industrial growth	91.33	2.64	0.63	2.59	2.81
Oil price	1.88	92.82	3.07	1.17	1.06
CPI	16.33	10.95	53.96	15.72	3.05
MMR	10.31	2.69	5.73	73.73	7.54
REER	0.82	3.47	9.91	35.30	50.49
Real oil price increase					
Industrial growth	92.70	1.84	0.48	2.35	2.63
Oil price	0.65	91.44	4.22	1.72	1.97
CPI	18.01	5.21	60.69	13.51	2.59
MMR	11.15	6.55	6.00	68.06	8.25
REER	1.76	4.24	13.33	25.75	54.93
Real oil price decrease					
Industrial growth	91.93	3.32	0.47	2.56	1.71
Oil price	3.48	92.94	0.74	0.87	1.97
CPI	16.62	13.00	51.90	15.81	2.67
MMR	11.09	4.65	3.46	74.43	6.37
REER	1.47	21.25	5.33	32.96	38.99
Net oil price increase over last four quarters (NOPI4)					
Industrial growth	93.90	0.88	0.35	3.03	1.85
Oil price	3.00	90.52	3.59	0.95	1.95
CPI	18.55	3.82	60.39	14.14	3.10
MMR	12.73	1.64	5.41	70.36	9.85
REER	1.26	0.20	12.85	30.35	55.34
Net oil price decrease over last four quarters (NOPD4)					
Industrial growth	92.28	4.03	0.32	2.09	1.28
Oil price	2.81	91.16	1.48	2.02	2.54
CPI	16.30	12.46	54.22	13.55	3.47
MMR	10.59	1.01	3.94	76.90	7.56
REER	0.56	5.88	8.57	38.81	46.18
Net oil price increase over last 12 quarters (NOPI12)					
Industrial growth	93.35	1.35	0.16	3.53	1.60
Oil price	3.49	91.07	3.67	1.14	0.63
CPI	18.47	2.51	63.15	12.71	3.16
MMR	17.06	3.14	4.00	66.55	9.25
REER	5.16	2.14	12.57	25.76	54.36
Net oil price decrease over last 12 quarters (NOPD12)					
Industrial growth	93.09	3.47	0.07	2.30	1.07
Oil price	3.13	91.96	0.42	1.96	2.53
CPI	16.85	6.88	58.88	14.21	3.19
MMR	10.20	0.98	4.47	76.68	7.67
REER	0.92	7.99	9.99	36.78	44.32
Scaled oil price change (SOPC)					
Industrial growth	91.57	2.42	0.56	2.66	2.79
Oil price	1.93	92.33	3.20	1.28	1.26
CPI	16.32	10.92	53.98	15.71	3.07

(continued)

Table 13.7 (continued)

Real oil price change	Industrial growth	Oil price	CPI	MMR	REER
MMR	9.89	2.93	5.69	74.32	7.16
REER	0.76	3.67	9.66	35.49	50.42
Scaled oil price increase (SOPI)					
Industrial growth	92.72	1.75	0.47	2.35	2.71
Oil price	0.61	90.90	4.32	1.93	2.24
CPI	17.94	5.16	60.77	13.55	2.58
MMR	10.80	6.38	6.04	68.66	8.12
REER	1.54	3.90	13.31	26.61	54.66
Scaled oil price decrease (SOPD)					
Industrial growth	92.03	3.27	0.40	2.63	1.67
Oil price	3.87	92.24	0.84	0.91	2.14
CPI	16.58	13.04	51.82	15.86	2.70
MMR	10.95	5.34	3.25	74.31	6.14
REER	1.42	21.70	5.13	32.47	39.29

Note: *CPI* consumer price index, *MMR* money market interest rate, *REER* real effective exchange rate. This table presents the results of the estimated variance decomposition at 12-period horizon
Source: Authors' calculations

Turning to variables other than industrial growth and REER, the results indicate that an oil price shock increases inflation and the short-term interest rate. These results are plausible and provide evidence of transmission mechanism – other than the exchange rate channel – playing the expected role.

13.5.2.2 Variance Decomposition Analysis

Table 13.7 presents the results of the forecast error variance decomposition for all specifications used in the study. The forecast error variance decomposition tells us the proportion of the movements in a sequence due to its *own* shocks versus shocks to the other variable. The variance decompositions suggest that oil shocks are a considerable source of volatility for many of the variables in the model. For industrial growth, oil prices together with short-term interest rate are the largest source of shock other than the variable itself. Innovations in the short-term interest rate represent monetary shocks in our model. The contribution of oil prices and the short-term interest rate to industrial growth variability is about 4% in the preferred model SOPI. REER exhibits a contribution to industrial growth variability of an approximate magnitude of 3%. Moreover, it is found that the movements in the short-term interest rate arise from changes in oil prices. For the SOPI model, the oil price variable contributes to industrial growth, inflation, short-term interest rate, and REER 1.75%, 5.16%, 6.38%, and 3.90% respectively. The contribution of oil prices to short-term interest rate variability can be interpreted as a reaction of monetary policy to oil price shocks.

13.6 Concluding Remarks

This chapter studies the oil price–macroeconomy relationship in the Indian economy by means of analyzing the impact of oil price shocks on the growth of industrial production over the period 1975Q1–2004Q3. Vector autoregressions are used to measure the impact of oil prices on the macroeconomic variables. We obtain a higher impact when oil prices are measured in Indian rupees (INR) than when they are expressed in US$. This could be due to the role of the exchange rate and variation in domestic prices. We also find that oil price shocks (especially an increase in real oil prices) Granger-cause the growth of industrial production.

It is found that increase in real oil prices negatively affects the growth rate of industrial production in linear and nonlinear specifications. For the Indian economy we find that a 100% increase in real oil prices reduced the growth of industrial production by 1%. This small impact of the growth of industrial production can be traced, among other factors, to depreciation in the real effective exchange rate. Furthermore, we find that the inflation rate and short-term interest rate are positively affected by the increase in real oil prices.

We also obtain evidence on the asymmetric relationship between oil prices and the growth of industrial production, confirming the relationship found in developed economies. Among all specifications used for oil prices, the one that turns out to be best performing from a statistical standpoint is the SOPI model. This implies that it is not just price changes, but also the environment in which the movements take place. An oil price shock in a stable environment has larger economic consequences than one in a volatile price environment.

The variance decomposition analysis shows that oil price shocks are a considerable source of volatility for the variables used in the study. For the growth of industrial production the oil price shocks combined with monetary shocks are the largest source of variation other than the variable itself; thus, the variance decomposition analysis puts the relationship between oil price and industrial growth into perspective, while the focus of the study is to analyze the impact of oil price shocks on the growth of industrial production.

Appendix

The quarterly data used in this study are mainly obtained from two sources: International Financial Statistics (IFS) CDROM and the Reserve Bank of India (RBI) Database of Indian Economy. The variable and source details are these:

Economic activity: The aggregate economic activity is proxied by Index of Industrial Production (IIP), since for India, the quarterly GDP series is available since 1996–1997 only. The series for IIP covers the period 1975Q1 to 2004 and is taken from IFS-CDROM.

Oil price variable: The world oil price measured in US$ for India is calculated as the average of UK Brent and Saudi Prices since India's oil imports are mainly based on the prices of these two markets. To convert these oil prices into real world prices we deflated the nominal prices by the world consumer price indices. Real oil prices measured in Indian rupees (INR) are calculated by converting world oil prices by the market rate of exchange and then deflating by the wholesale price indices (WPI) found in India. The series for oil price covers the period 1970Q1 to 2004Q4 and is taken from IFS-CDROM.

Inflation rate: Calculated from consumer Price Index (CPI) and taken from the IFS-CDROM for the period 1975Q1 to 2004Q3.

Short-term interest rate: Measured by the money market rate of interest (MMR) and obtained from RBI for the period 1975Q1 to 2004Q3. RBI provided the monthly estimated money market rate of interest. To convert the series into quarterly data we have taken the simple 3-month average.

Real effective exchange rate (REER): REER series is taken from the RBI for the period 1975Q1 to 2004Q3. RBI provided monthly estimates of the money market rate of interest. To convert the series into quarterly data we have taken the simple 3-month average. RBI constructs the 5-country trade-based nominal effective exchange rate (NEER) and REER on a daily basis. The countries chosen are the United States, Germany, Japan, United Kingdom, and France (G-5 countries). REER is defined as the weighted average of NEER adjusted by the ratio of the domestic inflation rate to foreign inflation rates. In terms of formula, $\mathrm{REER} = \prod_{i=1}^{5} \left[\left(\frac{e}{e_i}\right)\left(\frac{P}{P_i}\right)\right]^{w_i}$ where: e is the exchange rate of rupee against numeraire (SDRs) (i.e., SDRs per Rupee) (in index form), e_i: Exchange rate of currency i against the numeraire(SDRs) (i.e., SDRs per currency i) (in index form) (i = US Dollar, Japanese Yen, Deutsche Mark, Pound Sterling, French Franc), w_i: Weights attached to currency/country i in the index, P: India's wholesale price index (WPI) (in Index form), and P_i: Consumer Price Index (CPI) of country i (in Index form). The increase in the value of REER implies the appreciation of the currency and decline in the competitiveness of the country.

Chapter 14
Energy Prices and Induced Technological Progress

14.1 Introduction

Technological progress plays a crucial ameliorating role in reducing energy consumption for combating climate change. Energy economists often cite market-based instruments such as energy taxes for encouraging energy-saving technological progress. Energy policy interventions may change the constraints and incentives that affect technological change (TC). For instance, changes in current relative energy prices may induce substitution of energy by other factors of production, and changes in its long-run prices may induce development of new energy-saving technologies. The importance of relative prices as a stimulator of technological advancement is traceable to Hicks (1932).[1] The theory of induced innovation helps in measuring the impact of relative prices on the direction of technological change (Hayami and Ruttan, 1971).

In the earlier literature on energy and environmental policy models, technological change is incorporated as an exogenous variable; i.e., technological developments are autonomous and do not depend upon on policy or economic variables, and there is little empirical evidence for induced technological developments. However, recently some attempts have been made to model policy-induced technological changes in the climate-economy models.[2]

Most of the empirical studies on induced innovations are conducted using firm-level industrial data and measure technological progress either in terms of inputs (e.g., investments or research and development (R&D) spending in energy-saving

---

[1]Hicks argues that "...a change in relative prices of factors of production is itself a spur to invention, and to invention of a particular kind – directed to economizing the use of a factor which has become relatively expensive (pp. 124–125)."

[2]Special issues of the *Energy Journal* (2006), *Energy Economics* (2006) and *Ecological Economics* (2005) as evidence of recent attempts on modeling edogenous technological progress in the area. There are few studies that empirically measure the induced technological progress due to changes in environmental policy parameters, for example, Lichtenberg (1986, 1987), Lanjouw and Mody (1996), Jaffe et al. (1997), Newell et al. (1999), Nordhaus (1999), Popp (2002).

S. Kumar and S. Managi, *The Economics of Sustainable Development*,
Natural Resource Management and Policy 32,
DOI: 10.1007/978-0-387-98176-5_14,

innovations (ESI)) or outputs (e.g., the number of patents filed, granted, or cited in the area of ESI). But at the macro level, though data on energy R&D spending are collected in some countries, getting comparable data across countries is a daunting task. Moreover, the data collected by International Energy Agency (IEA) on energy R&D do not encompass deployment activities, which are essential components in the technological progress (Gallagher et al., 2006).

Popp (2002), using an output measure of ESI, analyzes the induced innovation hypothesis on a macroeconomic level. He uses U.S. patent data from 1970 to 1994 to estimate the effect of energy prices on innovations and finds a positive association between energy prices and energy-saving innovations. Patent data cannot be an appropriate measure of technological change because inventions might not be widely deployed (Basberg, 1987). For example, if a country produces a lot of useless patents that are never deployed, the country should not be rated as more innovative than another country with the same or smaller number of patents that are more useful.

Newell et al. (1999) also provide evidence of energy price induced innovations using a product-characteristics framework. Using the most tangible output metric of ESI, they find that energy prices have positively affected the energy efficiency of electrical appliances. Gallagher et al. (2006) point out that this output metric is again loaded with problems, since technologies are discrete and not often well defined, and ESIs relate to more energy-efficient system integration, which relies heavily on the accumulated knowledge of those doing the integration.

Technological change can be decomposed into two components – innovation and diffusion – and the transformation function[3] is best suited to measure technological change (Jaffe et al. 2003). The transformation function represents "best practice," i.e., what the economy would produce if all innovations made to date had fully diffused; therefore, the shift in transformation function captures innovations. The role of diffusion would then arise if some countries are not adopting "best practice" and operating at points inside the transformation frontier. The movement of these countries towards the frontier can be termed as "catch-up" effect or technological diffusion (TD).[4] The present study tends to extend the literature on induced technological progress by measuring both innovations and diffusion.

We use the directional distance function, which is a more general version of the transformation function, for measuring energy price induced technological change (TC). The directional distance function simultaneously seeks to expand output and contract inputs. It is particularly well suited to the task of providing a measure of technical efficiency in the full input-output space and satisfies all those properties, which are satisfied by the conventional representations of production technology.

---

[3]Transformation function describes a production possibility frontier, that is, a set of combinations of inputs and outputs that are technically feasible at a point in time.

[4]Directional distance function constitute the transformation function using the data of the countries under study, thus, it is a relative measure of technical inefficiency across countries. It can identify the practices adopted by most efficient country are diffused to other countries. This is not equivalent to saying that most efficient country uses only the latest innovations, i.e., directional distance function cannot say anything about the diffusion within a country.

There is considerable theoretical and empirical literature on the induced innovation hypothesis. See Hayami and Ruttan (1971), Binswanger (1974), Binswanger (1978), and Thirtle and Ruttan (1987) for a summary of this literature. That literature typically analyzes the inducement effect in the framework of the conventional representation of production technology, such as cost, production, or profit functions. Distinguishing between factor substitution and shift of transformation frontiers is problematic with the conventional representations. That is, in conventional representations the first-order comparative static optimization conditions cannot be followed since the direct derivatives of the demand and supply functions with respect to prices cannot be unambiguously signed, given the presence of the cross derivatives (Celikkol and Stefanou, 1999; Paris and Caputo, 2001).

We measure TC for a sample of 55 countries over the period 1974–2000 using macro variables. TC is similar in nature to any investment process, as it requires time and adjustment that is not instantaneous, and the choice of technology is influenced by long-term prices. Innovations are decomposed into two parts; namely, exogenous innovations (EI) and energy price induced innovations (PII). A time trend variable is used to measure exogenous innovation.[5] Similarly the inclusion of long-term energy prices as a sift factor in the transformation function is used for measuring the induced innovation effect. We use oil prices as proxy for energy prices.[6]

The chapter is organized as follows: Section 14.2 outlines the measurement of technological change. Section 14.3 presents the empirical model for the stochastic estimation of directional distance function, and the data is discussed in Section 14.4. Section 14.5 discusses the main results of the study. Summary and conclusions are presented in Section 14.6.

14.2 Measurement of Technological Change

We extend the Luenberger measure of productivity change or technological change,[7] introduced by Chambers et al. (1996a) and Chambers (2002), to a measure that also accounts for energy price induced innovations. The Luenberger productivity indicator is decomposed into two component measures: innovation and diffusion.

[5]Technological progress occurs both due to inducements and advancements in general science and technology. Therefore, a time trend is included as an argument in the transformation frontier to account for the impact of scientific innovation on the production technology (Lansink et al., 2000, p. 500, Footnote 1).

[6]In the energy consumption oil accounts for most of the consumption of hydrocarbons, although the use of natural gas has risen in the past decades or so and there is high positive correlation between oil and natural gas prices. Moreover, oil accounts for about 35% of global annual use of primary energy, with much of that oil coming from politically unstable regions (Gallagher et al., 2006), therefore, it is assumed that it is oil price volatility which induces technological progress which is energy saving.

[7]Productivity change is generally decomposed into technical change and efficiency change components. We use the term technological change in place of productivity change; technical change is termed as innovations and efficiency change is termed as technological diffusion or catch-up effect.

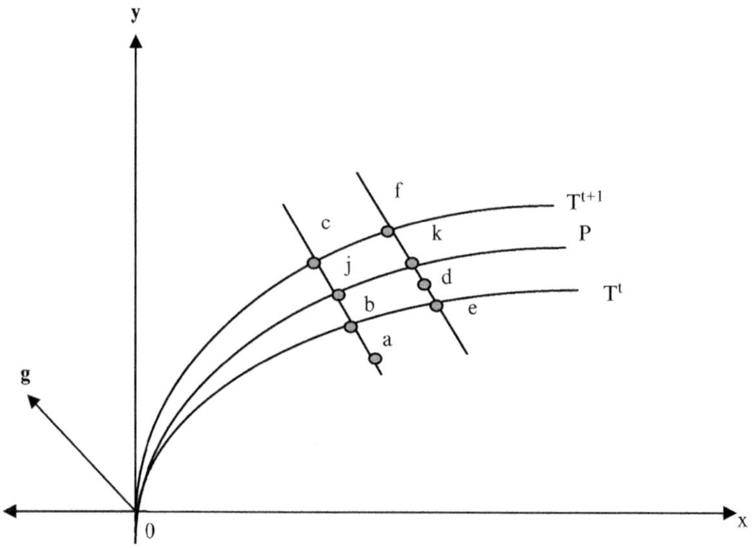

Fig. 14.1 Luenberger productivity indicators

We decompose innovation further into EI and PII. This can be illustrated through Fig. 14.1.

Suppose a country in the year t with input–output (x^t, y^t) vector is operating at point a, and in the year $(t + 1)$ with the input–output vector (x^{t+1}, y^{t+1}) is at d. The technologies at these two points of time are specified as T^t and T^{t+1}. The shift in technology from T^t to T^{t+1} is the combination of energy price induced and exogenous innovations, i.e., shift in the production technology from T^t to P is induced by the factors such as change in relative long-term energy prices and the shift from P to T^{t+1} is due to some external factors such as advancement in science and technology. Therefore we get

$$\text{Diffusion} = (b - a) - (f - d)$$
$$\text{Innovation} = 0.5((f - e) + (c - b)) = 0.5(((f - k) + (k - e)) + ((c - j) + (j - b)))$$

or

$$\text{Innovation} = 0.5((f - k) + (c - j)) = 0.5((k - e) + (j - b)) = EI + PII$$

Thus technological diffusion is measured by the distance of points a and d from the transformation functions T^t and T^{t+1}, respectively.[8]

[8]The reference (benchmark) technology may be of t or $t + 1$ period. In order to avoid choosing an arbitrary benchmark, we specify the technological change index or innovations index as the arithmetic mean of the two indexes.

To measure technological change, we use directional distance function. Directional distance function seeks to expand the desired output e.g., GDP and contract inputs such as labor, capital and energy, and inherits its properties from the production technology, T.[9] More formally the function is defined as:

$$D(x,y;g) = \max_{\beta} \{\beta : (y + \beta \cdot g_y, x - \beta \cdot g_x) \in T\} \tag{14.1}$$

where T=(x,y):x can produce y, and $y = (y_1, ..., y_M) \in \Re_+^M$, and $x = (x_1, ..., x_N) \in \Re_+^N$ are output and input vectors, respectively. The solution, β^* gives the maximum expansion and contraction of outputs and inputs, respectively. The vector $g = (g_y, -g_x)$ specifies in which direction an output–input vector, $(y,x) \in T$ is scaled so as to reach the boundary of the technology frontier at $(y + \beta^* \cdot g_y, x - \beta^* \cdot g_x) \in T$, where $\beta^* = D(x,y;g)$. This means that the producer becomes more technically efficient when simultaneously increasing outputs and decreasing inputs. The function takes the value of zero for technically efficient output–input vectors on the boundary of T, whereas positive values apply to inefficient output vectors below the boundary. The higher the value, the more inefficient is the input–output vector, i.e., the directional distance function is a measure of technical inefficiency.[10]

Moreover, directional distance function and profit function are dual to each other (Färe and Grosskopf, 2000) and the *dual Hotelling lemma*, i.e., the derivatives of directional distance with respect to output and input quantities, provide inverse supply and demand functions (Hudgins and Primont, 2004). The function also satisfies the translation property.[11]

Following Chambers (2002), the directional output distance function is parameterized using a (additive) quadratic flexible functional form. In our case, with one output, three inputs, time trend, and long-run relative energy prices, the particular form is

$$\begin{aligned} D^{kt}\left(x^{kt}, y^{kt}, b^{kt}; g, t, \bar{r}\right) &= \alpha_0 + \sum_{n=1}^{3} \alpha_n x_n^{kt} + \beta_1 y^{kt} + \gamma_1 t + \gamma_2 \bar{r}^{kt} + \frac{1}{2} \sum_{n=1}^{3} \\ &\times \sum_{n'=1}^{3} \alpha_{nn'} x_n^{kt} x_{n'}^{kt} + \sum_{n=1}^{3} \delta_{n1} x_n^{kt} y^{kt} + \sum_{n=1}^{3} \eta_{n1} x_n^{kt} t \\ &+ \sum_{n=1}^{3} \eta_{n2} x_n^{kt} \bar{r}^{kt} + \frac{1}{2} \beta_2 y^{kt} y^{kt} + \mu_1 y^{kt} t + \mu_2 y^{kt} \bar{r}^{kt} \\ &+ \frac{1}{2} \gamma_{11} t.t + \varphi t \bar{r}^{kt} + \frac{1}{2} \gamma_{22} \bar{r}^{kt}.\bar{r}^{kt} + \phi G \end{aligned} \tag{14.2}$$

---

[9] For properties of directional distance function see, Färe et al. (2005).

[10] Directional distance function can be used for the case of multiple outputs and multiple inputs. In our study the output is a scalar rather than a vector.

[11] The translation property may be stated as follows: $D(x, y + \alpha \cdot g_y, x - \alpha \cdot g_x; g) = D(x, y; g) - \alpha$, where α is a positive scalar, implying that if output is expanded by αg_y and inputs are contracted by αg_x, then the value of the distance function will be more efficient with the amount α.

with

$$\alpha_{nn'} = \alpha_{n'n}; \quad \beta_1 - \sum_{n=1}^{3} \alpha_n = -1; \quad \delta_{n1} - \sum_{n=1}^{3} \alpha_{nn'} = 0;$$

$$\beta_2 - \sum_{n=1}^{3} \delta_{n1} = 0; \quad n = 1, 2, 3.$$

where $g=(1,-1)$, 1 refers to g_y and -1 refers to $-g_b$; and t is a time-trend, $\bar{r}$ is the long-run energy prices and G is group dummy. The countries were grouped in two categories: developed and developing, based on per capita income following the World Bank classification.[12]

The specification of (14.2) allows for neutral and biased technological changes. The effect of neutral exogenous technological change is captured by the coefficients γ_1 and γ_{11} and the effect of neutral induced technological change is captured by the coefficients γ_2 and γ_{22}. The extent of input-biased exogenous and induced technological change are captured by the coefficients η_{n1} and η_{n2} respectively, and the effect of changes in output due to exogenous and induced factors (i.e., scale augmenting technological change) is captured by the coefficients μ_1 and μ_2 respectively. In addition, the interaction between exogenous and induced factors is captured by the coefficient ϕ.

We parameterize the directional distance function in quadratic form hence; it is possible to apply Diewert's (1976) *Quadratic Identity Lemma*.[13] Using this identity, changes in the directional distance function from one period to the next can be written as:

$$\begin{aligned}(D^t - D^{t+1}) = {} & 0.5\left[\frac{\partial D^t}{\partial y} + \frac{\partial D^{t+1}}{\partial y}\right].(y^{t+1} - y^t) + 0.5 \\ & \times \sum_{n=1}^{3}\left[\frac{\partial D^t}{\partial x_n} + \frac{\partial D^{t+1}}{\partial x_n}\right].(x_n^{t+1} - x_n^t) + 0.5\left[\frac{\partial D^{t+1}}{\partial t} + \frac{\partial D^t}{\partial t}\right] \\ & + 0.5\left[\frac{\partial D^{t+1}}{\partial \bar{r}} + \frac{\partial D^t}{\partial \bar{r}}\right].(\bar{r}^t - \bar{r}^{t+1}), \end{aligned} \tag{14.3}$$

[12] An important issue in efficiency studies is the credibility of the assumption that all production processes can actually reach the best practice production frontier. In the present study, when measuring technical efficiency it would be not be proper to assume that all countries included in the study have access to the best practice manufacturing frontier because currently, specialized journals, technological fairs, multinationals' global marketing strategies, etc., that guarantee new innovations are not readily available to all firms equally in all the countries. Therefore, to account for these differences across the nations we grouped the countries in two groups on the basis of per capita income according to World Bank classification and included one dummy in the estimation of directional distance function.

[13] Orea (2002) used the quadratic identity lemma for parametric decomposition of Malmquist productivity index using output distance function.

where D^t is short for $D(x^t, y^t; g, t, \bar{r})$. Technological change (TC) can be defined as:

$$\begin{aligned} \mathrm{TC} = & -0.5\left[\frac{-\partial D^{t+1}}{\partial y} + \frac{-\partial D^t}{\partial y}\right].(y^{t+1} - y^t) \\ & +0.5\sum_{n=1}^{3}\left[\frac{\partial D^{t+1}}{\partial x_n} + \frac{\partial D^t}{\partial x_n}\right].(x_n^{t+1} - x_n^t) \end{aligned} \tag{14.4}$$

Technological change can be broadly defined as the difference of the weighted average rates of change in outputs and inputs, where the weights are derivatives of directional distance function with respect to (negative) output and (positive) inputs respectively. Rearranging (14.4), TC can be decomposed as:

$$\begin{aligned} \mathrm{TC} = & \underbrace{(D^{t+1} - D^t)}_{\text{Diffusion}} \underbrace{- 0.5\left[\frac{\partial D^{t+1}}{\partial t} + \frac{\partial D^t}{\partial t}\right]}_{\text{EI}} \\ & \underbrace{-0.5\left[\frac{\partial D^{t+1}}{\partial \bar{r}} + \frac{\partial D^t}{\partial \bar{r}}\right].(\bar{r}^{t+1} - \bar{r}^t)}_{\text{PII}} \end{aligned} \tag{14.5}$$

Equation (14.5) provides a meaningful decomposition of TC into diffusion, exogenous innovations (EI), and energy price induced innovations (PII), respectively. Negative values of the derivatives of directional distance function with respect to time-trend and long-run energy prices imply positive change in EI and PII respectively. Therefore, the negative value of each component of productivity index implies positive change in technological change (TC).[14]

14.3 The Econometric Estimation

The function in (14.2) can be estimated using either linear programming (LP) or stochastic techniques.[15] Estimating distance functions stochastically has some advantages over the LP approach. Other than allowing for an appropriate treatment of measurement errors and random shocks, several statistical hypotheses can be tested: significance of parameters, separability between outputs and inputs, and monotonicity properties of distance functions.

Following Kumbhakar and Lovell (2000) and Färe et al. (2005), the stochastic specification of directional distance function takes the form

[14] In the discussion of results, for the sake of convention we have multiplied each of the component by minus one.

[15] The LP estimating procedure is adopted in Färe et al. (2001) and in Färe et al. (2005).

$$0 = D(x, y; -1, 1, t, \bar{r}) + \varepsilon \tag{14.6}$$

where $\varepsilon = v - \mu$with $v \sim N(0, \sigma_v^2)$ and μ (one-sided error term) is assumed to be exponentially distributed with θ as scale distribution parameters.

To estimate (14.6) we utilize the translation property of the directional output distance function. As in Färe et al. (2005), we choose the directional vector $g=(1,-1)$, where 1 refers to g_y and -1 refers to $-g_x$, (see Fig. 14.1). This choice of direction is consistent with profit maximization hypothesis. The translation property implies that

$$D(x - \alpha, y + \alpha, ; -1, 1, t, \bar{r}) + \alpha = D(x, y; -1, 1, t, \bar{r}) \tag{14.7}$$

By substituting $D(x - \alpha, y + \alpha; -1, 1, t, \bar{r}) + \alpha$ for $D(x, y; -1, 1, t, \bar{r})$ in (14.6) and taking α to the left hand side, we obtain

$$-\alpha = D(x - \alpha, y + \alpha; -1, 1, t, \bar{r}) + \varepsilon \tag{14.8}$$

where $D(x - \alpha, y + \alpha; -1, 1, t, \bar{r})$is the quadratic form given by (14.2) with α added to y and subtracted from x. Thus one is able to get variation on the left-hand side by choosing an α that is specific to each country. In our case it may be one of inputs, and we use capital input as α.[16]

The parameters of the quadratic distance function, as specified in (14.8), can be estimated either using corrected ordinary least square (COLS)[17] or maximum likelihood (ML) methods. The COLS approach is not as demanding as the ML method; The ML method requires numerical maximization of the likelihood function. This method is asymptotically more efficient than the COLS estimator, but the properties of two estimators in finite samples can be analytically determined. The finite sample properties of the half-normal frontier model were investigated in a Monte-Carlo experiment by Coelli (1995), who found the ML estimator to be significantly better than the COLS estimator when contribution to technical inefficiency effects to the total variance term is large.

Moreover, Greene (2000) shows that the gamma/exponential model has the virtue of providing a richer and more flexible parameterization of the inefficiency distribution in the stochastic frontier model than either of the canonical forms, half normal and exponential. Gamma/exponential specification enjoys essentially the same properties as normal/half-normal model, with the additional advantage of the flexibility of a two-parameter distribution. The primary advantage is that it does not require that the firm-specific inefficiency measures be predominately near zero (Greene, 1990). One can test down from the gamma to the exponential by testing

---

[16]Note the results were not affected by the choice of α. The parameters obtained alternatively with the other inputs as α showed little difference.

[17]For an application of COLS to the Shephard output distance function, see Lovell et al. (1994) and to the directional output distance function, see Färe et al. (2005)

if the shape parameter, P, equals 1.0. Gamma distribution is a generalization of the exponential distribution. The present study adopts the ML estimation approach and assumes exponential distribution for one-sided error term.[18]

14.4 Data

For measuring energy price induced technological change, the resource constraint consists of the net fixed standardized capital stock, labor force, measured by the number of employed workers and energy use measured in kilotons (kt) of oil equivalent. Real GDP (adjusted for 1996 prices) measured in $PPP is taken as an indicator of output. Data on the capital stock, labor, and real GDP are compiled from a recent data set in Marquetti (2002). World Development Indicators (World Bank) is the source for energy use. Crude oil prices, currency exchange rates, and country-specific consumer price indices are complied from International Financial Statistics (IMF) to create country-specific indices of relative oil prices as a proxy for country-specific energy prices. The annual panel data set includes 55 countries,[19] a mix of developed and developing countries for the period 1974–2000. The choice of countries and study period is constrained on the availability of the required information. The period of study starts just after the first oil shock.

The choice of oil price variables is difficult, and country-specific oil prices oil prices have been influenced by price controls, high and varying taxes on petroleum products, exchange rate fluctuations, and country-specific price index variations. All the differential characteristics which influence the effective oil price faced by each of the countries raise great difficulties in measuring the appropriate oil price variable for each country. Most of the empirical literature analyzing the effect of oil price shocks use either the $US world price of oil as a common indicator of the world market disturbances that affect all countries (see, e.g., Burbidge and Harrison, 1984) or this world oil price converted into each respective country's currency by means of the market exchange rate and adjusted by the domestic inflation (see, e.g. Mork et al., 1994, for OECD countries or Cunado and Gracia, 2005, for Asian countries). The main difference between the two variables is that only the second one takes into account the differences in the oil price that each of the countries faces

---

[18]The null hypothesis of gamma distribution of one-sided error term, $P = 1$ could not be rejected.

[19]We have grouped all the countries in two categories according to World Bank Classification on the basis of per capita income: developing and developed countries. The countries included in the study are: CAMEROON, COTE d'IVOIRE, EGYPT, EL SALVADOR, ETHIOPIA, GHANA, GUATEMALA, HONDURAS, INDIA, KENYA, NIGERIA, PAKISTAN, PARAGUAY, PHILIPPINES, SENEGAL, SRI LANKA, SYRIA, TANZANIA, TOGO, COLOMBIA, COSTA RICA, DOMINICAN REP., ECUADOR, GABON, INDONESIA, JAMAICA, JORDAN, MOROCCO, PERU, SOUTH AFRICA, TRINIDAD & TOBAGO, URUGUAY, VENEZUELA, ARGENTINA, CHILE, IRAN, MALAYSIA, MEXICO, THAILAND, TURKEY, AUSTRALIA, BOLIVIA, CANADA, DENMARK, GREECE, ICELAND, ISRAEL, JAPAN, KOREA REP. OF, NEW ZELAND, NORWAY, SWEDEN, SWITZERLAND, UNITED KINGDOM, USA.

due to its exchange rate fluctuations and its inflation levels. In the present study we use the second kind of oil price indices for analysis.[20] The oil price indices are created by taking the 1970 as the base year.

14.4.1 Long-Term Energy Prices

The notion of long-run prices serving as a stimulating factor to innovate is a critical component of the price-induced innovation model. Changes in current prices induce factor substitution where changes in long-run prices induce the development of new technologies leading to the shift of the technology frontier. Therefore, it is important to model long-run prices which depend on current and past price information as arguments in the production technology frontier to separate scarcity responses from biased ITC. Therefore, past country-specific prices of energy are included in the country-specific frontier function to measure ITC and are generated as a 3-year moving average of past energy prices. The choice of 3-year moving average is based on the assumption that firms use most recent years as having the greatest information content (Lansink et al., 2000). The choice of long-term energy prices is also consistent with an adaptive expectation model of prices, in which expected future prices depend on a weighted average of past prices (Popp, 2002).

14.5 Results

For the measurement of exogenous and energy price induced innovations, following Färe et al. (2005), we estimate directional distance function as specified in (14.2) using normalized values of inputs and outputs.[21] This normalization implies that $(x, y)=(1, 1)$ for a hypothetical country that uses mean inputs and produces mean output.

We estimated four specifications of directional distance function. In specification 1, we estimate the directional distance function only in input-output vectors; in specification 2, we include the trend variable as the shift parameter; and in specification 3 there are two shift parameters: time trend and long-run relative energy prices. As noted above, the sample consists 55 countries. We grouped the countries in two groups: developing and developed countries, and in the estimation we also included the group dummy (specification 4). The selection of model is done on the

[20] Ideally the oil price index should account for differences in taxes across countries, but due to nonavailability of energy taxes for all the countries, cross country variation in energy prices comes from exchange rate and domestic inflation.

[21] We normalized the data for each output and each input by their mean values before estimation.

Table 14.1 Tests of hypotheses for functional form of directional distance function

Null hypothesis	Log likelihood ratio test statistics (λ)	Critical χ^2	Decision value at 5%
H_0: $\gamma_1 = \gamma_{11} = \eta_{11} = \eta_{21} = \eta_{31} = \mu_1 = 0$	771.08	12.592	Reject
H_0: $\gamma_2 = \gamma_{22} = \eta_{12} = \eta_{22} = \eta_{32} = \mu_2 = \phi = 0$	167.51	14.067	Reject
H_0: $\psi = 0$	136.186	3.84	Reject

$\lambda = -2\{\text{Log(Likelihood } (H_0) - \text{Log(Likelihood } (H_1)\}$

Source: Authors' calculations

Table 14.2 Parameter estimates of mean normalized directional distance function

Name of variables/ parameters	Coefficient	*t* statistics	Name of variables/ parameters	Coefficient	*t* statistics
Constant (α_0)	−0.0244*	−4.0350	YX_3 (δ_{31})	0.0028*	2.8150
Y (β_1)	−0.4520*	−203.2570	Yt (μ_1)	−0.0001	−0.5520
X_1 (α_1)	0.0843*	17.6290	$Y\bar{r}$(μ_2)	−0.0048*	−4.6180
X_2 (α_2)	0.2984	–	$0.5X_1X_2$ (α_{12})	0.0077	–
X_3 (α_3)	0.1653*	16.3960	$0.5X_1X_3$(α_{13})	0.0199*	3.7830
T (γ_1)	−0.0009	−1.3860	$0.5X_2X_3$(α_{23})	0.0084	–
$\bar{r}$(γ_2)	0.0116*	2.5960	X_1t (η_{11})	−0.0011*	−6.2660
G (ψ)	0.0312*	8.1070	$X_1\bar{r}$(η_{21})	−0.0025	−0.9400
$0.5Y^2$(β_2)	0.0074*	12.2950	X_2t (η_{31})	0.0010	–
$0.5{X_1}^2$(α_{11})	−0.0179*	−19.0680	$X_2\bar{r}$(η_{12})	−0.0055	–
$0.5{X_2}^2$(α_{22})	−0.0213	–	X_3t (η_{22})	0.0001	0.2370
$0.5{X_3}^2$(α_{33})	−0.0255*	−12.4080	$X_3\bar{r}$(η_{23})	0.0032***	1.6350
$0.5t^2$(γ_{11})	0.0001**	2.4780	$t.\ \bar{r}$(ϕ)	−0.0003	−1.0230
$0.5\bar{r}^2$(γ_{22})	−0.0014	−1.4360	θ	15.4473*	39.731
$Y.X_1$ (δ_{11})	0.0097*	14.3840	σ_v	0.0172*	15.862
$Y.X_2$ (δ_{21})	−0.0051	–	Log likelihood function	2,233.901	

Note: Underlined parameters are calculated by applying the translation property of the directional distance function. Number of observation: 1,485. Y GDP, X_1 labor, X_2 Capital, X_3 Energy
*, **, *** implies level of significance at 1%, 5% and 10% respectively
Source: Authors' calculations

basis of log-likelihood ratio (LR) test. Table 14.1 provides the LR test statistics. On the basis of LR test statistics, specification 4 is finally selected for further analysis.

Table 14.2 provides the parameters estimate of directional distance function for specification 4. Most of the ML coefficients are accurately estimated. Technical inefficiency is correctly identified within the composed error term: (a) the LR test on the one-sided error is highly significant; (b) the share of technical inefficiency in total variance is high, i.e., 93% and (c) it appears to have an exponential distribution with $\theta = 15.45$.

A first look at the production technology parameters in Table 14.2 indicates that the first-order coefficients on output and inputs have expected signs regarding economic behavior. Looking at the signs of second-order parameters, it appears that they involve interesting results too; however, a more detailed analysis is necessary to measure their final influence. The resulting distance functions satisfy the regularity conditions of convexity on inputs and concavity on outputs for majority of observations.[22]

The parameters associated with time-trend and long-term energy price variables are of specific interest. Negative parameters indicate positive TC; a positive parameter indicates negative TC. The LR test statistics on these parameters allows us to reject the null hypotheses of no exogenous (EI) or energy price induced innovations (PII) (Table 14.2). We find absence of neutral EI as the coefficients γ_1 is statistically insignificant, although it has required sign, but the presence of biased or embodied EI as the coefficients of interaction terms between time-trend and output and time-trend and inputs are statistically significant. The coefficient γ_2 is positive and statistically significant, indicating regressive neutral PII. This observation is consistent with the literature on inverse relationship between oil prices and GDP growth. This is due to the classic supply-side effect, according to which rising oil prices are indicative of the reduced availability of a basic input to production, leading to a reduction of potential output. Consequently, there is a rise in cost of production, and the growth of output and productivity is slowed.[23] But the coefficients of interaction terms between output and energy prices, and inputs and energy prices indicate progressive embodied PII.

Moreover, the results reveal that TC varies considerably between countries. For instance, India in developing countries and Japan and the United States in developed countries observe larger technological change effects (Figs. 14.3–14.5). One explanation for this could be that the functional form used is only a local approximation, and the countries that differ significantly from the rest may be assigned extreme TC.[24]

14.5.1 Levels of Inefficiency in the Countries

Country-specific technological diffusion, innovations – exogenous and energy price induced, and technological change are generated for each year over the period of 1974–2000.[25] The pooled sample average value of directional distance function is 0.065, implying that the country that is operating at the average values of inputs and output has the potential to increase GDP and simultaneously decrease the quantities

[22] We find that the monotonicity conditions with respect to output is satisfied by all the observations, and with respect to inputs: labor, capital and energy these conditions are satisfied by 98.18%, 100%, and 100% observations respectively.

[23] ***See among others, Barro, 1984; Brown and Yücel, 1999; Abel and Bernanke, 2001.

[24] The size of these economies is quite large in comparison to other sample countries and they may be outlier in the sample.

[25] Country- and time specific inefficiency and components of technological change are not reported because of space restrictions. The results are available from the author upon request.

of inputs (labor, capital and energy) by 6.5% (Appendix). The level of inefficiency is higher in developed countries than in developing countries. Developed countries have the potential to increase GDP and reduce the consumption of inputs by about 9.6%, whereas this potential for developing countries is 5.3%. It is also observed that the level of inefficiency is increasing over time in developing countries. In developed countries the level of inefficiency was relatively low in the 1980s compared with the 1970s or 1990s. At the country level, we find that the lowest level of inefficiency is observed in Egypt; that is, Egypt is operating quite near to the frontier. The countries that observed an inefficiency of more than 10% are Nigeria (25.96%), Iran (13.24%), Thailand (18.13%), Canada (23.9%), Japan (20.99%), Republic of Korea (19%), United States (30.49%), Indonesia (11.76%), Ethiopia (10.85), and Australia (10.03%).

14.5.2 Technological Diffusion and Exogenous and Energy Price Induced Innovations

The components of technological change are presented in the Appendix. The world witnessed technological progress increasing by 0.1% per annum, and this is attributed to the growth of exogenous innovations since the technological diffusion effect was negative in a magnitude of −0.1% per annum.

During the study period, 25 countries observed positive TC and India experienced the highest growth rate of 3.72% per annum; about 97% of the growth can be attributed to innovations. In technological progress, India is followed by Japan (2.21%), United States (2.16%), and United Kingdom (1.18%). Korea and Nigeria experienced negative technological change of the magnitude of 1.66% and 1.07% per annum respectively.

The technological diffusion or catch-up effect is negligible across the groups, although it is positive in the developed countries and negative in developing countries. In the sample of 55 countries, Japan observed the highest technological diffusion effect of about 2.15% per year followed by the United States with 1.09%. On the other hand, Korea and Nigeria witnessed a decline in catch-up effect of 1.85% and 1.52% per annum, respectively, which explains the decline of technological change in these countries. Out of 55 countries, 11 tried to catch the world frontier and 44 countries observed a negative catch-up effect.

Innovations are decomposed into two categories: exogenous and energy price induced. It is found that developed countries witnessed higher exogenous innovations (EI) than did developing countries, and the gap between the groups in the growth of EI has narrowed down over time (Fig. 14.2). Fifty-four countries witnessed exogenous innovations (EI), and India observed the highest growth rate in EI of about 3.4% per annum followed by the United States (2.96%). Only Gabon experienced a decline in EI. This implies that although innovations have contributed positively to

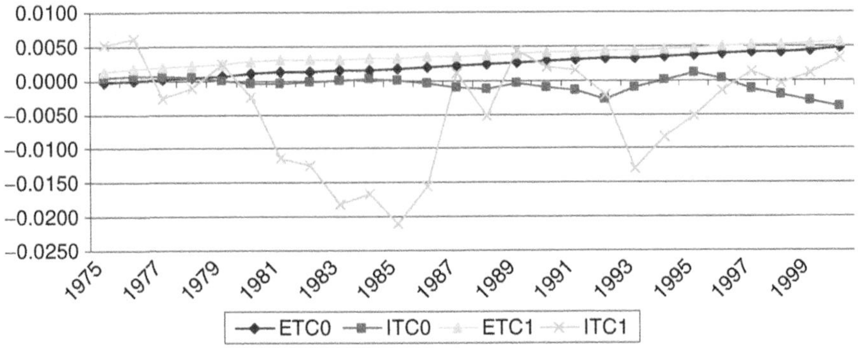

Fig. 14.2 Exogenous and induced innovations in developed and developing countries. *ETC0* exogenous innovations in developing countries, *ITC0* the induced innovations in developing countries, *ETC1* exogenous innovations in developed countries, and *ITC1* the induced innovations in developed countries
Source: Authors' calculations

growth for most countries, the pattern is very dissimilar, and developed countries have benefited more from exogenous innovations than developing countries.[26]

At the world level, on average we observe an absence of energy price induced innovations, and all progress in innovations can be attributed to progress in exogenous innovations. However, Fig. 14.2 reveals that developed countries observe substantial energy price induced innovations (PII) when long-term oil prices were rising, although the growth rate of PII is much volatile in these countries. In developing countries the magnitude of PII is negligible and is not associated with long-term changes in energy prices. This finding is consistent with the given level of energy consumption in the concerned economies. In the developed economies the per capita as well aggregate energy consumption is too high in comparison to developing economies so the expected magnitude of PII is expected to be higher. Developed countries account for more than half of the world total final consumption of energy (IEA, 2006). During the study period 22 countries observed an outward shift in the production frontier due to change in long-term oil prices, although the magnitude of progress was negligible.

Moreover, while analyzing the effect of the components of technological progress at the country level one should also keep in mind the nature of the economy and the level of energy use in production and consumption activities. The developing country group consists of most of the economies where state intervention is relatively high and market forces are allowed to play a limited role in economic activities in comparison to developed countries. Therefore to understand the implications of long-term oil prices the obvious way is to analyze the country-specific results. Due to space constraints, we present the analysis of results for three major

[26]The similar kind of trend is observed by Kumar and Russell (2002) using a sample of 55 countries for the period of 1965–1990. The countries taken in the studies are different.

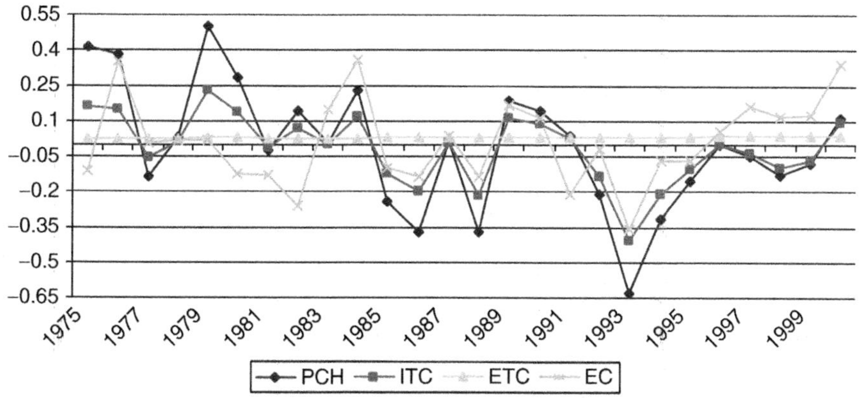

Fig. 14.3 Technological progress and oil price changes in USA. *PCH* long-term oil price changes, *ITC* the induced innovations, *ETC* exogenous innovations, and *EC* technological diffusion
Source: Authors' calculations

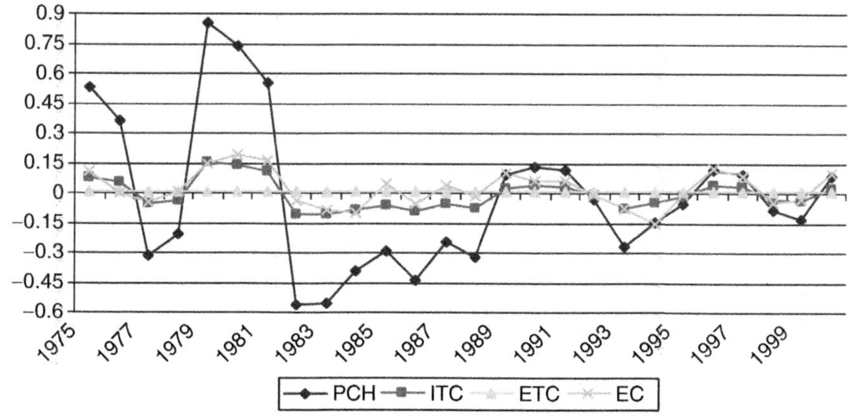

Fig. 14.4 Technological progress and oil price changes in Japan. *PCH* long-term oil price changes, *ITC* the induced innovations, *ETC* exogenous innovations, and *EC* technological diffusion
Source: Authors' calculations

economies, viz., United States, Japan, and India; the first two are developed and the third one is a major developing economy. We consider these three economies for further analysis because of their size and aggregate consumption of energy. Japan and the United States together account for about 75% of the estimated public sector spending in the area of energy research, development, and demonstration (ERD&D) by International Energy Agency (IEA) countries (Gallagher et al., 2006). Although there are no systematic and detailed data on public ERD&D spending in developing countries, spending in India is fairly large. India spent the equivalent of about 0.9 billion 2000 PPP$ in 1996–1997 (Sagar, 2002). The results for these three countries are presented in Figs. 14.3–14.5.

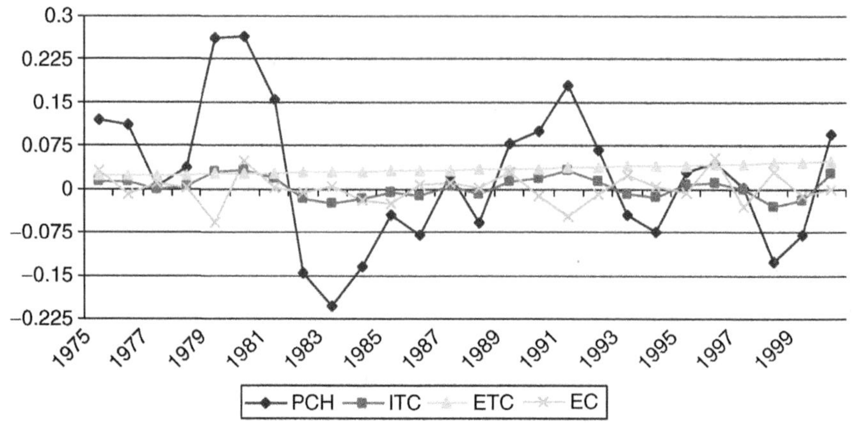

Fig. 14.5 Technological progress and oil price changes in India. *PCH* long-term oil price changes, *ITC* the induced innovations, *ETC* exogenous innovations, and *EC* technological diffusion
Source: Authors' calculations

In all three economies, we observe a stable growth path in exogenous innovations (EI). The U.S. economy experienced exogenous innovations of about 3% per annum, and it was 3.4% per annum for India. The annual growth rate of EI in Japan was 0.76%.[27] The path of technological diffusion is more volatile in the United States than in Japan and India. On average the contribution of diffusion in technological change is negligible; however, all three economies observed positive change in catch-up effect. Technological progress in Japan can be attributed mainly to the technological diffusion effect, whereas in the United States it is the function of both technological diffusion and exogenous innovations. In India, technological progress can be attributed mainly to exogenous innovations.

The annual growth path of energy price induced innovations (PII) is of particular interest. Figures 14.3–14.5 show that the path of PII is very volatile and is consistent with changes in long-term oil prices. All three countries observed a high growth rate in PII when oil prices were rising and decline in PII when oil prices were declining. It is observed that the growth rate was highest during the period when long-term oil prices were at their peak. This finding is consistent with the expenditure in ERD&D area in the United States and Japan. Public ERD&D in OECD countries showed a significant upward spike in the wake of the oil crises of the 1970s. These expenditures peaked in the early 1980s and then declined significantly (Gallagher et al. 2006). In developed economies it is not the public sector that spends for ERD&D technologies but the private sector that makes more investments in this area. Although the exact figures on private investments in ERD&D area are not available, some data certainly support this position. The National Science Foundation's

[27]This finding corroborate with Färe et al. (1994) and Kumar and Russell (2002), although their samples consist of the different groups of countries and different sample periods. They used the DEA technique for measuring exogenous technological changes.

annual survey of industrial R&D indicates that (public and private) funds for industrial energy R&D showed an almost continuous decline in the second half of the 1980s and 1990s; with the 1999 levels about a fifth of the peak value in 1980 in real terms.[28]

Moreover, it is also observed that the growth rate of energy price induced innovations was higher in the United States than in Japan. This finding is consistent with the dependence of the countries on imported oil and structural changes in energy consumption in the economies.[29]

During the study period, India observed positive growth in energy price induced innovations during the 1970s and 1980s, when the oil prices were at their peak and then showed positive changes in PII during 1995–1997 and in 2000 (Fig. 14.5), although the magnitude of PII was much lower in India in comparison to the United States and Japan.

14.6 Summary and Conclusions

Global climate change is linked to energy consumption. The reduction in energy consumption is possible with the innovations and diffusions of energy saving technologies. In this study, we have applied an analytical framework, developed by Robert Chambers and others, for estimating energy price-induced and exogenous technological change. A distinguishing feature of this framework is that it provides several pieces of information simultaneously: it describes the structure of production technology; it provides a measure of technological diffusion effect; and it provides the direction and pace of energy price induced as well exogenous innovations.

We used the directional distance function as an analytical tool in place of conventional representations of production technology, such as cost, production, or profit functions. The directional function simultaneously seeks to expand output and contract inputs. Conventional functions fail to distinguish between factor

[28] http:\\www.nsf.gov/statistics/iris/research_hist.cfm?index = 21 as quoted by Gallagher et al. (2006).

[29] Japan's dependence on imported crude oil is nearly 100%. Theoretically, the effect of a price rise on the Japanese economy should be extensive. However, the structural pattern of energy consumption in 1990s Japan differs from that of the first and second oil crises. During first and second crises, crude oil's share of energy consumption in Japan was nearly 80%, but this has fallen below 50% in late 1990s. Moreover, Japan has taken steps to cushion the impact of a rise in crude oil prices through a strategic stockpile and raising the efficiency of energy consumption. The ratio of energy consumption to GDP, calculated by dividing energy consumption by real GDP, declined rapidly from the middle of the 1970s to the first half of 1980 and mildly after that until the beginning of the 1990s. Since then, it has been flat or risen slightly. Between 1973 and 2000 it fell 33%, reflecting Japan's increased energy consumption efficiency (ono, 2005). However, the US dependence on oil in energy consumption has declined slightly from about 46% in 1973 to 38% in 2000.

substitution and shift of the production technology frontier. The shift in the production technology frontier with respect to long-run energy prices signals energy price induced innovations (PII).

We estimated the directional distance function for panel data on 55 countries over the period 1974–2000. The country estimates of directional distance function incorporate past energy prices as a factor inducing PII and a time trend to account for EI. This approach is also used to decompose technological changes into technological diffusion and innovations and to investigate input bias arising from EI and PII.

The application of this analytical framework to the macroeconomic data yields several important findings. First, the parameter estimates of directional distance function reveal the absence of neutral EI and the presence of biased innovations – either it is EI or PII. Second, the study provides an interesting descriptive look at innovations and diffusion across a wide range of countries. Third, in developed countries we observe larger PII in comparison to developing countries in the periods after first (1974), and second (1980) world oil crisis that caused substantial energy price increases. The time pattern of the PII effect in high-income countries also seems consistent with the economic theory and data that show most R&D activities occur in high-income countries, particularly in the United States and Japan.

Appendix: Average Annual Values of Luenberger Productivity Indicators

Country	INEFF	TD	EI	PII	EI + PII	TC
Argentina	0.0122	−0.0001	0.0025	0.0006	0.0031	0.0030
Australia	0.1003	−0.0009	0.0014	0.0002	0.0016	0.0007
Bolivia	0.0290	−0.0002	0.0005	0.0000	0.0005	0.0003
Canada	0.3174	−0.0095	0.0028	0.0003	0.0031	−0.0064
Switzerland	0.0363	−0.0005	0.0006	0.0001	0.0007	0.0002
Chile	0.0215	−0.0006	0.0007	0.0001	0.0008	0.0002
Cote d'Ivoire	0.0307	−0.0009	0.0005	−0.0002	0.0003	−0.0006
Cameroon	0.0342	−0.0010	0.0005	−0.0002	0.0003	−0.0007
Colombia	0.0113	−0.0003	0.0017	−0.0001	0.0015	0.0012
Costa Rica	0.0264	−0.0002	0.0002	−0.0001	0.0000	−0.0002
Denmark	0.0210	0.0006	0.0006	0.0001	0.0006	0.0012
Dominican Republic	0.0274	0.0000	0.0003	−0.0001	0.0001	0.0001
Ecuador	0.0313	−0.0010	0.0003	−0.0005	−0.0001	−0.0011
Egypt	0.0112	0.0003	0.0019	−0.0002	0.0017	0.0021
Ethiopia	0.1085	−0.0041	0.0020	−0.0003	0.0017	−0.0024
Gabon	0.0244	−0.0002	0.0000	−0.0004	−0.0004	−0.0006
United Kingdom	0.0542	0.0070	0.0050	−0.0003	0.0047	0.0118
Ghana	0.0401	−0.0016	0.0005	−0.0005	0.0000	−0.0017
Greece	0.0248	0.0003	0.0006	0.0001	0.0007	0.0011
Guatemala	0.0194	0.0000	0.0003	−0.0001	0.0003	0.0002
Honduras	0.0265	−0.0003	0.0001	−0.0001	0.0000	−0.0003

(*continued*)

(continued)

Country	INEFF	TD	EI	PII	EI + PII	TC
Indonesia	0.1176	−0.0097	0.0074	0.0018	0.0092	−0.0005
India	0.0389	0.0009	0.0339	0.0025	0.0363	0.0372
Iran	0.1324	−0.0069	0.0018	−0.0001	0.0017	−0.0052
Iceland	0.0134	−0.0001	0.0001	0.0000	0.0001	0.0000
Israel	0.0199	−0.0006	0.0005	0.0001	0.0006	0.0000
Jamaica	0.0301	−0.0002	0.0001	0.0000	0.0001	−0.0001
Jordan	0.0263	−0.0003	0.0001	−0.0001	−0.0001	−0.0003
Japan	0.2099	0.0215	0.0076	−0.0070	0.0006	0.0221
Kenya	0.0715	−0.0023	0.0011	−0.0001	0.0010	−0.0013
Korea, Rep. Of	0.1900	−0.0185	0.0021	−0.0002	0.0019	−0.0166
Sri Lanka	0.0325	−0.0005	0.0008	−0.0002	0.0006	0.0000
Morocco	0.0170	−0.0001	0.0009	−0.0002	0.0007	0.0006
Mexico	0.0648	−0.0045	0.0038	−0.0004	0.0034	−0.0011
Malaysia	0.0475	−0.0030	0.0009	−0.0002	0.0006	−0.0024
Nigeria	0.2596	−0.0152	0.0045	0.0000	0.0045	−0.0107
Norway	0.0482	−0.0004	0.0003	−0.0001	0.0003	−0.0001
New Zealand	0.0175	−0.0005	0.0004	−0.0001	0.0003	−0.0002
Pakistan	0.0925	−0.0025	0.0033	0.0000	0.0033	0.0007
Peru	0.0468	−0.0013	0.0018	0.0008	0.0026	0.0013
Philippines	0.0498	−0.0033	0.0026	0.0000	0.0026	−0.0007
Paraguay	0.0245	−0.0004	0.0002	−0.0003	−0.0001	−0.0005
Senegal	0.0318	−0.0008	0.0003	−0.0004	−0.0001	−0.0009
El Salvador	0.0291	0.0002	0.0005	0.0007	0.0012	0.0013
Sweden	0.0556	0.0005	0.0009	0.0000	0.0009	0.0014
Syria	0.0432	0.0002	0.0008	0.0006	0.0014	0.0016
Togo	0.0288	−0.0006	0.0001	−0.0003	−0.0002	−0.0008
Thailand	0.1813	−0.0094	0.0027	0.0000	0.0028	−0.0066
Trinidad & Tobago	0.0306	−0.0004	0.0001	0.0000	0.0001	−0.0003
Turkey	0.0346	−0.0054	0.0029	−0.0002	0.0027	−0.0027
Tanzania	0.0921	−0.0025	0.0010	−0.0002	0.0009	−0.0017
Uruguay	0.0223	0.0001	0.0002	0.0001	0.0003	0.0004
USA	0.3049	0.0109	0.0296	−0.0190	0.0106	0.0216
Venezuela	0.0760	−0.0029	0.0009	0.0000	0.0009	−0.0020
South Africa	0.0713	−0.0005	0.0022	0.0001	0.0023	0.0018
Average	0.065	−0.001	0.003	0.000	0.002	0.001

INEFF level of inefficiency, *TD* technological diffusion (catch-up effect), *EI* exogenous innovations, *PII* energy price induced innovations, *EI* + *PII* sum of exogenous and the induced innovations, and *TC* technological change (TD + EI + PII)

Source: Authors' calculations

Chapter 15
The Road Ahead

15.1 Findings

To gain a global economic perspective, India is one of the most important countries to understand. It is the world's largest democracy and the world's fourth largest economy in terms of purchasing power parity. However, about one-third of the total population of the country survives on less than US$1 per day. These two facts lead to degradation and depletion of the environment and natural resources. Similarly, the country has elaborate statutes, regulations, institutional frameworks and policies on almost every conceivable topic – from hazardous waste to public liability for forests and wildlife. However, monitoring and enforcement capabilities are weak.

This book contributes to the literature in several ways. First, the research provides a comprehensive analysis of these issues within the context of India's environmental and energy problems. The book's 14 chapters address key issues regarding economic development, environmental regulations, and technological change in the Indian context. Chapter 2 examines the magnitude of environmental problems, in addition to general economic performance in India. These contrasts raise a question about the sustainability of the present growth trajectory from both economic and environmental points of view. The first real impetus for developing a framework for environmental protection in India came after the UN Conference on the Human Environment in 1972. Environmental policy in the 1970s and 1980s recognized the need for an institutional identity for environmental policy making, resulting in the setting up of the government agency in India. Chapter 3 provides an explanation of environmental regulations and the state of the environment in India. The degree of compliance with environmental regulations in a state depends on the probability of detection of noncompliance and the severity of punishment if detected and convicted. Noncompliance with environmental standards is a problem, and possible causes are discussed.

Thorough analyses on environmental problems are provided in Chapter 4. We analyzed the role of intergovernmental fiscal transfers in achieving environmental sustainability. This study highlights the need for both lump-sum and earmarked

S. Kumar and S. Managi, *The Economics of Sustainable Development*, Natural Resource Management and Policy 32,
DOI: 10.1007/978-0-387-98176-5_15 © Springer Science+Business Media, LLC 2009

grants for internalizing spatial externalities. Earmarked grants are better suited for environmental clean-up activities and for financing ways in which human resources and built infrastructure can be improved to build resilience to environmental degradation. Lump-sum transfers are better suited for precautionary activities such as nature preservation and soil and water protection. The study also underscores the need to find appropriate biotic and abiotic indicators of environmental performance that constitute a link between environmental services and corresponding costs for their provision. These indicators would be used to modify the existing formulas of resource allocations for acknowledging environmental services provided by the states and local bodies. To understand the significance of intergovernmental fiscal transfers in internalizing environmentally positive externalities, the study provided an illustration, which demonstrated that inclusion of forest cover in the formula for lump-sum transfers benefits the poor states that contain ecological resources. Poor states with degraded environments can be compensated through grants-in-aid for clean-up activities.

Probably most important, technological advances continue to play an important role in facilitating global integration. A thorough understanding of the nature of technological change is essential for developing well-conceived policies that contribute to the long-term well-being of society. Chapter 5 illustrates the importance of understanding the process of technological change. We found that in the prereform period productivity had grown at the rate of 1.7% per year, while in the postreform era the corresponding growth rate was 3%. While the prereform growth rate in productivity was due to gains in technical efficiency, postreform growth was influenced by technical progress. Another interesting result of the present exercise is the nature of technical progress in Indian manufacturing. It was seen that the capital intensity of Indian firms has been increasing in recent years. Although regional differences in productivity persist, the variation has declined in the postreform period. The majority of states tried to be nearer the isoquant in the postreform era than in prereform years. Most of the states are also operating under increasing returns to scale, and the gain in productivity in the postreform era was due to gain in technical progress. In contrast, in the prereform period it was due to efficiency improvement. During the 1990s, capital intensity in the manufacturing sector seemed to have increased as technical progress was in favor of capital. The states which were exhibiting either neutral or labor-using technical bias in the prereform period also show capital-using technical change during the postreform era. It was also found that although there is a tendency toward convergence in terms of the productivity growth rate among Indian states during the postreform era, only those that were technically efficient at the beginning of reform remained innovative.

The Indian economy today is highly prone to industrial pollution and is making compliance decisions in order to meet environmental standards. Environmental regulations impose significant costs upon industry that are fairly high and that therefore require economic justification. This justification can be given by estimating the benefits associated with these costs. While the scientific rationale behind air quality preservation is well understood, its economic rationale for a developing country like India has to be verified. Chapter 6 estimates the economic value that people in an urban area in India. This chapter verifies the economic rationale of air quality preservation by the dose-response technique to derive a compensating variation type expression for marginal willingness to pay (WTP) for improved air quality. The estimated WTP using health and air pollution (PM10) data is a comparatively low figure that is less than 1% of the average monthly income. The study makes two contributions to the literature on nonmarket valuation of goods and services in developing countries: first, the paper presents, perhaps for the first time, estimates of the economic value to households of air quality improvements in a low-income developing country like India and second, the study suggests that the use of nonmarket valuation methods in developing countries can be both practical and feasible.

There are clear economic linkages between preservation efforts and the benefits derived from these efforts. We show that in a developing country like India, people's preferences are well formed to place values on air quality preservation. The benefits to local residents are estimated, and an estimate ranges about 2% of the average monthly income of the residents. The study makes two contributions to the literature on nonmarket valuation of goods and services in developing countries. First, it compares the estimated value to households of air quality

Chapter 7 examines the relationship between environmental regulations and production efficiency, which is a more general representation of production technology. The stochastic output distance function was estimated simultaneously with a model that explained the causes of inefficiency. The results of this study contradict the Porter hypothesis, i.e., that environmental regulations lead to production inefficiency. But according to Porter, it is not regulations as such; it is the intensity or stringency of regulations that encourages firms to adopt "pollution prevention methods." These methods restrict economic waste, and pollution is a manifestation of economic waste. If any pollution standards are satisfied in India, they are met through the "end-of-pipe" treatment. The average level of efficiency is only 0.58, and the coefficient of variation is quite high. Moreover, here it is found that there is a positive association between plant size and production efficiency and between the rate of capacity utilization and production efficiency. This reveals that the energy crisis in India can be resolved, to some extent, through better utilization of existing capacity.

The maintenance cost of water pollution abatement measures to Indian industry is estimated in Chapter 8. We estimate firm-specific shadow prices for pollutants. The estimates of production efficiency for water-polluting industries in India reported in this paper explain production efficiency with a joint production of good and bad outputs. For Indian water-polluting industries as a whole, the estimated efficiency index is approximately 90%. It means that by employing the same set of inputs, the good output can be further increased by 10%. Among industries for which an efficiency index is estimated, distillery has the lowest, while iron and steel has the highest efficiency in the sample of 60 firms from 17 water-polluting industries in India. We find that water-polluting industry has decreasing returns to scale. Estimates show that three industries, that is, fertilizers, refinery, and drugs, have increasing returns to scale, while others have decreasing returns to scale.

There is a positive correlation between the economies of scale and the turnover of a firm. Also, there is a positive association between pollution control and economies of scale (the higher the scale economies, the lower the effluent–influent quality ratio). The shadow prices of pollutants may be interpreted as the marginal costs of respective pollutants. The result – that there is a negative relationship between pollution load reductions and shadow prices across the firms found in this study – confirms the presence of scale economies in pollution abatement found in the earlier studies on industrial water pollution abatement in India.

Next, in Chapter 9, we estimate the effect of environmental regulation on the productive efficiency of water-polluting industries. Environmental regulation could provide incentives to firms for innovation and resource conservation in environmental management. Using firm-specific data, the Porter hypothesis is tested for the Indian water-polluting industry. The technical efficiency of firms increases with the intensity of environmental regulation and water conservation efforts. This result supports the Porter hypothesis about environmental regulation. Win–win opportunities from environmental regulation were found more in some industries than in others. Given the very high monitoring and enforcement cost of environmental regulation, this could result in the significant cost savings.

Chapter 10 investigates the structure of industrial water demand in India. We first show that there is high variability in the production efficiency of Indian manufacturing industries. They can produce the same level of output with less than half of the quantities of inputs that they are using on average. There are increasing returns to scale in our sample of firms, with an average of 1.42. Returns to scale is positively associated with turnover and water intensity. The estimated average shadow price of water is Rs. 7.21/kl. We observe a wide variation across industries and firms in these shadow prices. The shadow price varies from Rs.1.40 per kiloliter for petrochemicals to Rs. 30.54 per kiloliter for paper and paper products industry. We find that water is a complement to labor and materials and a substitute for capital. The price elasticity of water demand is about −0.902 (in conventional sense −1.11) at the sample mean. This high value is similar to what has been found by other researchers working on developing countries (for example, China and Brazil). Thus, given the high responsiveness of water demand to price, water charges may act as an effective instrument for water conservation.

Oil prices have sharply risen in recent years. India is the seventh largest consumer of oil in the world. Chapter 13 studies the oil price–macroeconomy relationship in the Indian economy by analyzing the impact of oil price shocks on the growth of industrial production. We find an increase in real oil prices negatively affects the growth rate of industrial production. For the Indian economy we find that a 100% increase in real oil prices reduced the growth of industrial production by 1%. This small impact of the growth of industrial production can be traced, among other factors, to depreciation in the real effective exchange rate. Furthermore, we find that the inflation rate and short-term interest rate are positively affected by the increase in real oil prices. We also obtain evidence on the asymmetric relationship between oil prices and the growth of industrial production, confirming the relationship found in developed economies. The variance decomposition analysis shows that oil price shocks are a considerable source of volatility for the variables used in the study. For the growth of industrial production, oil price shocks combined with monetary shocks are the largest source of variation other than the variable itself; thus, the variance decomposition analysis puts the relationship between oil price and industrial growth into perspective, while the focus of the study is to analyze the impact of oil price shocks on the growth of industrial production.

Extremely rapid economic growth brought serious environmental problems. Whether pollution abatement technologies are utilized efficiently is crucial in the analysis of environmental management because they influence the cost of alternative production and pollution abatement technologies, at least in part. In Chapter 11, we show that overall environmental productivity decreases over time. At present, existing environmental management is not sufficient to bring sustainable development. However, once we disaggregate the pollutants to specific pollution by SO_2, NO_2, and SPM, we find environmental productivity recently increases in SO_2. The results for NO_2 and SPM are the main causes of the productivity reduction over the study periods. Furthermore, we analyze the determinants of environmental productivity and find an EKC-type relationship exists between environmental productivity and income. However, environmental productivities in general decline more in high-income states than in low-income states. Panel analysis results show that the scale effect is negative and dominant over the positive technique effect. Therefore, a combined effect of income on environmental productivity is negative, which answers the puzzle of why productivity has declined faster in developed states than in their underdeveloped counterparts. We conclude that if the ongoing pace of industrialization is not met with effective environmental management, there will be untoward consequences in India. Indian society is required to introduce environmental practices based on incentives for industries to perform well in environmental management and to simultaneously formulate economic and environmental policies that achieve a sustainable growth process.

Chapters 12 and 14 analyze the position of India in a more global context. Chapter 12 provides insight into the sources of productivity growth to estimate an adjusted rate of productivity growth while accounting for CO_2 emissions minimization activities in the world. Through an asymmetrical treatment of good and bad outputs, the productivity index is decomposed into efficiency and technical changes. This index provides a common dialog of different perspectives on the climate change debate by expanding the basic economic concept of productivity to identify the combined role of technological innovation and adoption and green accounting. The productivity index is calculated using 41 countries consisting of 21 Annex-I countries and 20 Non-Annex-I countries. In the components of productivity, technical and technical efficiency changes, the null hypothesis of whether the different indexes are same when emissions are ignored and when they are accounted for cannot be accepted for either of the groups of countries. Out of 41 countries, only six – Iceland, Hong Kong, Japan, Luxembourg, Netherlands, and Switzerland – were innovators. None of the developing countries was shifting the frontier under either scenario. We also find that the environmentally sensitive measure of productivity is higher in those countries, which have higher GDP per capita. We also find a presence of convergence hypothesis, and the energy intensity of production is negatively related to the environmentally sensitive measure of productivity. However, the conventional measure of productivity remains unaffected by the composition of output growth. The openness of a country increases its productivity.

Technological progress plays a crucial ameliorating role in reducing energy consumption for combating climate change. Energy economists often cite market-based instruments such as energy taxes for encouraging energy-saving technological progress. Energy policy interventions may change the constraints and incentives that affect technological change. The reduction in energy consumption is possible with innovations and diffusions of energy-saving technologies. In Chapter 14, we estimate energy price-induced and exogenous technological change for a panel data of 55 countries. We the parameter estimates of directional distance function reveal the absence of neutral exogenous innovations (EI) and the presence of biased innovations, either EI or long-run energy prices signal energy price induced innovations (PII). Also, the study provides an interesting descriptive look at innovations and diffusion across a wide range of countries. In developed countries we observe larger PII in comparison to developing countries in the periods after the first (1974) and second (1980) world oil crises that caused substantial energy price increases. The time pattern of the PII effect in high-income countries also seems consistent with economic theory and data that show most R&D activities occur in high-income countries, particularly in the United States and Japan.

Technological change plays a key role in maintaining standards of living in economies with increasingly stringent environmental goals. Successful environmental policies can contribute to efficiency by encouraging, rather than inhibiting, technological innovation. Over time, economists have greatly improved our understanding of the role of technological change in economic growth and of the constituents of technological change. We have progressed from confessions of ignorance based on mere observations that productivity increases over time to an increasingly sophisticated understanding of the mechanisms that drive technological change and empirical measures of the various components of technological change.

However, little research to date has focused on analyses of environmental regulations that encourage technological progress or on ensuring productivity improvements in the face of the increasing stringency of environmental regulations. In particular, almost no previous studies analyze the case in India. Using case studies in India, we believe our analyses shed light on several different perspectives on environmental problems in India.

15.2 Climate Change Policy

Finally, we briefly discuss climate change and India policy and consider the global significance of Indian policy. Climate change is a long-run global problem and requires global efforts. It is a classic case of stock externalities. It is due to historical accumulation of greenhouse gas (GHG) emissions and to a large extent due to anthropogenic activities. The IPCC Fourth Assessment Report shows that climate has been changing due to human activities and that strong actions are required to ensure that the world does not face excessive risks from the global warming. The recently released National Action Plan on Climate Change recognizes the need for a national strategy, while simultaneously acknowledging the need for keeping engaged with the international community. There are two reasons for the need to collectively and cooperatively deal with the problem of climate change: first, to adapt to climate change and second, to enhance the ecological sustainability of the development path. The plan restates India's stand on the problem, which is that India would remain engaged "actively in multilateral negotiations in the UN Framework Convention on Climate Change, in a positive, constructive, and forward looking manner."

India emits only about 4% of annual global GHG emissions, and its contribution to the accumulated atmospheric concentration of emissions is only about 2% versus the contribution of about 30%, 27%, and 7.3% by the United States, EU-25, and China, respectively. According to former Secretary, Ministry of Environment and Forests, and a member of the Council on Climate Change, Government of India, "if India were to eliminate all its [greenhouse gas] emissions, essentially by going back to the Stone Age, it would hardly matter for the climate change impacts on India, or indeed, anywhere else!" Then the question is, Why India should be concerned about climate change? India has been concerned with the climate change phenomenon due to its substantial adverse impacts on agriculture, sea level rise leading to submergence of coastal areas, and increased frequency of extreme events. Therefore, a global climate policy that helps in reducing future vulnerability by getting the developed countries to reduce their emissions is very important for India.

In 1997, the Kyoto Protocol (KP) was formalized to take action against the problem of climate change. The principle governing the KP was common but differentiated responsibility and relative capabilities, and was enshrined in the UN Framework Convention on Climate Change (UNFCCC). It was also decided that the targeted emission cuts could be realized through emissions trading, joint implementation (JI), and CDM so that the total costs of meeting the targets are minimized. CDM, similar to JI, is a project-based mechanism; it enables countries with a specific emission reduction target to obtain credit for implementing abatement projects in developing countries. In return, developing countries are expected to get financial and technological assistance.

The main weakness of the KP is that it has failed to promote the participation of the largest emitter in the world, the United States. As a result of this nonparticipation, the potential benefits of the flexibility mechanism are lost, and the total costs of the KP mitigation effort are going to be higher than they would be with U.S. participation. Second, the protocol contains no provision to deal with a country that fails to comply with committed emission cuts. For example, in 2005, the level of GHG emissions in Canada was about 55% higher than the level of emissions in 1990, and it seems to be very difficult for Canada to comply with the target. Third, CDM is also considered as a major source of technological transfer and diffusion.

Moreover, the NAPCC states that India does not escape from its responsibilities and is ready to make its contribution to the solution of problem, provided that the developed countries accept their responsibilities for the legacy of the problem and fulfill their commitment regarding financial and technology transfers under the UNFCCC. Further, it emphasizes the need for an equitable and efficient solution that allows equal allocation of global environmental space to all human beings.

Note that the NAPCC does not set any concrete numerical targets for emissions cuts or even for increasing energy efficiency; it suggests measures that promote development objectives, simultaneously yielding "cobenefits" for addressing climate change. The plan offers a list of eight technological missions that help in moving towards a sustainable development path. It emphasizes research and development of solar energy and improving energy efficiency in the economy. It asks for mandated reductions in consumption by energy-intensive industries, improved urban planning, and a new building code. But it fails to give details on how these objectives would be realized. Questions about how government policies would help Indian industry be competitive in a carbon-constrained world are yet to be answered. (That is, crafting proper public policies that promote carbon-efficient technological progress is a moot question before the Indian government.)

The NAPCC stresses that India's per capita emissions would not exceed the average global emissions of the developed countries. India's special envoy on climate change has interpreted this statement to mean that India will set limits to its emissions according to the limits that developed countries are ready to establish. "The more ambitious they are, lower the limit that India would be prepared to accept. Thus there is an inbuilt mutuality of incentives." In this way, the Indian stand may help break the deadlock between developed and developing countries at the international climate negotiations.

Though presently per capita emissions in India are about one tonne, they are bound to increase under the business-as-usual scenarios and given the projections of energy requirements in the Integrated Energy Policy (Government of India, 2006). About 600 million people do not have access to electricity – even for lighting. The Integrated Energy Policy projects that in the next 25 years, electricity generation in the country will increase by 7–8 times, which involves a 4–5 times increase in coal and 9–10 times increase in natural gas. Aggregate emissions would be higher by 4–5 times, and per capita emissions would be about 2.8–4.0 tons by 2030 (Government of India, 2006).

If the promise made in the NAPCC is taken together with what scientists conjecture are the requirements for avoiding disastrous impacts of climate change, then India would have tight constraints with respect to emissions. If the world were to agree on emissions reductions of 80% below 1990 levels by 2050, one likely scenario is that world emissions would have to peak by 2015 before declining to less than 20 billion tons of CO_2 emissions by 2030. By that time, according to UN projections, the world population would be about 8 billion or more, and if the allowed emissions are shared equally, the per capita limit comes out to be around 2.5 tons of CO_2 emissions. This necessitates the need to plan for a low-carbon Indian society along with the given objectives in the NAPCC.

India needs financial and technology transfers from developed countries for reducing carbon emissions in a way that simultaneously realizes the objectives of climate protection and sustainable development. The Bali Road Map pledges developing countries to consider nationally appropriate mitigation actions in the context of sustainable development, supported and enabled by technology, financing, and capacity building in a measurable, reportable, and verifiable manner.

The post Kyoto climate regime should continue with features like CDM. Technology transfers are associated with the size of projects. The CDM design should encourage exploitation of economies of scale that can be realized by designing sector-specific CDM. The sector-specific approach could lower the transaction costs and also garner a large flow of funds and technology.

As for technological diffusion and absorption by transferring from developed countries to developing countries, it was assumed that promarket institutional reforms alone would encourage rapid technology transfer and industrial growth. Leapfrogging is considered to be possible through these reforms, provided domestic entrepreneurs and foreign investors are given free rein to exploit market opportunities at home and abroad. However, this assumption turned out to be false because technological transitions in industries entail conscious effort by firms, governments, and the network of actors involved in transition management.

15.3 What to Do?

The adjustment to external landscape pressures does not happen in a stylized manner. Instead, it happens through negotiations, power struggles, and shifting coalitions of different actors. A change in the social network is often important to start a transformation process because incumbent regime actors initially tend to downplay the need for transformation. The neo-laissez-faire scene failed to transform these power relations, coalitions, and networks and consequently was unable to effectively restart or sustain the engines of growth. Therefore, improvements of productivity increase came about slowly at best, and there were unsatisfactory incentives to encourage local firms to invest in the capabilities needed to learn, upgrade technologies, and transform production systems.

The prospects for sustainability transitions through future environment and development policies depend on how many actual incentives sustainable transitions exist in each policy. In summary, it is essential to account for the power of different shareholders in shaping the environmental quality of value chains. Future studies in this field are essential to understand a society for sustainable development.

Reliable baseline forecasts and responses to different policy actions of production and pollution in India are critical for the formation of sound technology, energy, and environmental policy. Improved understanding of the role of technology and environmental policy in demand, supply, and productivity changes will lead to improvements in decision making and the design of environmental regulations. A detailed policy scenario study provides a quantitative assessment of the potential cost and benefits, indicating the significance of the potential benefits of environmental policy that reduce emissions and encourage innovation. For example, Khanna and Zilberman (1999, 2001) construct a microeconomic framework to analyze the impact of alternative mixes of policy reforms and develop a framework to explore the implications of trade and domestic policy distortions for the magnitude of carbon emissions and for the welfare costs of abating these emissions for India. Future studies need to estimate and understand how the development of sound policy, such as taxing or emission trading, requires an improved understanding of the nature and role of technological change in shaping future living standards.

Bibliography

Abel A.B., Bernanke B.S. (2001) Macroeconomics. Addison-Wesley Longman, Boston, MA.

Acemoglu D., Johnson S., Robinson J. (2002) Reversal of fortune: geography and institutions – the making of the modern world income distribution. Quarterly Journal of Economics 117: 1231–1294.

Afriat S.N. (1972) Efficiency estimation of production function. International Economic Review 13: 568–598.

Aghion P., Howitt P. (1998) Endogenous growth theory. MIT, Cambridge, MA.

Aghion P., Burgess R., Redding S., Zilibotti F. (2003) The unequal effects of liberalization: theory and evidence from India. London School of Economics, London.

Ahluwalia I.J. (1991) Productivity and growth in Indian manufacturing. Oxford University Press, Delhi.

Ahluwalia I.J. (1995) India: industrial development review. The Economist Intelligence Unit/ UNIDO, London/Vienna.

Ahmad J.K., Goldar B.N., Jakariya M., Misra S. (2002) Willingness to pay for arsenic free, safe drinking water in rural Bangladesh: methodology and results. Report, Water and Sanitation Programme: South Asia, Dhaka, Bangladesh.

Aigner D.J., Chu S.F. (1968) Estimating the industry production function. American Economic Review 58: 826–839.

Aigner D.J., Lovell C.A.K., Schmidt P. (1977) Formulation and estimation of stochastic frontier production function models. Journal of Econometrics 6: 21–37.

Alberini A., Krupnick A. (2000) Cost of illness and willingness to pay estimates of the benefits of improved air quality: evidence from Taiwan. Land Economics 76: 37–53.

Alberini A., Cropper C., Fu T., Krupnick A., Liu J., Shaw D., Harrington W. (1997) Valuing health effects of air pollution in developing countries: the case of Taiwan. Journal of Environmental Economics and Management 34: 107–126.

Altaf M.A., Hughes J.A. (1994) Measuring the demand for improved urban sanitation services: results of a contingent valuation study in Ouagadougou, Burkina Faso. Urban Studies 31(10): 1763–1776.

Altaf M.A., Whittington D., Jamal H., Smith V.K. (1993) Rethinking rural water supply in the Punjab, Pakistan. Water Resources Research 29(7): 1943–1954.

An M.Y. (2000) A semiparametric distribution for willingness to pay and statistical inference with dichotomous choice contingent valuation data. American Journal of Agricultural Economics 82: 487–500.

Anand R.B., Perman R. (1999) Preferences, inequality and entitlements: some issues from a CVM study of water supply in Madras, India. Journal of International Development 11(1): 27–46.

Antle J.M. (1984) The structure of U.S. agricultural technology: 1910–78. American Journal of Agricultural Economics 66: 414–421.

Antweiler W., Copeland B., Taylor S. (2001) Is free trade good for the environment? American Economic Review 91(4): 877–908.

Arbues F., Garcýa-Valiñas M.Á., Martýnez-Espiñeira R. (2003) Estimation of residential water demand: a state-of-the-art review. Journal of Socio-Economics 32: 81–102.

Arellano M., Bond S. (1991) Some tests of specification for panel data: Monte Carlo evidence and an application to employment equations. Review of Economic Studies 58: 277–297.

Arellano M., Bover O. (1995) Another Look at the Instrumental Variable Estimation of Error-components Models. Journal of Econo-metrics, 68, pp. 29–52

Arrazola M., De Hevia J. (2004) More on the estimation of the human capital depreciation rate. Applied Economics Letters 11: 145–148.

Arrow K.J. (1962) The economic implications of learning by doing. Review of Economic Studies 29(3): 155–173.

Arrow K.J., Solow R., Portney P.R., Learmer E., Radner R., Schuman H. (1993) Report of the NOAA panel on contingent valuation. National Oceanic and Atmospheric Administration, US Department of Commerce, Washington, DC.

Arrow K.J., Dasgupta P., Mäler K.G. (2003) Evaluating projects and assessing sustainable development in imperfect economies. Environmental and Resource Economics 46(4): 647–685.

Arrow K.J., Dasgupta P., Goulder, L., Daily G, Ehrlich P.R., Heal G.M., Levin, Maler K.-G., Schneider S., Starrett D.A., Walker B. (2004) Are We Consuming Too Much? Journal of Economic Perspectives, 18: 147–172.

Arrow K.J., Dasgupta P., Goulder L., Daily G., Ehrlich P., Heal G., Levin S., Mäler K.G., Schneider S., Starrett D., Walker B. (2005) Are we consuming too much? Journal of Economic Perspectives 18: 147–172.

Atkinson A.B., Stern N. (1974) Pigou, taxation and public goods. Review of Economic Studies 41: 119–128.

Atkinson S.E., Dorfman R.H. (2002) Bayesian measurement of productivity and efficiency in the presence of undesirable outputs: crediting electric utilities for reducing air pollution. Department of Economics, University of Georgia, Athens, GA.

Auty R.M. (2007) Natural resources, capital accumulation and the resource curse. Ecological Economics 61: 627–634.

Ayres R.U. (2008) Sustainability economics: Where do we stand? Ecological Economics. 67(2): 281–310.

Babin F., Willis C., Allen G. (1982) Estimation of substitution possibilities between water and other production inputs. American Journal of Agricultural Economics 64: 148–151.

Balakrishnan P., Pushpagandan K. (1994) Manufacturing industry: a fresh look. Economic and Political Weekly October: 2028–2035.

Balk B.M. (1998) Industrial price, quantity, and productivity indices: the micro-economic theory and an application. Kluwer, Boston, MA.

Balke N.S., Brown S., Yucel M. (1998) Crude oil and gasoline prices: an asymmetric relationship? Federal Reserve Bank of Dallas Economic Review First Quarter: 2–11.

Balke N.S., Brown S.P.A., Yucel M.K. (2002) Oil price shocks and the U.S. economy: where does the asymmetry originate? Energy Journal 23(3): 27–52.

Ball E.V., Lovell C.A.K., Nehring R., Somwaru A. (1994) Incorporating undesirable outputs into models of production: an application to U.S. agriculture. Cahiers d'Economie et Sociologie Rurales 31: 60–74.

Barbara A.J., McConnell V.D. (1990) The impact of environmental regulations on industry productivity: direct and indirect effects. Journal of Environmental Economics and Management 18: 50–65.

Barro R.J. (1984) Macroeconomics. Wiley, New York.

Basberg B. (1987) Patents and the measurement of technological change: a survey of the literature. Research Policy 16: 131–141.

Bateman I., Willis K.G. (1999) Valuing environmental preferences: theory and practice of the contingent valuation method in the US, EU, and developing countries. Oxford University Press, New York.

Battese G.E., Coelli T.J. (1993) A stochastic frontier production function incorporating a model for technical inefficiency effects. Working Papers on Econometrics and Applied Statistics, No. 69, Department of Econometrics, University of New England, Australia.

Battese G.E., Coelli T.J. (1995a) A model for technical inefficiency effects in a stochastic frontier production function for panel data. Empirical Economics 20: 325–332.

Battese G.E., Coelli T.J. (1995b) Prediction of firm-level technical efficiencies with a generalised frontier production function and panel data. Journal of Econometrics 38: 387–399.

Baumol W.J. (1986) Productivity growth, convergence and welfare: what the long-run data show. American Economic Review 76(5): 1072–1085.

Baumol W.J., Oates W.E. (1988) The theory of environmental policy. Cambridge University Press, Cambridge.

Baumol W.J., Blackman S.A.B., Wolff E.N. (1989) Productivity and American leadership: the long view. MIT, Cambridge.

Bernanke B.S., Gertler M., Watson M. (1997) Systematic monetary policy and the effects of oil price shocks. Brookings Papers on Economic Activity 1: 91–157.

Bernstein M.A., Feldman S.L., Schinnar A.P. (1990) Impact of pollution controls on the productivity of coal-fired power plants. Energy Economics 12: 11–17.

Bhalla G.S., Hazell P., Kerr J. (1999) Prospects for India's cereal supply and demand to 2020. 2020 Brief No. 63.

Bhatia R., Rogers P., Briscoe J., Sinha B., Cestti R. (1994) Water conservation and pollution control in Indian industries: how to use water tariff, pollution charges and fiscal incentives. Currents, UNDP–World Bank Water and Sanitation Program.

Binswanger H.P. (1974) A cost function approach to the measurement of elasticities of factor demand and elasticities of substitution. American Journal of Agricultural Economics 56: 377–386.

Binswanger H.P. (1978) Measured biases of technical change: the United States. In Binswanger H.P., Ruttan V.W. (eds.) Induced innovation: technology, institutions, and development. John Hopkins University Press, Baltimore, MD.

Binswanger H.P. (1974) The measurement of technical change biases with many factors of production. American Economic Review 64: 964–976.

Blackorby C., Russell R.R. (1989) Will the real elasticity please stand up? American Economic Review 79: 882–888.

Blundell R., Bond S. (1998) Initial conditions and moment restrictions in dynamic panel data models. Journal of Econometrics 87: 115–143.

Boggs R.L. (1997) Hazardous waste treatment facilities: modeling production with pollution as both an input and an output. Unpublished Ph.D. Dissertation, University of North Carolina, Chapel Hill.

Bolt K., Matete M., Clemens M. (2002) Manual for calculating adjusted net savings. World Bank, Environment Department Publication, Washington, DC.

Boussemart J.P., Briec W., Kerstens K., Poutineau J.C. (2003) Luenberger and Malmquist productivity indices: theoretical comparisons and empirical illustration. Bulletin of Economic Research 55: 391–405.

Boyd G.A., McClelland J.D. (1999) The impact of environmental constraints on productivity improvement in integrated paper plants. Journal of Environmental Economics and Management 3: 121–142.

Brandon C., Hommann K. (1995) The cost of inaction: valuing the economy-wide cost of environmental degradation in India. Asia Environmental Division, World Bank.

Breton A. (1965) A theory of government grants. Canadian Journal of Economics and Political Science 31(2): 175–187.

Breton A. (1996) Competitive governments. Cambridge University Press, Cambridge.

Briscoe J., de Castro P., Griffim C., North J., Olsen P. (1990) Toward equitable and sustainable rural water supplies: a contingent valuation study in Brazil. World Bank Economic Review 4: 115–134.

Brown B.W., Walker M.B. (1995) Stochastic specification in random production models of cost-minimising firms. Journal of Econometrics 66: 175–205.

Brown S.P.A., Yucel M.K. (1999) Oil prices and US aggregate economic activity. Federal Reserve Bank of Dallas Economic Review Second Quarter: 16–23.

Brown S.P.A., Yucel M.K. (2002) Energy prices and aggregate economic activity: an interpretative survey. Quarterly Review of Economics and Finance 42(2): 193–208.

Bruno M., Sachs J. (1982) Input price shocks and the slowdown of economic growth: the case of U.K. manufacturing. Review of Economic Studies 25: 459–484.

Buchanan, J.M. (1950) Federalism and fiscal equity. American Economic Review 40: 583–599.

Burbidge J., Harrison A. (1984) Testing for the effects of oil price rises using vector autoregression. International Economic Review 25: 459–484.

Burgess R., Venables A.J. (2003) Towards a microeconomics of growth. Mimeo, London School of Economics, London.

Caves D.W., Christensen L.R., Diewert W.E. (1982) The economic theory of index numbers and the measurement of input, output, and productivity. Econometrica 50: 1393–1414.

Celikkol P., Stefanou S. (1999) Measuring the impact of price induced innovation on technological progress: application to the U.S. food processing and distribution sector. Journal of Productivity Analysis 12: 135–151.

Central Electricity Authority (CEA) (1993) Performance review of thermal power stations 1991–92. Generation Data Analysis Directorate, CEA, New Delhi.

Central Pollution Control Board (1995, 1998, 2003) National air quality monitoring programme report. Ministry of Environment and Forest, New Delhi.

Central Pollution Control Board (2001) Annual report. Ministry of Environment and Forest, New Delhi.

Central Statistical Organization (1995, 2004) Annual survey of industries. Ministry of Planning and Programme Implementation, Government of India, Calcutta.

Chambers R.G. (2002) Exact nonradial input, output and productivity measurement. Economic Theory 20: 751–765.

Chambers R.G., Chung Y., Färe R. (1996a) Benefit and distance functions. Journal of Economic Theory 70: 407–419.

Chambers R.G., Färe R., Grosskopf S. (1996b) Productivity growth in APEC countries. Pacific Economic Review 1: 181–190.

Chambers R.G., Chung Y., Färe R. (1998) Profit, directional distance functions, and Nerlovian efficiency. Journal of Optimization Theory and Applications 98: 351–364.

Chapple W., Harris R. (2003) Accounting for solid waste generation in measuring of regional productivity growth. ICCSR Research Paper Series No. 01-2003. Nottingham University Business School, Nottingham.

Chhabra S.K., Pande J.N., Joshi T.K., Kumar P. (2001) Air quality and health. Paper presented at the Workshop on Land Use, Transportation and the Environment. Pune, India (available at http://www.deas.harvard.edu/TransportAsia/workshop_papers/ChhabraetalFinal2.pdf).

Choe K., Whittington D., Luria D.T. (1996) The economic benefits of surface water quality improvements in developing countries: a case study of Davao, Philippines. Land Economics 72: 519–537.

Chopra K., Kadekodi G.K., Murty M.N. (1990) Participatory development. Sage, New Delhi.

Chung Y., Färe R., Grosskopf S. (1997) Productivity and undesirable outputs: a directional distance function approach. Journal of Environmental Management 51: 229–240.

Coase R.H. (1960) The problem of social cost. Journal of Law and Economics 3: 1–44.

Coelli T. (1994) A guide to frontier 4.1: a computer program for stochastic frontier production and cost function estimation. Department of Econometrics, University of New England, Australia.

Coelli T. (1995) Estimators and hypothesis test for a stochastic frontier function: a Monte Carlo analysis. Journal of Productivity Analysis 6: 247–268.
Coelli T. (2000) Econometric estimation of the distance function representation of a production technology. CEPA, University of New England, Mimeo.
Coelli T., Perelman S. (1996) Efficiency measurement, multiple output technologies and distance functions: with application to European railways. CREP Working Papers 96/05 University Liege de.
Coggins J.S., Swinton J.R. (1996) The price of pollution: a dual approach to valuing SO2 allowances. Journal of Environmental Economics and Management 30: 58–72.
Cohen M., Fenn S., Naimon J. (1995) Environmental and financial performance. IRRC, Washington, DC.
Conrad K., Morrison C.J. (1989) The impact of pollution abatement investment on productivity change: an empirical comparison of the US, Germany and Canada. Southern Economic Journal 55: 684–698.
Cormier D., Magnan M., Morard B. (1993) The impact of corporate pollution on market valuation: some empirical evidence. Ecological Economics 8: 133–155.
Courant P., Porter R. (1981) Averting expenditures and the cost of pollution. Journal of Environmental Economics and Management 8: 321–329.
Cropper M.L., Oates W.E. (1992) Environmental economics: a survey. Journal of Economic Literature 30(2): 675–740.
Cropper M.L., Simon N.B., Alberini A., Sharma P.K. (1997a) The health effects of air pollution in Delhi, India. Policy Research Working Paper No. 1860, World Bank, Washington, DC.
Cropper M.L., Simon N.B., Alberini A., Arora S., Sharma P.K. (1997b) The health benefits of air pollution control in Delhi. American Journal of Agricultural Economics 79(5): 1625–1629.
Cunado J., Gracia P. (2005) Oil prices, economic activity and inflation: evidence for some Asian countries. Quarterly Review of Economics and Finance 45: 65–83.
Curnally A. (2002) Environment and rehabilitation. India Infrastructure Report 2002, 3i Network, pp. 96–137.
Dahlby B. (1996) Fiscal externalities and the design of intergovernmental grants. International Tax and Public Finance 3: 397–411.
Darby M.R. (1982) The price of oil and world inflation and recession. American Economic Review 72(4): 738–751.
Das D.K. (2003) Quantifying trade barriers: has trade protection declined substantially in Indian manufacturing. Working Paper No. 105, Indian Council for Research on International Economic Relations, New Delhi.
Dasgupta P. (1993) An inquiry into well-being and destitution. Clarendon, Oxford.
Dasgupta P. (2001) Human well-being and the natural environment. Oxford University Press, Oxford.
Dasgupta P. (2007a) Nature and the economy. Journal of Applied Ecology 44: 475–487.
Dasgupta P. (2007b) Nature in economics. Environmental and Resource Economics 39: 1–7.
Dasgupta P., Mäler K.G. (2000) Net national product, wealth, and social well-being. Environment and Development Economics 5(2): 69–93.
Dasgupta S., Laplante B., Mamingi N. (2001) Pollution and capital markets in developing countries. Journal of Environmental Economics and Management 42: 310–335.
Datt D., Garg S.C., Kadekodi G.K., Narang K.K., Sharma D., Singh J.P. (2004) Environmental fiscal reform in India: issues and some steps forward. Draft, TERI, New Delhi.
Datt G., Ravallion M. (1998) Why have some Indian states done better than others at reducing rural poverty. Economica 65(1): 17–38.
Debreu G. (1951) The coefficient of resource utilization. Econometrica 22: 14–22.
DeRooy Y. (1974) Price responsiveness of the industrial demand for water. Water Resources Research 10: 403–406.
Dev S.M., Ravi C. (2007) Poverty and inequality: all-India and states, 1983–2005. Economic and Political Weekly February 10: 519–520.

Dhryms P.J., Kurz M. (1964) Technology and scale in electricity generation. Econometrica 32: 287–315.

Diewert W.E. (1976) Exact and superlative index numbers. Journal of Econometrics 4: 115–145.

Dinda S. (2004) Environmental Kuznets curve hypothesis: a survey. Ecological Economics 49(1): 431–455.

Dohner R.S. (1981) Energy prices, economic activity and inflation: survey of issues and results. In Mork K. (ed.) Energy prices, inflation and economic activity. Ballinger, Cambridge, MA.

Dupont D.P., Renzetti S. (2001) The role of water in manufacturing. Environmental and Resource Economics 18: 411–432.

Dur R., Staal K. (2008) Local public good provision, municipal consolidation and national transfers. Regional Science and Urban Economics 38: 160–173.

Economic Survey of India (1998–1999) Ministry of Finance, Economic Division, Government of India, New Delhi.

ESPASSA (Ecosystem Services and Poverty Alleviation Study in South Asia) (2008) A situation analysis for India and the Hindu Kush Himalayan region. The Energy and Resource Institute, New Delhi.

Etkins P., Folke C., Costanza R. (1994) Trade, environment and development: the issues in perspective. Ecological Economics 9: 1–12.

Färe R. (1988) Fundamentals of production theory: lecture notes in economics and mathematical systems, vol. 311. Springer, Berlin.

Färe R., Grosskopf S. (2000) Theory and applications of directional distance functions. Journal of Productivity Analysis 13: 93–103.

Färe R., Primont D. (1990) A distance function approach to multi-output technologies. Southern Economic Journal 56: 879–891.

Färe R., Primont D. (1995) Multi-output production and duality: theory and applications. Kluwer, Boston, MA.

Färe R., Grosskopf S., Lovell C.A.K. (1985) The measurement of efficiency of production. Kluwer–Nijhoff, Boston, MA.

Färe R., Grosskopf S., Pasurka C. (1986) Effects on relative efficiency in electric power generation due to environmental controls. Resources and Energy 8: 167–184.

Färe R., Grosskopf S., Lovell C.A.K., Pasurka C. (1989) Multilateral productivity comparisons when some outputs are undesirable: a non-parametric approach. Review of Economics and Statistics 71: 90–98.

Färe R., Grosskopf S., Nelson J. (1990) On price efficiency. International Economic Review 31: 709–720.

Färe R., Grosskopf S., Lovell C.A.K., Yaisawarng S. (1993) Derivation of shadow prices for undesirable outputs: a distance function approach. Review of Economics and Statistics 75: 375–380.

Färe R., Grosskopf S., Lovell C.A.K. (1994a) Production frontiers. Cambridge University Press, Cambridge.

Färe R., Grosskopf S., Norris M., Zhang Z. (1994b) Productivity growth, technical progress, and efficiency change in industrialized countries. American Economic Review 84(1): 66–83.

Färe R., Grosskopf S., Lee W.F. (1995) Productivity and technical change: the case of Taiwan. Mimeo, Southern Illinois University, Carbondale, IL.

Färe R., Grifell-Tatje E., Grosskopf S., Lovell C.A.K. (1997) Biased technical change and Malmquist productivity index. Scandinavian Journal of Economics 99: 119–127.

Färe R., Grosskopf S., Pasurka C. (2001a) Accounting for air pollution emissions in measures of state manufacturing productivity growth. Journal of Regional Science 41(3): 381–409.

Färe R., Grosskopf S., Weber W. (2001b) Shadow prices of Missouri public conservation land. Public Finance Review 26(6): 444–460.

Färe R., Grosskopf S., Noh D., Weber W. (2005) Characteristics of a polluting technology: theory and practice. Journal of Econometrics 126: 469–492.

Farrell M.J. (1957) The measurement of productive efficiency. Journal of the Royal Statistical Society Series A 120: 253–290.
Ferderer J.P. (1996) Oil price volatility and the macroeconomy. Journal of Macroeconomics 18(1): 1–26.
Feres J., Reynaud A. (2003) Industrial water use, cost structure and environmental policy in Brazil (available at http://www.sbe.org.br/ebe25/056.pdf).
Fisher A., Zeckhauser R. (1976) Averting behavior and external diseconomies. Manuscript.
Forest Survey of India (2005) State of forest report 2005. Forest Survey of India, Ministry of Environment and Forest, Government of India, Dehradun.
Foster A.D., Rozenzweig M.R. (2003) Agricultural development, industrialization and rural inequality. Harvard University, Cambridge, MA.
Frederick K.D., Vandenberg T., Hanson J. (1997) Economic values of freshwater in the United States. Discussion Paper No. 97-03, Resources for the Future, Washington, DC.
Freeman A.M. (1993) The measurement of environmental and resource value: theory and methods. Resources for the Future, Washington, DC.
Fuss M.D., McFadden D., Mundlak Y. (1978) A survey of functional forms in the economic analysis of production. In Fuss M.D., McFadden D. (eds.) Production economics: a dual approach to theory and applications. North-Holland, Amsterdam.
Gallagher K.S., Holdren J.P., Sagar A.D., (2006) Energy-technology innovation. Annual Review of Environment and Resource 31: 193–237.
Gerking S., Stanley L.R. (1986) An economic analysis of air pollution and health: the case of St. Louis. Review of Economics and Statistics 68: 115–121.
Goldar B.N. (1986) Productivity growth in Indian industry. Allied Publishers, New Delhi.
Goldar B.N. (2003) Water use in Indian industry: estimates of value and price elasticity of demand. In Chopra K., Hanumantha Rao C.H., Sengupta R. (eds.) Resources, sustainable livelihoods and eco-system services. Concept Publishing Company, New Delhi.
Goldar B.N. (2004) Productivity trends in Indian manufacturing in the pre and post reform periods. Working Paper No. 137, Indian Council for Research on International Economic Relations, New Delhi.
Goldar B.N., Pandey R. (2001) Water pricing and abatement of industrial water pollution: study of distilleries in India. Environmental Economics and Policy Studies 4: 95–113.
Goldman Sachs (2007) The N-11: more than an acronym. Global Economics Paper No. 153.
Gollop F.M., Roberts M.I. (1983) Environmental regulations and productivity growth: the case of fossil-fuelled electric generation. Journal of Political Economy 91: 654–674.
Government of India (GoI) (1995) Annual report 1994–95. Ministry of Environment and Forests, New Delhi.
Government of India (2004) Report of the twelfth finance commission. Finance Commission, Government of India, New Delhi.
Government of India (2006) Integrated energy policy report. Planning Commission, New Delhi.
Government of India (2007) Terms of reference of the thirteenth finance commission. Finance Commission, Government of India, New Delhi.
Government of India (2008) Economic Survey 1997–1998, p. 28; Government of India, Economic Survey.
Grafton R.Q., Jotzo F., Wasson M. (2004) Financing sustainable development: country undertaking and rights for environmental sustainability CURES. Ecological Economics 51: 65–78.
Grandstaff S., Dixon J. (1986) Evaluation of Lumpinee park in Bangkok. In Dixon J., Hufschmidt M. (eds.) Economic valuation techniques for the environment: a case study workbook. Johns Hopkins University Press, Baltimore, MD.
Gray W.B., Shadbegian S.J. (1995) Pollution abatement cost regulation and plant level productivity. Working Paper No 4964, National Bureau of Economic Research, Washington, DC.
Grebenstein C., Field B. (1979) Substituting for water inputs in U.S. manufacturing. Water Resources Research 15: 228–232.

Greene W.H. (1990) A gamma-distributed stochastic frontier model. Journal of Econometrics 46: 141–163.

Greene W. (1993a) Frontier production functions, Working Paper Series EC-9320, Department of Economics, New York University.

Greene W. (1993b) The econometric approach to efficiency analysis. In Fried H.O., Lovell C.A.K., Schmidt S.S. (eds.) The measurement of productive efficiency. Oxford University Press, Oxford.

Greene W.H. (2000) Simulated likelihood estimation of the normal-gamma stochastic frontier function. Stern School of Business, New York University, New York.

Griffin C.C., Briscoe J., Singh B., Ramasubban R., Bhatia R. (1995) Contingent valuation and actual behavior: predicting connections to new water systems in the state of Kerala, India. World Bank Economic Review 9(1): 373–395.

Griliches Z., Mairesse J. (1995) Production functions: the search for identification. Working Paper No. 5067, National Bureau of Economic Research, Washington, DC.

Griliches Z., Ringstad V. (1971) Economies of scale and form of the production function. North-Holland, Amsterdam.

Groot W. (1998) Empirical estimates of the rate of depreciation of education. Applied Economics Letters 5: 535–538.

Grosskopf S. (1986) The role of the reference technology in measuring productive efficiency. Economic Journal 96: 499–513.

Grosskopf S. (1996) Statistical inference and non-parametric efficiency: a selective survey. Journal of Productivity Analysis 7: 161–176.

Grosskopf S., Hayes K. (1993) Local public sector bureaucrats and their input choices. Journal of Urban Economics 33: 151–166.

Grosskopf S., Hayes K., Hirschberg J. (1995a) Fiscal stress and the production of public safety: a distance function approach. Journal of Public Economics 57: 277–296.

Grosskopf S., Margaritis D., Valdmonis V. (1995b) Estimating output substitutability of hospital service: a distance function approach. European Journal of Operational Research 80: 519–533.

Grossman G.M., Krueger A.B. (1993) Environmental impacts of a North American free trade agreement. In Garber P.M. (ed.) The U.S.-Mexico free trade agreement. MIT, Cambridge MA, pp. 13–56.

Grossman G.M., Krueger A.B. (1995) Economic growth and the environment. Quarterly Journal of Economics 110(2): 353–377.

Gupta D.B., Murty M.N., Pandey R. (1989) Water conservation and pollution abatement in Indian industry – a case study of water tariff. Mimeo, National Institute of Public Finance and Policy, New Delhi.

Gupta S. (2001) Environmental policy and federalism in India. Working Paper, Delhi School of Economics, University of Delhi, Delhi.

Gupta S., Goldar B. (2005) Do stock markets penalize environment-unfriendly behaviour? Evidence from India. Ecological Economics 52: 81–95.

Haab T., McConnell K. (1997) A simple method for bounding willingness to pay using a Probit or Logit model. Working Paper, University of Maryland, College Park.

Hadkar N., Sharma S., David A., Muraleedharan T.T., Geetha S., Babu P. (1995) Are people in developing countries willing to pay for natural resource preservation? Evidence from a CV of the Borivali National Park, Bombay. Indira Gandhi Institute of Development Research, Bombay.

Hailu A., Veeman T.S. (2000) Environmentally sensitive productivity analysis of the Canadian pulp and paper industry, 1959–1994: an input distance function approach. Journal of Environmental Economics and Management 40: 251–274.

Hajkowicz S. (2007) Allocating scarce financial resources across regions for environmental management in Queensland, Australia. Ecological Economics 61: 208–216.

Hakuni M. (1994) Efficiency in the Finnish sulphate pulp industry in 1972–90. Finnish Forest Research Report No. 517, Research Institute.

Hamilton J.D. (1983) Oil and macroeconomy since World War II. Journal of Political Economy 91(2): 228–248.

Hamilton J.D. (1988) A neoclassical model of unemployment and the business cycle. Journal of Political Economy 96(3): 593–617.

Hamilton J.D. (1996) This is what happened to oil price-macroeconomy relationship. Journal of Monetary Economics 38: 215–220.

Hamilton J.D. (2003) What is an oil shock? Journal of Econometrics 113: 363–398.

Hamilton K. (2002) Sustaining per capita welfare with growing population: theory and measurement. Presented at Second World Congress of Environmental and Resource Economists, Monterey, California, June 27–30.

Hamilton K., Clemens M. (1999) Genuine saving in developing countries. World Bank Economic Review 13(2): 333–356.

Hamilton K., Hassan R. (2003) Measuring development prospects by greening national accounts. Working Paper, Initiative for Policy Dialogue, Columbia University.

Hanemann W.M. (1984) Welfare valuation in contingent valuation experiments with discrete responses. American Journal of Agricultural Economics 66: 332–341.

Hanemann W.M. (1989) Welfare evaluation in contingent valuation experiments with discrete response data: reply. American Journal of Agricultural Economics 71: 1051–1061.

Hanley N., Mourato S., Wright R.E. (2001) Choice modeling approaches: a superior alternative for environmental valuation? Journal of Economic Surveys 15(3): 435–462.

Harford J. (1984) Averting behavior and the benefits of reduced soiling. Journal of Environmental Economics and Management 11: 296–302.

Hartwirk J.M. (1977) Intergenerational equity and the investing of rents from exhaustible resources. American Economic Review 67(5): 972–974.

Hayami Y., Ruttan V.W. (1971) Agricultural development: an international perspective. John Hopkins University Press, Baltimore, MD.

Haynes K.E., Ratick S., Bowen W.M., Cummings-Saxton J. (1993) Environmental decision models: U.S. experience and a new approach to pollution management. Environment International 19: 261–275.

Haynes K.E., Ratick S., Cummings-Saxton J. (1994) Toward a pollution abatement monitoring policy: measurements, model mechanics, and data requirements. The Environmental Professional 16: 292–303.

Hazra K.S., Rakhe P.B., Gajbhiye D. (2008) Issues before the Thirteenth Finance Commission: correction of horizontal and vertical imbalances. Economic and Political Weekly March 22: 89–95.

Hetemaki L. (1996) Essays on the impact of pollution control on a firm: a distance function approach. Finnish Forest Research Institute, Helsinki.

Hettige H., Lucas R.E.B., Wheeler D. (1992) The toxic intensity of industrial production: global patterns, trends, and trade policy. AEA Papers and Proceedings, pp. 478–481.

Hicks J.R. (1932) The theory of wages. St. Martins, New York.

Hicks J.R. (1946) Value and capital, 2nd edition. Oxford University Press, Oxford.

Himanshu (2007) Recent trends in poverty and inequality: some preliminary results. Economic and Political Weekly February 10.

Holtz–Eakin D., Selden T.M. (1995) Stoking the fires? CO2 emissions and economic growth. Journal of Public Economics 57: 85–101.

Hooker M. (1996) What happened to the oil price-macroeconomy relationship? Journal of Monetary Economics 38: 195–213.

Hooker M. (1999) Oil price and macroeconomy revisited. Mimeo, Federal Reserve Board.

Hooker M. (2002) Are oil shocks inflationary? Asymmetric and non-linear specifications versus change in regime. Journal of Money, Credit and Banking 34: 540–561.

Hossain M.A., Karunaratne N.D. (2004) Trade liberalization and technical efficiency: evidence from Bangladesh manufacturing industries. The Journal of Development Studies 40(3): 87–114.

Hudgins B.L., Primont D. (2004) Directional technology distance functions: theory and applications. Working Paper 10, Department of Economics, Southern Illinois University, Carbondale, IL.

Huntington H.G. (1998) Crude oil prices and U.S. economic performance: where does the asymmetry reside? Energy Journal 19(4): 107–132.

IEA (2006) Energy Technology Perspectives: Scenarios and Strategies to 2050, IEA, Paris.

Jaffe A.B., Palmer K. (1997) Environmental regulation and innovation: a panel data study. Review of Economics and Statistics 79: 610–619.

Jaffe A.B., Peterson S.R., Portney P.R., Stavins R.N. (1995) Environmental regulation and the competitiveness of U.S. manufacturing: what does the evidence tell us? Journal of Economic Literature 33: 132–163.

Jaffe A.B., Palmer K., (1997) Environmental regulation and innovation: a panel data study. Review of Economics and Statistics 79: 610–619.

Jaffe A.B., Newell R.G., Stavins R.N. (2003) Technological change and the environment. In Mäler K., Vincent J. (eds.) Handbook of environmental economics. North-Holland, Amsterdam, pp. 462–516.

James A.J., Murty M.N. (1996) Water pollution abatement: a taxes-standards approach for Indian industry. Working Paper No. E/177/96, Institute of Economic Growth, Delhi.

Jasanoff S. (1986) Managing India's environment: new opportunities, new perspectives. Environment 28(8): 12–16, 31–38.

Jha R., Whalley J. (1999) The environmental regime in developing countries. Paper presented at Distributional and Behavioural Effects of Environmental Policy, NBER/FEMI Conference, pp. 11–12.

Jha V. (1999) Investment liberalization and environmental protection: conflicts and compatibilities in the case of India. UNCTAD.

Jimenez-Rodriguez R., Sanchez M. (2005) Oil price shocks and real GDP growth: empirical evidence for some OECD countries. Applied Economics 37: 201–228.

Johansson R.C., Tsur Y., Roe T.L., Doukkali R., Dinar A. (2002) Pricing irrigation water: a review of theory and practice. Water Policy 4: 173–199.

Jones D.W., Leiby P.N. (1996) The macroeconomic impacts of oil price shocks: a review of literature and issues. Oak Ridge National Laboratory, Oak Ridge, TN.

Jones D.W., Leiby P.N., Raik I.K. (2004) Oil price shocks and the macroeconomy: what has been learned since 1996. Energy Journal 25(2): 1–32.

Jorgenson D., Wilcoxen P.J. (1990) Environmental regulation and US economic growth. RAND Journal of Economics 21: 314–340.

Kahn G., Hampton R. (1990) Possible monetary policy responses to the Iraqi oil shock. Federal Reserve Bank of Kansas City Economic Review 2: 19–32.

Kalirajan K.P., Sankar U. (2001) Agriculture in India's economic reform program. Journal of Asian Economics 12: 383–399.

Kammen D.M. (2006) The rise of renewable energy. Scientific American, Special issue on Energy's Future Beyond Carbon. September: 52–61.

Khanna M., Zilberman D. (1999) Adoption of energy efficient technologies and carbon abatement: the electricity generating sector in India. Energy Economics 23(6): 637–658.

Khanna M., Zilberman D. (2001) Freer markets and the abatement of carbon emissions: the electricity-generating sector in India. Resource and Energy Economics 21(2): 125–152.

Khanna M., Quimio W.R.H., Bojilova D. (1998) Toxics release information: a policy tool for environmental protection. Journal of Environmental Economics and Management 36: 243–266.

Kohlin G. (2001) Contingent valuation in project planning and evaluation: the case of social forestry in Orissa, India. Environment and Development Economics 6(2): 237–258.

Köllner T., Schelske O., Seidl I. (2002) Integrating biodiversity into intergovernmental fiscal transfers based on cantonal benchmarking: a Swiss case study. Basic and Applied Ecology 3: 381–391.

Kopp G. (1998) Carbon dioxide emissions and economic growth: a structural approach. Journal of Applied Statistics 25(4): 489–515.

Krupnick A., Harrington W., Ostro B. (1990) Ambient ozone and acute health effects: evidence from daily data. Journal of Environmental Economics and Management 18(1): 1–18.

Kumar S. (1999) Economic evaluation of development projects: a case analysis of environmental health implications of thermal power projects in India. Ph.D. Thesis, Jawaharlal Nehru University, New Delhi, India.

Kumar S., Russell R.R., (2002) Technological change, technological catch-up, and capital deepening: relative contributions togrowth and convergence. American Economic Review 92: 527–548.

Kumar S. (2006) Environmentally sensitive productivity growth: a global analysis using Malmquist–Luenberger index. Ecological Economics 56(2): 280–293.

Kumar S., Managi S. (2008) Energy price induced and exogenous technological changes: assessing economic and environmental outcomes. Working Paper, http://papers.ssrn.com/sol3/papers.cfm?abstract_id=1203202, 2008.

Kumar S., Rao D.N. (2001) Valuing benefits of air pollution abatement using health production function: a case study of Panipat thermal power station, India. Environmental and Resource Economics 20: 91–102.

Kumbhakar S.C., Lovell C.A.K. (2000) Stochastic frontier analysis. Cambridge University Press, Cambridge.

Lall P., Featherstone A.M., Norman D.W. (2002) Productivity growth in the Western Hemisphere (1978–94): the Caribbean in perspective. Journal of Productivity Analysis 17: 213–231.

Lanjouw J.O., Mody A., (1996) Innovation and the international diffusion of environmentally responsive technology. Research Policy 25: 549–571.

Lansink A.O., Silva E., Stefanou S. (2000) Decomposing productivity growth allowing efficiency gains and price-induced technical progress. European Review of Agricultural Economics 27(4): 497–518.

Lee K., Ni S., Ratti R.A. (1995) Oil price shocks and the macroeconomy: the role of price variability. Energy Journal 16: 39–56.

Lilien D. (1982) Sectoral shift and cyclical unemployment. Journal of Political Economy 90: 777–793.

List J.A., Bulte E.H., Shogren J.E. (2002) Beggar thy neighbour: testing for free riding in state-level endangered species expenditure. Public Choice 111(3): 303–315.

Litchtenberg F.R., (1986) Energy prices and induced innovation. Research Policy 15: 67–75.

Litchtenberg F.R., (1987) Changing market opportunities and the structure of R&D investment: the case of energy. Energy Economics 9: 154–158.

Loungani P. (1986) Oil price shocks and the dispersion hypothesis. Review of Economics and Statistics 72: 53–78.

Lovell C.A.K., Richardson S., Travers P., Wood L. (1994) Resources and functioning: a new view of inequality in Australia. In Eichhorn W. (ed.) Models and measurement of welfare and inequality. Springer, Berlin, pp. 787–807.

Luenberger D.G. (1992a) Benefit functions and duality. Journal of Mathematical Economics 21: 461–481.

Luenberger D.G. (1992b) New optimality principles for economic efficiency and equilibrium. Journal of Optimization Theory and Applications 75: 221–264.

Maddala G.S. (1983) Limited dependent and qualitative variables in econometrics. Cambridge University Press, Cambridge.

Mäler K.G. (2007) Wealth and sustainable development: the role of David Pearce. Environmental and Resource Economics 37: 63–75.

Managi S. (2006) Are there increasing returns to pollution abatement? Empirical analytics of the environmental Kuznets curve in Pesticides. Ecological Economics 58(3): 617–636.

Managi S., Opaluch J.J., Jin D., Grigalunas T.A. (2005) Environmental regulations and technological change in the offshore oil and gas industry. Land Economics 81(2): 303–319.

Mandal S., Rao M.G. (2005) Overlapping fiscal domains and effectiveness of environmental policy in India. Working Paper, National Institute of Public Finance and Policy, New Delhi.

Manoharan T.R. (2000) Natural resource accounting: economic valuation of intangible benefits of forests. RIS Discussion Paper 04/2000, Research and Information System for the Non-aligned and other Developing Countries, New Delhi.

Markandya A., Murty M.N. (2000) Cleaning up Ganges: the cost benefit analysis. Oxford University Press, New Delhi.

Marquetti A. (2002) Extended Penn World Tables 2.0 Data set (available at http://homepage.newschool.edu/~foleyd/epwt).

May P.H., Veiga Neto F., Denardin V., Loureiro W. (2002) Using fiscal instruments to encourage conservation: municipal responses to the ecological value added tax in Parana and Minas Gerais, Brazil. In Pagiola S., Bishop J., Landell-Mills N. (eds.) Selling forest environmental services: market based mechanisms for conservation and development. Earthscan, London, pp. 173–199.

McConnell J., Ducci J. (1988) Valuing environmental quality in developing countries: two case studies. Paper presented at the Applied Social Science Association, Atlanta, GA.

McConnell K.E. (1995) Issues in estimating benefits with non-market methods. Office of the Chief Economist, Working Paper Series 308, Inter-American Development Bank, Washington, DC.

McFadden D. (1994) Contingent valuation and social choice. American Journal of Agricultural Economics 76: 689–708.

McKenzie D., Ray I. (2004) Household water delivery options in urban and rural India. Paper prepared for the 5th Stanford Conference on Indian Economic Development, June 3–5.

Mehta S., Mundel S., Shankar U. (1995) Controlling pollution: incentives and regulation. Sage, Delhi.

Menkhaus S. (1994) Measurement of economic and other benefits of wildlife preservation: a case study of Keoladeo National Park, Bharatpur, India. Institute of Economic Growth, Delhi.

Millennium Ecosystem Assessment (2005) Ecosystems and human well-being. Island Press, Washington, DC.

Ministry of Environment and Forest (2001) Annual report. Government of India, New Delhi.

Ministry of Finance (1993) Public sector commercial banks and financial sector reform: rebuilding for a better future. Discussion Paper, Government of India, New Delhi.

Mishan E.J. (1974) What is the optimal level of pollution? Journal of Political Economy 82: 1278–1279.

Misra S. (1996a) Accounting for costs of water pollution abatement: a case study of Nandesari Industrial Area. Working Paper E/179/96, Institute of Economic Growth, Delhi.

Misra S. (1996b) Measuring benefits from industrial water pollution abatement: use of contingent valuation method in Nandesari Industrial Area of Gujarat in India. Working Paper No. E/185/97, Institute of Economic Growth, Delhi.

Misra S. (1999) Water pollution abatement in small scale industries: an exploration of collective action possibilities in Handesari industrial area in Gujarat. Ph.D. Thesis, University of Delhi, Delhi.

Mitchel R.C., Carson R.T. (1989) Using surveys to value public goods: the contingent valuation method. Resources for the Future, Washington, DC.

Mohan R. (2006) Economic reforms in India: where are we and where do we go? RBI Bulletin, December.

Mohan R. (2008) Growth record of the Indian economy, 1950–2008: a story of sustained savings and investment. Economic and Political Weekly May 10: 61–71.

Morgenstern R.D., Pizer W.A., Shih J. (1997) Are we overstating the real economic cost of environmental protection. Discussion paper 97-36, Resource for the Future, Washington, DC.

Mork K. (1989) Oil and the macro-economy, when prices go up and down: an extension of Hamilton's results. Journal of Political Economy 97(3): 740–744.

Mork K. (1994) Business cycles and the oil market. Energy Journal 15: 15–38.

Mork K., Olsen O., Mysen H.T. (1994) Macroeconomic responses to oil price increases and decreases in seven OECD countries. Energy Journal 15(4): 19–35.

Murty M.N. (2007) India environmental outlook. OECD, Paris.
Murty M.N. (2008) Policy responses to ecosystem conservation in Hindu Kush Region. Discussion Paper, Ecosystem services and poverty alleviation study in South Asia (ESPASSA): a situation analysis for India and the Hindu Kush Himalayan Region, Annex 3, TERI, New Delhi.
Murty M.N., Gulati S.C. (2007) Measuring cost of environmentally sustainable industrial development and designing pollution taxes: a case study of thermal power generation in India. Indian Economic Journal 54(2): 135–144.
Murty M.N., Kumar S. (2002) Measuring cost of environmentally sustainable industrial development in India: a distance function approach. Environment and Development Economics 7: 467–486.
Murty M.N., Kumar S. (2004) Environmental and economic accounting for industry. Oxford University Press, New Delhi.
Murty M.N., Kumar S. (2006) Measuring productivity of natural capital. In Tendulkar S.D., Mitra A., Narayanan K., Das D.K. (eds.) India: industrialisation in a reforming economy: essays for K. L. Krishna. Academic Foundation, New Delhi.
Murty M.N., Kumar S., Paul M. (2006) Environmental regulation, productive efficiency and cost of pollution abatement: a case study of sugar industry in India. Journal of Environmental Management 79(1): 1–9.
Murty M.N., James A.J., Misra S. (1999) Economics of water pollution: the Indian experience. Oxford University Press, New Delhi.
Murty S., Russell R.R. (2002) On modeling pollution generating technologies. University of California, Riverside, CA.
Myers J.G., Nakamura L. (1980) Energy and pollution effects on productivity: a putty-clay approach. In Kenderick (ed.) New developments in productivity measurement and analysis. University of Chicago Press, Chicago, IL.
Nordhaus W. (1999) Modeling induced innovation in climate-change policy. In: A. Grubler, N. Nakicenovic and W. Nordhaus. (Eds.) Technological Change and the Environment, Washington, DC: Resources for the Future Press. Popp, D., 2002. Induced innovation and energy prices. American Economic Review 92: 160–180.
Newell R.G., Jaffe A.B., Stavins R.N. (1999) The induced innovation hypothesis and energy-saving technological change. Quarterly Journal of Economics 109: 941–975.
Oates W.E. (1972) Fiscal federalism. Harcourt, Brace and Jovanovich, New York.
Oates W.E. (1998) Environmental policy in the European community: harmonization or national standards? Empirica 25: 1–13.
Oates W.E. (1999) An essay on fiscal federalism. Journal of Economic Literature 37(3): 1120–1149.
Oates W.E. (2001) A reconsideration of environmental federalism. Discussion Paper 01-54, Resource for the Future, Washington, DC.
OECD (1998) Future liberalisation of trade in environmental goods and services: ensuring environmental protection as well as economic benefits. Joint Working Party on Trade and Environment, COM/ENV(98)37/FINAL.
OECD (2006) Environmental compliance and enforcement in India: rapid assessment. OECD, Paris.
Olson M., Jr. (1969) The principle of fiscal equivalence: the division of responsibilities among different levels of government. American Economic Review 59: 479–487.
Olson M. (1996) Big bills on the side walk: why some nations are rich, and others poor. Journal of Economic Perspectives 10(2): 3–34.
Onjala J. (2001) Industrial water demand in Kenya: industry behaviour when tariffs are not binding. Mimeo, Roskilde University, Denmark (available at http://www.environmental-economics.dk/papers/waterkenya.pdf).
Ono M. (2005) Rising crude oil prices affect the Japanese economy. Economic Research Department (available at http://www.jetro.go.jp/en/market/trend/special/pdf/jem0406-1e.pdf).

Orea L. (2002) Parametric decomposition of a generalized Malmquist productivity index. Journal of Productivity Analysis 18: 5–22.

Ostro B. (1990) Estimating the health effects of air pollutants: a method with an application to Jakarta. Policy Research Working Paper No. 1301, World Bank, Washington, DC.

Pachauri R.K. (2004) The future of India's economic growth: the natural resources and energy dimension. Futures 36: 703–713.

Pallikara R. (2004) India – national reports. In Asian Productivity Organization (eds.) Total factor productivity growth: survey report. APO Survey on Total Factor Productivity, Tokyo.

Palmer K., Oates W.E., Portney P.R. (1995) Tightening environmental standards: the benefit-cost or no cost paradigm. Journal of Economic Perspectives 9: 119–132.

Pandey R. (1999) Pollution taxes for industrial water pollution control. Mimeo, National Institute of Public Finance and Policy, New Delhi.

Papapetrou E. (2001) Oil price shocks, stock market, economic activity and employment in Greece. Energy Economics 23(5): 511–532.

Pargal S., Wheeler D. (1996) Informal regulation of industrial pollution in developing countries: evidence from Indonesia. Journal of Political Economy 104(6): 1814–1827.

Parikh J. (2004) Environmentally sustainable development in India (available at http://scid.stanford.edu/events/India2004/JParikh.pdf).

Parikh J., Parikh K. (2001) Environmentally adjusted GDP, Report to the United Nations University.

Paris Q., Caputo M.R. (2001) Price-induced technical progress and comparative statics. Economics Bulletin 15(8): 1–8.

Pearce D.W., Atkinson G. (1993) Capital theory and the measurement of sustainable development: an indicator of weak sustainability. Ecological Economics 8(2): 103–108.

Pearce D.W., Markandya A., Barbier E. (1989) Blueprint for a green economy. Earthscan, London.

Pearce D.W., Whittington D., Moran D., Georgiou S. (1996) Economic values and the environment in the developing world. Edward Elgar, London.

Perrings C., Gadgil M. (2003) Conserving biodiversity: reconciling local and global public benefits. In Kaul I., Conceicao P., Le Goulven K., Mendoza R.U. (eds.) Providing global public goods: managing globalization. Oxford University Press, Oxford, pp. 532–555.

Pezzey J. (2004) Sustainability policy and environmental policy. Scandinavian Journal of Economics 106: 339–359.

Pierce J.L., Enzler J.J. (1974) The effects of external inflationary shocks. Brookings Papers on Economic Activity 1: 13–61.

Pittman R.W. (1981) Issues in pollution interplant cost differences and economies of scale. Land Economics 57: 1–17.

Pittman R.W. (1983) Multilateral productivity comparisons with undesirable outputs. Economic Journal 93: 883–891.

Planning Commission (1994) Annual report on the working of state Electricity boards and electricity departments. Planning Commission, New Delhi.

Porter M.E. (1990) The competitive advantages of nations. Free Press, New York.

Porter M.E. (1991) America's green strategy. Scientific American 264: 168.

Porter M.E., van der Linde C. (1995) Towards a new conception of the environment competitiveness relationship. Journal of Economic Perspectives 9: 97–118.

Pradhan G., Barik K. (1999) Sustainability of output growth in Indian manufacturing: a decomposition analysis of selected industries. Paper presented at the Annual Meeting of the Australian Econometric Society, University of Technology, Sydney.

Priyadarshini K., Gupta O.P. (2003) Compliance to environmental regulations: the Indian context. International Journal of Business and Economics 2(1): 9–26.

Raghbendra Jha (2004) Alleviating Environmental Degradation in the Asia-Pacific Region: International cooperation and the role of issue-linkage. Departmental Working Papers 2005–01, Australian National University, Economics RSPAS.

Rangarajan C., Srivastava D.K. (2008) Reforming India's fiscal transfer system: resolving vertical and horizontal imbalances. Economic and Political Weekly June 7: 47–59.

Rao C.H.H. (2008) Growing regional disparities in development in India: post-reform experience and challenges ahead. In Chopra K., Rao C.H.H. (eds.) Growth, equity, environment and population: economic and sociological perspectives. Sage, New Delhi, pp. 198–207.

Rao M.G. (2000) Fiscal decentralization in Indian federalism. Working Paper, Institute for Social and Economic Change, Bangalore.

Rao Vijayendra (2005) Symbolic public goods and the coordination of collective action : a comparison of local development in India and Indonesia. Policy Research Working Paper Series 3685, The World Bank.

Rao M.G. (2006) Fiscal federalism in planned economies. In Ahmad E., Brosio G. (eds.) Handbook of fiscal federalism. Edward Elgar, London.

Rao M.G., Mandal S. (2008) Resource endowment, fiscal flows and regional equity in Indian federalism. In Rao M.G., Shah A. (eds.) States' fiscal management and regional equity. Oxford University Press, New Delhi.

Ray A.S. (1997a) India's economic reforms in progress: an introductory note. Journal of Asian Economics 8: 421–423.

Ray S.C. (1997b) Regional variation in productivity growth in Indian manufacturing: a non-parametric analysis. Journal of Quantitative Economics 13: 73–94.

Ray S.C. (2002) Did India's economic reforms improve efficiency and productivity? A non-parametric analysis of the initial evidence from manufacturing. Indian Economic Review 37: 23–57.

Ready R.C., Malzubris J., Senkane S. (2002) The relationship between environmental values and income in a transition economy: surface water quality in Latvia. Environment and Development Economics 7(1): 147–156.

Rees J. (1969) Industrial demand of water: a study of South East England. Weidenfeld and Nicolson, London.

Reinhard S., Lovell C.A.K., Thijssen G. (1999) Econometric estimation of technical and environmental efficiency: an application to Dutch dairy farms. American Journal of Agricultural Economics 81(1): 44–60.

Renzetti S. (1988) An econometric study of industrial water demands in British Columbia, Canada. Water Resources Research 24: 1569–1573.

Renzetti S. (1992) Estimating the structure of industrial water demands: the case of Canadian manufacturing. Land Economics 68: 396–404.

Renzetti S. (1993) Examining the differences in self and publicly supplied firm's water demands. Land Economics 69: 181–188.

Renzetti S. (2002) The economics of industrial water use: the management of water resources. Edward Elgar, London.

Repetto R., Rothman D., Faeth P., Austin D. (1996) Has environmental protection really reduced productivity growth? World Resource Institute, Washington, DC.

Reserve Bank of India (2004) Report on currency and finance, 2002–03. Mumbai.

Reynaud A. (2003) An econometric estimation of industrial water demand in France. Environmental and Resource Economics 25: 213–232.

Ring I. (2002) Ecological public functions and fiscal equalization at the local level in Germany. Ecological Economics 42: 415–427.

Ring I. (2008a) Compensating municipalities for protected areas: fiscal transfers for biodiversity conservation in Saxony, Germany. GAIA 17(Suppl. 1): 143–151.

Ring I. (2008b) Biodiversity governance: adjusting local costs and global benefits. In Sikor T. (ed.) Public and private natural resource governance: a false dichotomy? Earthscan, London, pp. 107–126.

Ring I. (2008c) Integrating local ecological services into intergovernmental fiscal transfers: the case of ecological ICMS in Brazil. Land Use Policy 25(4): 485–497.

Rodrik D., Subramanian A. (2004) From 'Hindu growth' to productivity surge: the mystery of the Indian growth transition. IMF Working Paper WP/04/77, International Monetary Fund, Washington, DC.

Rosen S. (1976) A theory of life earning. Journal of Political Economy 84: S45–S67.

Rowe R.D., Chestnut L.G. (1985) Valuing changes in morbidity: WTP versus COI measures. Energy and Resource Consultants, Inc., Boulder.

Sachs J., Warner A. (1995) Natural resources and economic growth. Mimeo, HIID, Cambridge, MA.

Sagar A.D., (2002) India's energy and energy R&D landscape: a brief overview, Belfer Center for Science and International Affairs Working Paper 2002–08, Energy Technology Innovation Project, Harvard University.

Salim R.A., Kalirajan K.P. (1999) Sources of output growth in Bangladesh food processing industries: a decomposition analysis. Developing Economies 37: 355–374.

Samuelson P. (1961) The evaluation of social income: capital formulation and wealth. In Lutz F.A., Haguae D.C. (eds.) The theory of capital. Macmillan, London.

Sankar U. (1998) Laws and institutions relating to environmental protection in India. Working Paper, Madras School of Economics, Chennai.

Seldon T.M., Song D. (1994) Environmental quality and development: is there a Kuznets curve for air pollution emissions? Journal of Environmental Economics and Management 27(2): 147–162.

Sengupta R. (2006) Inclusive growth and Indian energy scene: energy transition. Initiative for Policy Dialogue, Columbia University.

Shephard R.W. (1953) Theory of cost and production functions. Princeton University Press, Princeton, NJ.

Shephard R.W. (1970) Theory of cost and production functions. Princeton University Press, Princeton, NJ.

Shephard R.W. (1974) Indirect production functions. Mathematical systems in economics, No. 10, Verlag Anton Hain, Meisenheim am Glan.

Shephard, R.W. (1974) Indirect Production functions. Meisenheim am Glan,Verlag Anton Hain.

Shibata H., Winrich J.S. (1983) Control of pollution when the offended defend themselves. Economica 50: 425–438.

Shrestha R.M., Acharya M.P. (1992) Electricity planning and environmental planning in selected Asian countries. Natural Resources Forum 192–201.

Shyamsundar P., Kramer R. (1993) Does contingent valuation work in non-market economies? Center for Resource and Environmental Policy, Duke University, Durham, NC.

Sigman H. (2005) Transboundary spillovers and decentralization of environmental policies. Journal of Environmental Economics and Management 50: 82–101.

Simar L., Wilson P.W. (2007) Estimation and inference in two-stage, semi-parametric models of production processes. Journal of Econometrics 136: 31–64.

Simpson R.D., Bradford R.L. (1996) Taxing variable cost: environmental regulation as industrial policy. Journal of Environmental Economics and Management 1996: 282–300.

Sims C. (1980) Macroeconomics and reality. Econometrica 48: 1–48.

Singh B., Ramasubban R., Bhatia R., Briscoe J., Griffin C., Kim C. (1993) Rural water supply in Kerala, India: how to emerge from a low-level equilibrium trap. Water Resources Research 29(7): 1931–1942.

Singh H.P. (1998) Sustainable development of the Indian desert: the relevance of the farming systems approach. Journal of Arid Environments 39: 279–284.

Singh J. (1991) Plant size and technical efficiency in the Indian thermal power industry. Indian Economic Review 26: 239–252.

Smith K.R. (2000) National burden of disease in India from indoor air pollution. PNAS 97(24): 13286–13293.

Smith V.K. (1993) Non-market valuation of environmental resources: an interpretive appraisal. Land Economics 69: 1–26.

Solow R.M. (1974) The economics of resources or the resources of economics? American Economic Review 64(2): 1–14.

Srinivasan T.N. (2006a) China and India: economic performance, competition and cooperation: an update. Journal of Asian Economics 15: 613–636.

Srinivasan T.N. (2006b) India's economic growth and global integration: experience since reforms and future challenges. Lecture delivered for Jackson Hole Symposium, August 24.

Stern D.I. (1998) Progress on the environmental Kuznets curve? Environment and Development Economics 3: 173–196.

Stern D.I. (2004) The rise and fall of the environmental Kuznets curve. World Development 32: 1419–1439.

Sundram K. (2007) Employment and poverty in India: 2000–2005. Working Paper No. 155, Centre for Development Economics, Department of Economics, Delhi School of Economics, University of Delhi, Delhi.

Taskin F., Zaim O. (2001) The role of international trade on environmental efficiency: a DEA approach. Economic Modelling 18: 1–17.

Tatom J.A. (1993) Are there useful lessons from the 1990–91 oil price shock? Energy Journal 14(4): 129–141.

TERI (2001) Delhi water situation: new sources of water vital for Delhi's health. Issue no. 35.

Thirtle C.G., Ruttan V.W. (1987) The role of demand and supply in the generation and diffusion of technical change. In Scherer F.M. (ed.) Fundamentals of pure and applied economics, vol. 21 in the Economics and Technological Change Section, Harwood Academic, New York.

Thompson P., Taylor T.G. (1995) The capital–energy substitutability debate: a new look. Review of Economics and Statistics 77: 565–569.

Tiebout C.M. (1956) A pure theory of local expenditure. Journal of Political Economy 64: 416–424.

Torero M., Chowdhury S.K., Galdo V. (2003) Willingness to pay for the rural telephone service in Bangladesh and Peru. Information Economics and Policy 15: 327–361.

Turnovsky S. (1969) The demand for water: some empirical evidence on consumers' response to a commodity in uncertain supply. Water Resources Research 5: 350–361.

Unel B. (2003) Productivity trends in India's manufacturing sectors in the last two decades. IMF Working Paper, WP/03/22, International Monetary Fund, Washington, DC.

United Nations (1993) Integrated environmental and economic accounting. United Nations, New York.

Varela-Ortega C., Sumpsi J., Garrido A., Blanco M., Iglesias E. (1998) Water pricing policies, public decision making and farmers' response: implications for water policy. Agricultural Economics 19: 193–202.

Vaughan W.J., Russell C.S., Rodriguez D.J., Darling A.H. (1999) Willingness to pay: referendum contingent valuation and uncertain benefits. Sustainable Development Department Technical Papers Series, Inter-American Development Bank, Washington, DC.

Vincent J.R. (1997) Resource depletion and economic sustainability in Malaysia. Environment and Development Economics 2: 19–37.

Vincent J.R., Castaneda B. (1997) Economic depreciation of natural resources in Asia and implications for net savings and long-run consumption. Development Discussion Paper No. 614, Harvard Institute for International Development, Harvard University.

Vouvaki D., Xepapadeas A., (2008) Changes in social welfare and sustainability: Theoretical issues and empirical evidence, Ecological Economics, vol. 67(3), pp. 473–484

Wang H., Lall S. (2002) Valuing water for Chinese industries: a marginal productivity analysis. Applied Economics 34: 759–765.

Weber W.L., Domazlicky B.R. (1999) Total factor productivity growth in manufacturing: a regional approach using linear programming. Regional Science and Urban Economics 29: 105–122.

Weber W., Domazlicky B. (2001) Productivity growth and pollution in State manufacturing. Review of Economics and Statistics 83(1): 195–199.

Weitzman M.L. (1976) On the welfare significance of national product in a dynamic economy. Quarterly Journal of Economics 90(1): 156–162.

Whittington D., Swarna V. (1994) The economic benefits of potable water supply projects to households in developing countries. Economics Staff Papers 53, Asian Development Bank, Manila.

Whittington D., Mujwahuzi M., McMahon G., Choe K. (1988) Willingness to pay for water in Newala District, Tanzania: strategies for cost recovery. Water and Sanitation for Health Project Field Report No. 246, USAID, Washington, DC.

Whittington D., Briscoe J., Mu X., Barron W. (1990) Estimating the willingness to pay for water services in developing countries: a case study of the use of contingent valuation surveys in Southern Haiti. Economic Development and Cultural Change 38(2): 293–311.

Whittington D., Lauria D.T., Wright A., Choe K., Hughes J.A., Swarna V. (1993) Household demand for improved sanitation services in Kumasi, Ghana: a contingent valuation study. Water Resources Research 29(6): 1539–1560.

Whittington D., Pinheiro A.C., Cropper M. (1996) The economic benefits of malaria control: a contingent valuation study in Marracuene, Mozambique. World Bank, Washington, DC.

Williams M., Suh B. (1986) The demand for urban water by customer class. Applied Economics 18: 1275–1289.

World Bank (1999) Greening industry: new roles for communities, markets, and governments. World Bank and Oxford University Press, New York.

World Bank (2006) Where is the wealth of nations: measuring capital for the 21st Century. The World Bank, Washington, DC.

World Bank (2007) India: strengthening institutions for sustainable growth. Country Environmental Analysis, South Asia Environment and Social Development Unit South Asia Region.

Yaisawarng S., Klein D.J. (1994) The effects of sulphur dioxide controls on productivity change in the US electric power industry. Review of Economics and Statistics 76: 447–460.

Yoo S.H., Chae K.S. (2001) Measuring the economic benefits of the ozone pollution control policy in Seoul: results of a contingent valuation survey. Urban Studies 38(1): 49–60.

Young A.T., Higgins M.J., Levy D. (2003) Sigma convergence versus beta convergence: evidence from U.S. county-level data. Department of Economics Working Paper, Emory University, Atlanta, GA.

Zaim O., Taskin F. (2000) A Kuznets curve in environmental efficiency: an application on OECD countries. Environmental and Resource Economics 17: 21–36.

Zellner A., Kamenta J., Drèze J. (1966) Specification and estimation of Cobb-Douglas production function models. Econometrica 34: 784–795.

Zhengfei G., Lansink O.A. (2006) The source of productivity growth in Dutch agriculture: a perspective from finance. American Journal of Agricultural Economics 88: 644–656.

Ziegler J., Bell S. (1984) Estimating demand for intake water by self-supplied firms. Water Resources Research 20: 4–8.

Zofio J.L., Prieto A.M. (2001) Environmental efficiency and regulatory standards: the case of CO2 emissions from OECD countries. Resource and Energy Economics 23: 63–83.

Index

G

H

I

J

L

M

GPSR Compliance

The European Union's (EU) General Product Safety Regulation (GPSR) is a set of rules that requires consumer products to be safe and our obligations to ensure this.

If you have any concerns about our products, you can contact us on ProductSafety@springernature.com

In case Publisher is established outside the EU, the EU authorized representative is:

Springer Nature Customer Service Center GmbH
Europaplatz 3
69115 Heidelberg, Germany

Batch number: 10755284

Printed by Printforce, the Netherlands